Issued under the

HOME OFFICE

EMERGENCY PLANNI

CW01663940

MANUAL OF FIREMANSHIP

A Survey of the Science of Fire-fighting

Part 6c

Practical Firemanship—III

London: HMSO

Eighth impression 1995

Preface to the 2nd edition

THE first edition of this part of the *Manual* was published in 1962 and, now that reprinting is necessary, the opportunity has been taken of bringing up to date much of the information without recourse to a complete revision. In Section 2, 'Fire involving Explosives,' the new fire classes 7 and 8 have been added, and in Section 5, 'Fuels', details of the bulk storage of liquefied petroleum gases and of pipe-lines have been included. Some revisions have also been made to Section 11, 'Radioactive Materials.' Finally, the information contained in Section 16, 'Dangerous Chemicals', has been re-checked.

Following the publication of British Standard 4422, 'Glossary of terms associated with fire' in which the use of the word '*inflammable*' is deprecated in favour of '*flammable*', the change has been made throughout this Part of the *Manual*. It must not be forgotten, however, that there is legislation in force in which the word '*inflammable*' is used, and in any legislative context it is, of course, necessary to use the authorised word.

ISBN 0 11 340168 X

Metrication

List of SI units for use in the Fire Service

Quantity and basic or derived SI unit and symbol	Approved unit of measurement	Conversion factor
Length metre (m)	kilometre (km) metre (m) millimetre (mm)	1km = 0.621 mile 1m = 1.093 yards = 3.279 feet 1mm = 0.039 inch
Area square metre (m^2)	square kilometre (km^2) square metre (m^2) square millimetre (mm^2)	$1km^2 = 0.386$ $mile^2$ $1m^2 = 1.196$ $yards^2$ $= 10.764$ $feet^2$ $1mm^2 = 0.002$ $inch^2$
Volume cubic metre (m^3)	cubic metre (m^3) litre (l) ($= 10^{-3}m^3$)	$1m^3 = 35.7$ $feet^3$ 1 litre = 0.22 gallon
Volume, flow cubic metre per second (m^3/s)	cubic metre per second (m^3/s) litres per minute (l/min)	$1m^3/s = 35.7$ $feet^3/s$ 1l/min = 0.22 gall/min
Mass kilogram (kg)	kilogram (kg) tonne (t)	1kg = 2.205 lbs 1t = 0.984 ton
Velocity metre per second (m/s)	metre per second (m/s) international knot (kn) (= 1.852km/h) kilometre per hour (km/h)	1m/s = 3.281 feet/second 1km/h = 0.621 mile/hour
Acceleration metre per second^2 (m/s^2)	metre/second^2 (m/s^2)	$1m/s^2 = 3.281$ feet/second^2 = 0.102‘g’

Quantity and basic or derived SI unit and symbol	Approved unit of measurement	Conversion factor
Force		
newton (N)	kilonewton (kN) newton (n)	1kN = 0.1 ton force 1N = 0.225lb force
Energy, work		
joule (J) (= 1Nm)	joule (J) Kilojoule (kJ) Kilowatt/hour (kW/h)	1kJ = 0.953 British Thermal Unit 1J = 0.738 foot lb force
Power		
watt (W) (= 1J/s = 1Nm/s)	kilowatt (kW) watt (W)	1kW = 1.34 horsepower 1W = 0.735 foot lb force/second
Pressure		
newton/metre²(N/m²)	bar (= 10^5N/m²) millibar (mbar) (= 10^2N/m²) metrehead (= 0.0981 bar)	1bar = 0.991 atmosphere = 14.5lb force/in² 1mbar = 0.0288 inch Hg 1 metrehead = 3.28 foot head
Heat, quantity of heat		
joule (J)	joule (J) kilojoule (kJ)	1kJ = 0.953 British Thermal Unit
Heat flow rate		
watt	watt (W) kilowatt (kW)	1W = 3.41 British Thermal Units/hour 1kW = 0.953 British Thermal Unit/Second
Specific energy, calorific value, specified latent heat		
joule/kilogram (J/kg) joule/m³ (J/m³)	kilojoule/kilogram (kJ/kg) kilojoule/m³ (kJ/m³) megajoule/m³ (MJ/m³)	1kJ/kg = 0.43 British Thermal Unit/lb 1kJ/m³ = 0.0268 British Thermal Unit/ft³
Temperature		
degree Celsius (°C)	degree Celsius (°C)	1 degree Celsius = 1 degree Centigrade

Foreword

THE object of this Part of the *Manual* is to provide an insight into some of the special or unusual risks which the fireman or fire officer may be called upon to face in the course of his duties. It cannot, within the space of one volume, attempt to deal with all the risks of this kind which modern industry offers, and a selection has been made of those in which prior knowledge of the nature of the risk and the methods of fire-fighting applicable to it is particularly valuable.

In most cases each Section of this Chapter deals with a particular industrial material (metal, rubber, resin, etc.) and then describes the processes and hazards associated with the industries in which the material is principally used or encountered. A few Sections, however, (such as 'Dusts' and Radioactive Materials'), are not based in this way upon a single material but describe risks of a special kind which are liable to be found in many aspects throughout industry as a whole.

Finally, in Section 16, a catalogue is provided for reference purposes giving details of over one hundred of the principal chemical substances of interest to the fireman because of their dangerous qualities. The list is not confined to substances of particular flammability, but includes many which are toxic, explosive or corrosive. Again, however, this list can only be a selection from the many hundreds of substances which have some dangerous properties, and it is suggested that students of the *Manual* may care to supplement the catalogue for themselves by means of notes in similar form when information about additional substances comes to their notice.

Contents of Part 6c

Chapter 45

Section *Page*

Part 6c, Chapter 45, Section 1

Fires in dusts

It is a characteristic of almost every combustible material, including some which are not by any means readily combustible when normally encountered, that when they are in the form of a finely divided powder they present special dangers to those who work with them and to the fireman. These dangers are associated on the one hand with the possibilities which they contain of extremely rapid combustion, and on the other with the characteristics of very slow combustion indeed: at one end of the scale is the dust explosion, while at the other is the phenomenon of smouldering. Both these require special attention from fire-fighting personnel, and the whole range of forms of combustion in between offers a variety of causes of fire and of ways in which a fire can spread. The factors which decide whether a dust burns very slowly or with explosive violence are, firstly, the nature of the materials, (and materials classed as high explosives offer a good example of the effect of this), and secondly the readiness with which air can reach the individual particles of the dust.

Dust explosions

1. NATURE AND BEHAVIOUR

The greatest danger associated with combustible dusts is the dust explosion. This is the term commonly used to denote the very rapid flame propagation which can occur when particles of a finely divided combustible solid, suspended in air or in a gas which will support combustion, are ignited. The almost instantaneous combustion which is characteristic of a dust explosion is due to the very high specific surface area of the particles, *i.e.* the very high proportion of surface area exposed to the air by each particle in relation to its mass. Each particle burns very readily, due to the ample supply of oxygen available, and its small mass is consumed in a fraction of a second; the heat of combustion, however, manages to bring the next nearest particles to their ignition temperature, and the process spreads with the rapidity of an explosion. A dust cloud containing the right proportions of a combustible solid of the right particle size and air is capable of completely wrecking the building in which it occurs, and, in some cases, has razed whole industrial plants to the ground: many lives have been lost, including those of firemen, in such disasters.

If the dust particles are not of ideal size, or if the degree of concentration of the particles in the cloud is not favourable, the explosion is either inhibited altogether or occurs comparatively slowly and 'mildly'; in some cases the explosive force of a mild explosion has been equal to that of a severe one, but, because the force has been developed slowly, the relatively gradual rise in pressure has been neutralised by the escape of gases through windows, doorways, vents, etc., and damage has been reduced. The degree of concentration of the dust particles is critical, both to the likelihood of the explosion occurring and to the severity of the detonation: too high a concentration prevents the free access of oxygen to the particles and inhibits combustion, while too low a concentration prevents the heat of combustion of one particle being carried to the next, and makes continuation of the explosion impossible. Much study has been undertaken of the subject, and many other factors have been discovered and assessed; the results of research are steadily incorporated year by year into industrial plants, and into the codes of practice by which they are guided, with a view to lessening the dangers of handling dusts.

A notable feature of this type of fire is the freakish effects and behaviour which it frequently exhibits. It will be realised that the degree of concentration which is reached by a dust cloud in different parts of a room or building is largely a matter of pure chance; it is quite common, therefore, to find that an explosion has taken place in one room, apparently 'passed through' an adjacent room quite mildly, and completely devasted a room beyond. One assumes from this that the concentration in the intervening room was either unfavourable throughout, or was confined possibly to a narrow belt of suitable concentration which 'carried' the explosion through the room. It is not common for a dust explosion to produce a severe fire within a short time, except in those industrial premises in which flammable materials are present in large quantities and are readily ignited, or where the combustible dust itself is lying in large quantities throughout the building. More commonly, the explosion lasts too short a time to bring large masses of combustibles up to their ignition temperature, and only leaves a number of small fires in flimsy or easily flammable materials. These fires, however, will soon reach serious proportions if not dealt with immediately, and indeed may give rise to further dust explosions as dust clouds are raised by further falls of debris, etc.

2. CAUSES

Dust explosions most commonly occur in grinding and pulverising mills and similar machinery, and in the dust extraction plants and 'cyclones' in which dusts are collected (Fig. 1). They are very frequently caused by the accidental inclusion of foreign bodies—

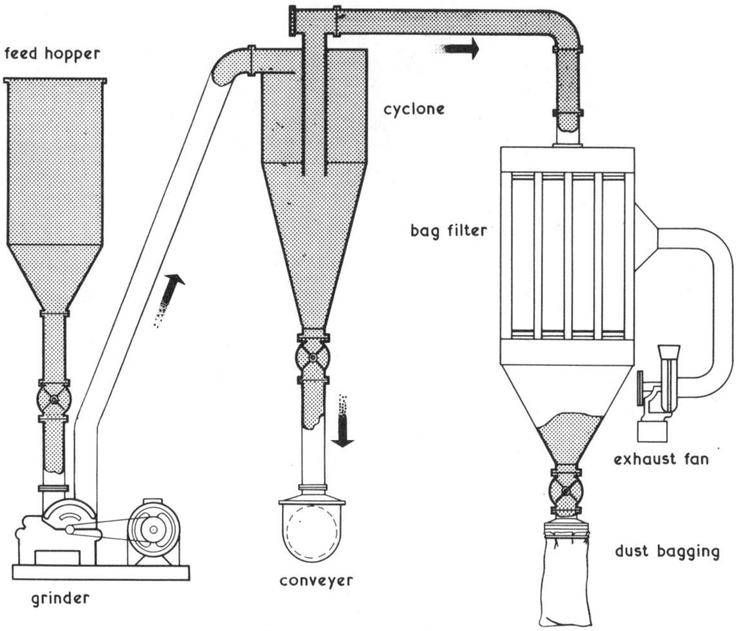

Fig. 1. Layout of a grinding and dust extraction plant.

particles of steel, perhaps, or stone—in the grinding or pulverising process, resulting in heat or sparks; these sparks frequently occur only too often in a favourable atmosphere for explosion, since the plant is usually of the enclosed type and thus may easily contain the correct degree of concentration. In dust extraction ducts and cyclones the cause has frequently been established as originating in overheating somewhere within the enclosed machinery, either by friction or by the breakdown of electric motors or wiring. Another frequent cause is the spark which follows an accumulation of static electricity; it is unfortunately true that the passage of hot, dry dusts through trunking and ducts is itself conducive to the generation of static electricity, and, if the most careful and effective means of earthing such charges are not installed, there is a grave risk that a discharge will take place in a favourable atmosphere for ignition. Since, however, all types of dust handling plant should be provided with explosion vents and releases (Fig. 2), it is common for a fire from any of these causes to be confined within the plant, and to subside, or cease, once the explosion is over.

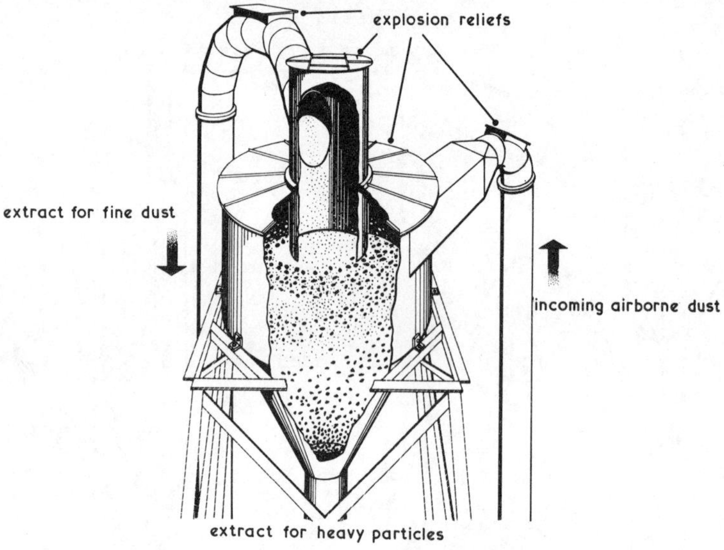

Fig. 2. Diagram of a dust cyclone showing the explosion reliefs at the top. There are no moving parts.

In many industrial premises, in which combustible dusts are apt to be present, and in which the building is unsuitable or the 'housekeeping' poor, a much more dangerous though less common cause of explosions is prevalent. In such places, a coating of the flammable dust slowly collects on horizontal surfaces on the tops of walls and machinery, among roof girders, on floors and furniture, on window sills and in the niches of old brickwork (Fig. 3). An occasion may arise when some of this dust is disturbed and becomes suspended in the presence of a heating appliance, a source of sparks or a naked flame. When conditions are right, this can give rise to a 'pilot' dust explosion, igniting a relatively small cloud of the right concentration. This 'pilot' explosion stirs up the dust lying on exposed surfaces, and the same source of ignition can then produce an explosion of devastating power, capable of destroying the building

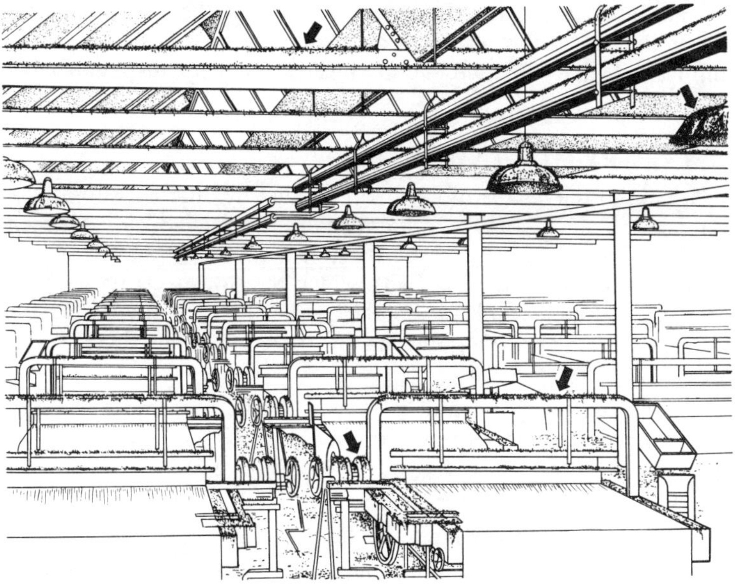

Fig. 3. Diagram showing how dust can accumulate on horizontal surfaces and ledges.

and of killing or severely burning the occupants. On numerous occasions, employees have afterwards described how a flash and a slight detonation have been noticed, and how this has been followed by the explosion proper. A special and persistent danger of the kind described above is inherent in coal mines, where it is additionally complicated by the risk of firedamp.

Spontaneous combustion must not be overlooked as a possible indirect cause of dust explosion, though it can never be the direct one, because spontaneous ignition of a dust cloud itself can only occur in a limited number of hazardous substances. Fires have been caused in dust heaps, however, by spontaneous combustion, and this has led, through disturbance of the burning dust, to the ignition of a dust cloud and consequent explosion. Fortunately, the number of substances in which spontaneous combustion of the dry dust has been known to occur is small: coal, rubber and zinc sulphide are the only examples usually quoted.

In time of war an additional cause of dust explosions has been recorded, arising from the dropping of high explosive bombs on or near mills or factories. A bomb can score a direct hit, or a near miss, on such a building, and raise clouds of flammable dust within a fraction of a second; the hot fragments of the bomb, or some secondary means of ignition occurring as a result of the bomb damage, can produce a dust explosion such a short time after the fall of the bomb itself that the two detonations are hardly distinguishable from each other.

3. HAZARDOUS INDUSTRIES

A guide to the degree of hazard presented by different substances is obtained from a study of a number of materials which have been tested and classified by the Ministry of Labour and National Service according to the ease with which their dusts can be ignited. (Ministry of Labour and National Service, 'Flammability of Dusts', Form No. 830 (Revised) published by H.M.S.O.). The three classes of dusts are as follows:

I Dusts which ignite and propagate flame readily, the source of heat required for ignition being small.

II Dusts which ignite readily but require a larger source of ignition.

III Dusts which do not ignite in the tests applied.

The following table of familiar materials has been extracted from the Ministry of Labour's very comprehensive list, and demonstrates clearly that most organic substances, and many others, offer a dust explosion hazard:

Class I

Aluminium	Dextrine
Bronze	Ebonite
Cellulose acetate	Ferromanganese
Chicory	Flax
Coal	Flour
Cocoa	Grass, dried
Cork	Grains
Cornflour	Magnesium
Cotton, ground	Malt

Class I—continued

Oat-husks	Starch
Paper	Sugar
Phenol-formaldehyde resin	Sulphur
Polythene	Tea
Potato	Titanium
Rice bran	Tobacco
Rubber	Wheat
Shellac	Wood-dust

Class II

Barley	Prussian blue
Castor oil meal	Seaweed
Herring meal	Shoddy
Meal, yellow	Steel grindings

Class III

Bonemeal	Ivory and mineral black
Charcoal	Kelp
Detergent powder	Lamp black
Graphite	Plumbago
Insect powder (mainly pyrethrum)	

Among the industrial plants in which particular care has to be taken against this hazard are flour mills, starch works, coal pulverising plants and metal grinding mills: it is perhaps true to say that the greatest difficulties from the fireman's point of view are attached to the metal dusts, particularly aluminium, magnesium, bronze powder and their alloys. This is partly because their particles are relatively heavy, and dust collections tend to remain in place about the premises, and partly because of the well-known difficulty of extinguishment of such metals. Whereas water, in suitable form, can be used for the extinguishment of most organic dusts, it is wrong and dangerous to use it on burning metal powders (*see* paragraph 5, page 9). In some places, the hazardous dust is not itself the product of a grinding process, but exists incidentally to the handling and storage of some commodity; this applies to cotton mills and grain storages, for example, and it usually means that the dust is not confined to the particular machine or plant or part of the factory where it is being processed, but may be found anywhere. Table I gives an analysis of over one thousand dust explosions (which, it should be noted, is not necessarily the same thing as fires in dusts) reported to the N.F.P.A. of America from 1900 to 1959:

TABLE I

Classification	*No. of explosions*	*No. killed*	*No. injured*
Grain dust	205	130	463
Wood dust	146	40	142
Feed and cereal dusts	137	69	309
Flour dust	88	8	69
Metal dusts	81	108	199
Coal dust (pulverised)	65	31	41
Starch and starch products dusts	55	145	169
Miscellaneous dusts	52	24	84
Cork dust	40	6	29
Plastic dusts	35	19	71
Sulphur dust	35	2	40
Fertiliser dust	28	7	21
Sugar dust	28	12	33
Malt dust	23	2	18
Cotton dust	18	6	12
Bark dust	15	—	2
Rubber dust	15	13	2
Coffee and spice dusts	12	5	13
Seed and seed products dusts ..	11	9	6
Paper dust	10	—	1
Phonograph record dust	6	—	1
Pitch and resin dusts	5	12	47

(*Extracted from NFPA Handbook of Fire Prevention, Twelfth Edition*).

4. PREVENTIVE MEASURES

It is not the function of this Section to go into the details of the steps which can be taken to reduce the risk of a serious dust explosion. Suffice it to say that these steps are directed on the one hand towards preventing an explosion from occurring, and on the other towards minimising the effects which would result if an explosion did occur: the former object may be secured by good housekeeping, suitable adaptation or construction of buildings and plant, efficient dust extraction systems, the use of inert gases inside enclosed plants and the injection of inert dust into hazardous dust clouds while the key to the latter object lies in adequate venting and pressure relief combined with good fire prevention. The fire officer who examines an industrial building to assess its fire risk need not concern himself overmuch with the elaborate systems of compartmenting, separation, venting and other safety measures with which dust-handling premises should be equipped; his most pressing questions will usually be answered adequately by a study of the general tidiness and dust-free condition of the rooms and buildings and a knowledge of the properties of the commodities which are handled in them. As will be seen in later Sections of this Chapter, the majority of buildings in which foodstuff dusts and many other organic dusts are handled incorporate sprinkler systems, and the importance of this needs no emphasis.

5. PRECAUTIONS WHEN FIGHTING FIRES

When fire-fighting is in progress in any place in which flammable dusts or powders might be encountered in any quantity, the dangers of an explosion will be much reduced if:

(**a**) The fireman is aware that he is in circumstances which offer a special risk, and of what that risk consists.

(**b**) He avoids the use of jets, or even of sprays, which strike harshly against piles or layers of dust, and uses only a fine spray or any low-velocity, gentle form of water application. He should use even this with scrupulous care to avoid the stirring up of any cloud of dust in the presence of flame or heat.

(**c**) In the case of metal powders, he uses extinguishers such as powdered talc, powdered asbestos, graphite, soda ash or dry sand, and avoids the use of water, foam or chemical extinguishers except where necessary for the protection of surrounding risks, (avoiding the use of vaporising liquid extinguishers under any circumstances).

(**d**) He avoids handling bins, cartons, drums or open containers of any flammable dust or powder in such a way that a dust cloud could be produced by his action or by inadvertence. Such containers, if free of fire, should be covered before removal; if they may still contain fire, they should be carefully drenched or flooded before handling by means of open ends or sprays or, if metal powders are involved, treated correspondingly with extinguishing materials as in (**c**) above.

(**e**) Consideration could also be given to the use of high expansion foam in certain circumstances.

Note: For further details of the extinguishment of metal fires see *Section* 7.

Other characteristics of fires in dusts

Dusts and powders create special conditions for the fireman, apart from the risk of explosion. Deposits of finely divided solids, even when they do not become suspended in air but remain spread on horizontal surfaces, can cause fire to spread through a building with exceptional speed; this applies particularly to dust fires inside trunking and ducts, and for this reason it is vital to see that extractors, exhaust systems or any machinery creating a draught in trunking systems are shut off at once. In cases where a flammable dust is being fed from, produced in or passed through machinery, it is also urgent to see that the product supply is cut off, since cases have been known of the supply lines continuing to feed a fire while it was being fought. Where fire has been caused by a dust explosion, it is very likely that small pockets of the dust may still be alight on all sorts of horizontal surfaces, whether visible or hidden, and these should be searched for with a hose reel spray of diffuser branch unless metal powders are involved; it is especially important that all communicating compartments and those linked by conveyors, however remote, should be thoroughly searched.

Smouldering fires in dusts

When dusts or powders, of organic materials particularly, are lying in heaps or layers, it is quite common for a smouldering fire to start in or under them; such a fire may have been originated by some small source of ignition, such as a cigarette end, or by spontaneous combustion. When the layer or pile is thin, the fire will be easily detected; in a large or deep heap, however, it has been shown that a fire can smoulder for as long as a week or more with little external evidence. It is particularly important to treat such fires with special care, since in many cases a slight disturbance is capable of causing the upper layer to fall into the cavity which has been burned away, and this might give rise to a minor explosion (*see* Fig. 4); this might in turn disturb further dust and create a major one. If, however, any

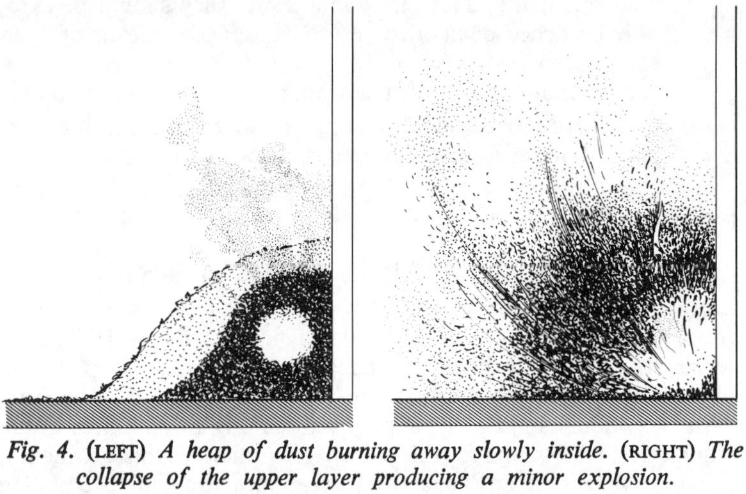

Fig. 4. (LEFT) *A heap of dust burning away slowly inside.* (RIGHT) *The collapse of the upper layer producing a minor explosion.*

approach to such a fire is always made with a spray branch, and care is taken to avoid disturbance of the heap, it is virtually certain that explosive conditions will be avoided. Water would be incorrect, of course, if applied to a metal powder, as has been said earlier, but such powders would not normally be found smouldering; they can be relied upon to burn fiercely in all circumstances.

Part 6C, Chapter 45, Section 2

Fires involving explosives

OF all the commonly encountered materials which may be involved in fire, it is likely that explosives offer the greatest and most immediate danger to life and property. However, because of this very fact, the highest degree of care and precaution is exercised by those responsible for the manufacture, handling and transport of explosives, and a very substantial and complex system of legal safeguard controls them, and the materials used in their manufacture, at every stage of their use. For this reason it is very rare indeed for large fires to occur in explosives installations, especially in peace-time, although a number of minor outbreaks or incidents not calling for outside help is reported annually by H.M. Inspectors of Explosives. For reasons which are easy to imagine, and which will become evident in this Section, it is likely that fire brigade personnel will always act under the guidance, if not the control, of those responsible at the explosives installation whenever they are engaged in fire-fighting at such premises, and will not therefore be exposed to any unforeseen risk from the dangerous materials surrounding them; it is desirable, none the less, that they should have a general understanding of explosives and their nature, and of the principles which have been laid down for the conduct of fire-fighting operations among or in the vicinity of explosives.

Nature and properties of explosives

Explosives are substances which can be transformed into gases at high temperatures. For purposes of this Section, this includes the commonly accepted commercial and military explosives, together with coloured fires, fog signals, flares, fireworks, rockets, fuses, percussion caps, detonators and all such devices: additionally, paragraphs will be found at the end of the Section dealing with the hazards of manufacture and storage of matches which are not classed as explosives.

Some explosives are simple chemical compounds, such as fulminate of mercury; others are mixtures of commonplace ingredients *e.g.* gunpowder is a mixture of saltpetre, sulphur and charcoal. The explosives industry handles an enormous variety of mixtures, some fundamentally different in character from others and some only differing in very slight details of their composition: these, in turn, are processed, packed or manufactured into very many different forms of product. For the purposes of the *Explosives Act* 1875, Orders in Council Nos. 1 and 1A divide explosives into seven classes as under:

Class 1 Gunpowder class;

,, 2 Nitrate-mixture class, consisting of preparations other than gunpowder formed by the mechanical mixture of a nitrate with any form of carbon or carbonaceous substance not possessed of explosive properties;

,, 3 Nitro-compounds class, comprising chemical compounds of explosive properties which are produced by the chemical action of nitric acid upon any carbonaceous substance;

,, 4 Chlorate mixture class, consisting of explosives containing a chlorate;

,, 5 Fulminate class, comprising compounds or mixtures which, because of their extreme sensitivity or instability, are especially suitable for initiating explosions, or are especially dangerous.

Class 6 Ammunition class, comprising explosives of any of the foregoing classes when made up or prepared or enclosed as a cartridge or a charge for a weapon, etc.;

„ 7 Firework class, comprising firework compositions and manufactured fireworks.

These classes, or the majority of them, are in turn split into divisions for the purpose of grouping together products which have features in common. Nitroglycerine itself, when not forming part of a nitro-compound under Class 3, Division 1, is treated separately and constitutes an unnumbered class of its own. It does not leave the factory unmixed.

It is most important to realise that the classification of explosives referred to above is founded on the composition of the explosives themselves, and not on their behaviour in fire: for this purpose a separate system of grouping exists, based upon the behaviour of different kinds of explosives in fire and the fire-fighting procedure to be adopted in respect of each fire-fighting class. This code of rules is of paramount importance to the fireman and is dealt with fully in later pages of this Section under the heading 6 '*Fire-fighting Classification*'.

There are other classifications of explosives: it is, for example, a common practice to differentiate between 'high' and 'low' explosives. The difference lies in their reaction speed: in the former case reaction can be so nearly instantaneous that the process of explosion is termed 'detonation' while low explosives have a slower development in their explosion, and the term 'deflagration' (literally, 'flaming off') is used to describe this. However, it is possible for the same material to detonate or deflagrate, or it may even burn without explosion, for whether it is brought to detonation or not depends upon the character of the initiation.

The following are some of the main characteristics and uses of common high explosives:

Nitroglycerine

A very powerful and sensitive explosive with qualities that make it difficult to use in unmixed form, but excellent for many purposes in combination with other explosives. It very commonly forms the basis of dynamites and blasting gelatines.

Dynamites

These are not an individual substance, as is popularly supposed: dynamites are a very considerable group of powerful explosives, embodying nitroglycerine and absorbents, often with gelatines and other ingredients, like ammonium nitrate.

Nitrates

Ammonium nitrate is one of the commonest constituents of nitrate mixtures although it is not classed as an explosive itself. When used in combination with selected other substances it has a wide variety of civil and military uses. The well-known 'amatol' for example combines ammonium nitrate with T.N.T. (trinitrotoluene).

Cellulose nitrate, nitrocellulose (guncotton, pyrocotton, nitrocotton). Burns very rapidly and fairly harmlessly (except in very large masses) but is a very powerful explosive when compressed and suitably initiated. Its principal uses are in propellant mixtures for shotgun and small arms ammunition and in nitroglycerine explosives.

T.N.T. (Trinitrotoluene)

Generally included in mixtures for blasting and bursting charges for some types of ammunition. They are nevertheless relatively safe to handle; in bulk may burn to detonation.

Chlorates

Usually as the salts of sodium or potassium. Not used in blasting explosives; limited use in cap and detonator compositions; its principal use is firework compositions.

Fulminates, Azides, Styphnates

Not explosive alone but form very sensitive mixtures with sulphur, phosphorus and carbonaceous materials. Are very sensitive and dangerous; the duration of their explosion is short and violent. They are commonly used with other substances as initiators in the form of caps or detonators.

The principal member of the low explosives group is *gunpowder* (black powder, black blasting powder). From a fire point of view, this is one of the most noteworthy of explosives, since it is extremely sensitive to ignition from sparks, heat and friction: it burns very violently even when quite free and uncompressed, and releases volumes of smoke on exploding. It is used in blasting fuse and pyrotechnic compositions.

Industrial processes and hazards of explosives

1. EXPLOSIVES AND FIREWORKS MANUFACTURE AND FILLING

Every factory for the manufacture of explosives (including fireworks) must be licensed by the Secretary of State under the *Explosives Act* 1875; the licence covers both manufacturing operations and storage. The conditions of each licence are presented for the protection of the employees and the general public.

The Report of Her Majesty's Inspectors of Explosives for 1968 shows that there were at that time 66 factories for manufacturing and filling explosives including ammunition and one Toy Firework factory. These figures do not include a number of factories operated by Government departments or agents for those departments.

A very large variety of processes is employed. The scope of manufacture ranges from the production of the raw materials themselves, which is often a process of little hazard, through a whole series of operations such as grinding, mixing, curing, melting, rolling and drying, to the actual filling and packing processes from which the finished products emerge.

Throughout the processes the hazards will be widely different according to the materials being handled. They will vary from a fire risk to an explosion risk. The protective measures are prescribed accordingly under the Explosives Act, by the terms of the licence, by H.M. Inspectors of Explosives and by the managements.

Factories normally take the form of a large number of separate buildings or sheds in order to ensure that the results of an accident will affect as far as possible only the building in which it occurs.

Where there may be a risk of explosive blast, buildings may be of light construction throughout or have light roofs or blow-out panels. Usually these buildings are wholly or partly surrounded by protective mounds or barriers. In certain processes involving explosives which are sensitive to ignition by static discharge, the building temperature and humidity may be controlled (for example, in processes for detonators and their compositions), or in special cases (*e.g.*, dry nitrocellulose) the floors may be wetted.

2. EXPLOSIVES STORAGE

In 1968, 211 magazines, the majority for quantities exceeding 1814 kg were operated under licences issued by the Secretary of State. In addition local authorities issued licences for several thousand explosives stores for quantities between 70–1815 kg as

Plate 1. An explosion in an aluminium powder factory, which flung burning dust some distance away and started a fire in another building. (Section 1).

Plate 2. Complete devastation of a building in which nitro-cotton blocks were being dried, resulting in the deaths of three people. Note the earthworks round the building. (Section 2).

Plate 3. The earth traverse round this small explosive storehouse has the sides sloping at the natural angle of rest. The building in the background is untraversed. (Section 2).

Plate 4. Interior of an explosives storehouse showing one method of storing ammunition by assembling it on pallets in 'unit loads'. (Section 2).

Plate 5. Firemen in danger from burning solvent at the British Cocoa Mills fire. A number of explosions occurred during this fire, in which several people lost their lives. (Section 3).

PHOTO: *Associated Newspapers Ltd.*

Plate 6. Plant used in the production of hydrogen by electrolysis for the manufacture of margarine. (Section 3).

Plate 7. Paraffin wax solidified by water from branches after flowing from a fire. This caused considerable inconvenience by blocking the drains. (Section 3).

well as registering premises (about 50,000) for small quantities, many of them in respect of fireworks. The technique of these storages is again a specialised and skilled matter, demanding rigorous adherence to detailed safety code, partly statutory. Storage is done in specialised types of building or sometimes in excavations either above or below ground (*see* Fig. 5), or even in moored ships. These storage places must maintain certain safety distances to houses, highways, railways, public places, etc. In some cases protective mounds are prescribed (*see* Plate 3). Special requirements as to access, ventilation for temperature control, dryness, security, cleanliness and tidiness are imposed. All these requirements apply in principle whether the storage is associated with a large civil or military installation or the small explosives store of a quarry or the registered premises of a fireworks shop.

3. TRANSPORT

The packing and conveyance of explosives by road (including the construction of the vehicle) are controlled by regulations and conditions made under the Explosives Act. By rail the control is by the Railway Bye-laws, and by sea or canal by the Ministry of Transport. The fact that very large quantities are involved every year with an extremely low accident rate is a measure of the enforcement and effectiveness of precautions.

4. CAUSES OF FIRE AND METHODS OF PROTECTION

Almost without exception, fires recorded as occurring in the use, manufacture or storage of explosives including fireworks, originate from the inadvertent ignition of a material or component containing explosives. There may be a rapid intense fire of short duration or a more steady fire of variable persistence. Often there will be an explosion or series of explosions of more or less violence, with projection of debris which may start other fires. Damage may vary between minor effects within the building itself to complete destruction by fire or explosion. Personnel may suffer from burns or injuries which in some cases prove fatal. Fortunately, really serious outbreaks of fire or conflagrations are rare. Table II on page 19 of accidents in the year 1968 is extracted from the information given in the Report of H.M. Inspectors of Explosives for that year. The originating cause of ignitions during manufacture is often impossible to determine, but may be due to the inadvertent inclusion of foreign bodies, grit or dirt in the materials themselves or in the processing machinery, variations in the composition, too low

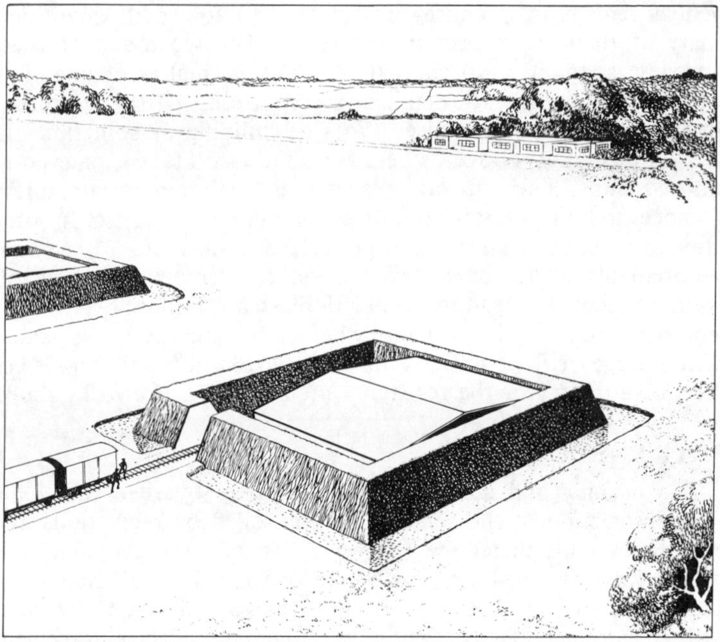

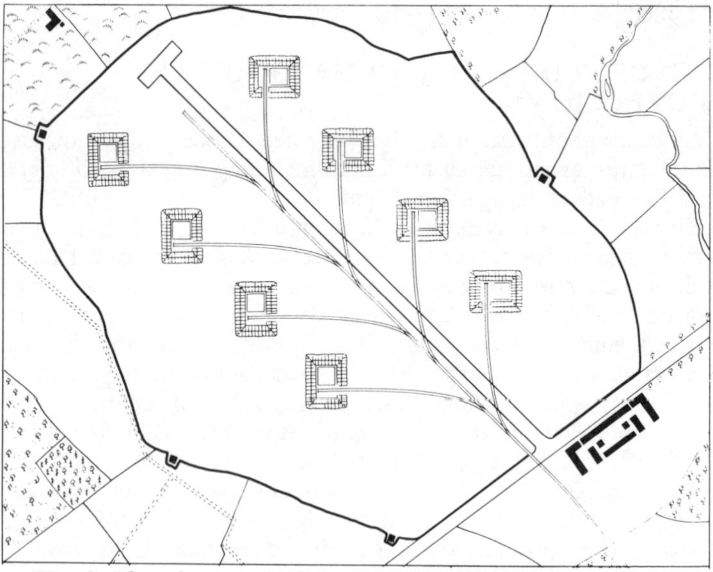

Fig. 5. Plan showing the lay-out of a modern explosives store. (TOP) The protective earthworks and moat round the building.

TABLE II

Summary of explosive accidents for the year 1968.

	No. of accidents	No. of persons killed	No. of persons injured
Manufacture	33	3	20
Illegal manufacture	9	1	10
Keeping	1	—	2
Conveyance	—	—	—
Blasting	3	2	1
Tampering with Service explosives	—	—	—
Tampering with explosives other than Service explosives	2	—	2
Explosives in coal	2	—	2
Miscellaneous	6	—	5
TOTALS ..	56	6	42

Note: These figures do not now include accidents reported to the Mines Department.

atmospheric humidity, friction, mishandling whereby explosives are accidentally dropped or struck, or to other factors associated with the processes. As in many other industries of exceptionally high fire risk, the incidence of accidents due to flagrant neglect of fire precautions is very low.

(The foregoing remarks are only intended to apply to legally authorised manufacture, storage and use: a substantial proportion of the accidents reported are due to experimenting with home-made mixtures, careless keeping of and mishandling or tampering with explosives, particularly fireworks, by the public, when inadequate safety precautions are usually observed. In such cases the quantities involved are almost invariably small but serious injuries do arise nevertheless.)

Protective procedures constitute a large part of the technique of explosives: they are elaborate and far-reaching. They depend on a very high standard of control, cleanliness and fire prevention and on the exclusion of sources of ignition by the use of special clothing and footwear, flooring, tools and other equipment. They also include the provision of first-aid fire-fighting appliances and, in the larger establishments, the usual resources of a factory fire brigade. Few fixed fire-fighting installations of the type normally provided against fire outbreaks are provided, since in most cases the action of these would be too late to be of any benefit and in others might itself produce very dangerous conditions. In many of the processes, however, automatic explosion protection and suppression is incorporated. Means of escape plays a larger part than any fire extinguishing measures in ensuring the safety of workpeople.

5. FIRE-FIGHTING PROCEDURE

It will be evident from the foregoing paragraphs that in no circumstances can fire-fighting among explosives be carried out in accordance with the practice and techniques adopted for ordinary risks. The key to correct fireground tactics and procedure in these circumstances lies in an intimate knowledge of the behaviour of different types: the staff of explosives establishments are experts in this field, and it is therefore indispensable that the officer-in-charge of the fire brigade units present should be guided by and act in close co-operation with the appropriate officer of the explosives establishment. To ensure that this co-operation works smoothly and effectively, the *Inter-Departmental Explosives Storage and Transport Committee* of the Government has drawn up a code of safety conditions for fire-fighting in government explosives establishments, and this, while only mandatory in such establishments, forms an invaluable guide for fire brigades in the majority of circumstances where explosives are involved. The code lays down the responsibilities of the different officers who will be concerned and the procedure for co-ordinating the efforts of the establishment's own personnel, the fire brigade and the police, and for pre-fire planning generally.

The code then classifies explosives into nine 'Fire Classes' (separate and quite distinct from the classification of types of explosive mentioned in the paragraphs headed 'Nature and Properties of Explosives' of this Section) and proceeds to detail the fire-fighting procedure called for by each of these Fire Classes. The Fire Classes which are laid down, the symbols used to identify storage of the different classes (*see* Plate 31) and the fire-fighting procedure for each class are quoted below in later paragraphs, but it is emphasised that the administrative and fire prevention procedures in the code, which are not quoted here, are also of vital importance to the fire brigade officers to whom they are appropriate and should be studied and implemented by them.

Fire officers attending fires or other accidents in explosives buildings should endeavour to disturb the site and the remains of the buildings and their contents as little as possible consistent with ensuring general safety. Inquiries into incidents can be seriously hampered or rendered abortive by avoidable disturbance of remains, for the clue to the causes may depend on establishing the nature and position of all loose articles present before the accident.

6. FIRE-FIGHTING CLASSIFICATION

CLASS 1 explosives are those which must be expected to explode EN MASSE VERY SOON AFTER FIRE REACHES THEM.

CLASS 2 explosives are those which are readily ignited and burn with great violence without necessarily exploding.

CLASS 3 explosives are those which may explode EN MASSE but, compared with Fire Class 1 explosives, may be exposed to a fire for some time before exploding. There will be a blast and fragment hazard.

CLASS 4 explosives are those which burn fiercely and give off dense smoke, with, in some instances, toxic effects. THERE IS NO RISK OF MASS EXPLOSION.

CLASS 5 explosives are those containing toxic substances.

CLASS 6 explosives are those which may be exposed to a fire for some time before exploding. The risk of mass explosion is not involved, but small sporadic explosions will occur with increasing frequency as the fire takes hold. There will be a fragment hazard, but not a serious blast risk.

CLASS 7 explosives are those which involve a combined flammable, toxic and corrosive hazard. These may be exposed to a fire for some time before exploding. THE RISK OF MASS EXPLOSION is not involved but explosions will occur with increasing frequency as the fire takes hold. There will be a fragment hazard arising from pressure bursts but not a serious blast risk. Personnel involved in fire-fighting must wear full acid-proof protective clothing and self-contained breathing apparatus of a type to withstand fuming nitric acid.

CLASS 8 explosives are those which, for fire-fighting purposes under storage conditions, involve a radiological hazard which may exceed the explosive hazard when involved in a fire. Explosives in which the presence of a radioactive substance does not materially increase the fire hazard associated with the explosive substance are not included in this class but are classified on the explosive hazard and do not require the display of the trefoil symbol (*see* Plate 31) indicating a radioactivity hazard.

Class 8 explosives are required to be stored under conditions which will limit mass explosion to the contents of one cell. Compared with Class 1 explosives they may be exposed to a fire for some time before exploding. There will be a blast and fragment hazard in addition to the radiological hazard.

CLASS MP comprises substances of the non-explosive dangerous goods class containing metallic powders such as magnesium, aluminium or zinc powders either in ammunition or in bulk when these are held in explosives establishments and provision for fire-fighting has, therefore, to be included.

The symbols denoting each of the Fire Classes are shown in Plate 31.

7. FIGHTING FIRES IN ABOVE-GROUND SITES

(a) For the purpose of fire-fighting 'above-ground sites' are those where the explosives are above ground level. It includes those where part of the explosives may be below ground.

(b) Success in fighting fires involving explosives is largely dependent on action being taken before the fire can develop and the availability of copious* water supplies. Where petrol is present, foam is required instead of water. For fires involving metallic powders, *see* par. (*m*), page 26 of this section.

(c) When a fire occurs at an open storage site, tarpaulins, other than those of fireproof material, will be removed from all accessible stacks.

(d) *Class* 1 *explosives fires.* If detected early, the prompt application of fire-fighting first-aid measures may prevent the development of a serious fire. The fire alarm must be operated immediately and all non-essential personnel evacuated to a safe position.

On arrival of the fire-fighting service, action should be directed towards preventing the explosives from becoming involved, fire-fighting operations being carried out from behind *substantial cover* such as that afforded by intervening traversed buildings or suitable natural features. The officer in charge is responsible for advising on the safety aspect of this operation and the fire brigade officer must defer to his advice.

If the fire reaches the explosives an explosion is to be expected, the external effects of which will depend upon the quantity and type of explosive and the design of the building and its protecting traverses. Thereafter, in such cases, action is to be confined to preventing the fire involving adjacent buildings or stacks.

The roofs and walls of buildings sufficiently near the burning building to feel the effects of thc heat or be within range of burning debris or sparks should be thoroughly sprayed with water from the outset.

The officer in charge will decide whether the explosives contained in these buildings or stacks are also to be sprayed. If the number or

* *Note:* The minimum amount of water which should be available on the site where explosives are stored should be 113,000 litres with hydrants available so that a pressure of at least 3·45 bar should be immediately available.

size of fragments falling on such buildings is large, no time should be lost in doing this. If drenchers or sprinklers are fitted in buildings they should be operated directly the risk is apparent.

At open storage sites the adjacent stacks are to be dealt with in a similar manner.

(e) *Class* 2 *explosives fires*. When a fire occurs in a building of an explosives factory containing these explosives, it will spread rapidly and burn fiercely. There is no risk of mass explosion. The fire-fighting service should, therefore, be employed in subduing the fire and in preventing its extension to adjacent buildings in the manner prescribed for dealing with Class 1 explosives fires.

When a fire occurs in a filling factory or an above-ground storage site, fire-fighting first-aid measures should be operated and an alarm sounded. The fire-fighting services can fight the fire before and after the explosives become involved as there is no great risk of explosion. The fire is likely to be fierce.

The spread of the fire to neighbouring buildings is to be dealt with as for Class 1 explosives fires.

(f) *Class* 3 *explosives fires*. If detected early, the prompt application of the fire-fighting first-aid measures will prevent the development of a serious fire. The fire alarm should be operated immediately and all non-essential personnel evacuated from the vicinity.

Once the explosives have become involved the fire-fighting services should only continue to operate when substantial cover such as that afforded by intervening mounded buildings or high ground is available.

The spread of the fire to neighbouring buildings should be dealt with as for Class 1 explosives fires.

(g) *Class* 4 *explosives fires*. If detected early, the prompt application of fire-fighting first-aid measures may prevent the development of the fire. The fire alarm should be operated immediately.

The fire-fighting services may fight the fire, whether the substances are or are not involved, as there is no risk of mass explosion. When the substances are involved dense smoke will be given off and, as there may be some risk of toxic effects, the use of self-contained breathing apparatus is essential. When incendiary fillings are involved a fierce fire, with intense heat, must be expected.

Action to prevent the fire spreading to other buildings or stacks should follow the lines prescribed for Class 1 explosives fires.

If phosphorus fillings have been involved in the fire special precautions are necessary in disposing of residue, as loose phosphorus will ignite spontaneously when it dries out. The affected packages and loose residue should, therefore, be removed to a safe spot, split up into small isolated groups, and allowed to dry. If spontaneous ignition again occurs, the groups should be split still further and allow contaminated material to burn out.

When the fire has been subdued, the building should be strewn with wet sand and thoroughly cleaned, all residue being removed to a safe place and allowed to burn out. The building should be dried out and carefully inspected in the dark for signs of phosphorus before being used again.

Special care should be exercised by fire-fighting personnel to avoid contact between phosphorus and fire-fighting equipment, or clothing, or their person. Phosphorus causes severe flesh burns which require medical treatment.

(h) *Class 5 explosives fires.* If detected early, the prompt application of fire-fighting first-aid measures may prevent the development of the fire. The fire alarm should be operated immediately.

The fire-fighting services may fight the fire, whether the chargings are involved or not, as THERE IS NO RISK OF MASS EXPLOSION. The toxic effects may, however, be very serious. The bursting charges of the weapons are liable to explode in a fire and scatter the toxic fillings, consequently, fire-fighting personnel must be equipped to deal with liquid contamination and toxic effects in the form of a cloud of lethal gas. They must have protective clothing and Service respirators or self-contained breathing apparatus. The fire should be fought from the windward side. A general alarm must be sounded to put the police and civilian population on the alert and enable them to proceed in accordance with the pre-arranged scheme. All personnel within the enclosed area should don gas masks. During fire-fighting operations personnel in close proximity to the fire should be sprayed with water to prevent their protective clothing becoming ignited.

(i) *Class 6 explosives fires.* If detected before the explosives are involved, prompt application of fire-fighting first-aid measures will prevent the development of the fire. The alarm should be operated immediately.

Once the explosives have become involved, all non-essential personnel should be evacuated from the vicinity, but the fire-fighting services should continue to operate from behind light cover such as brick buildings. Fires should be fought with copious quantities of water but the use of hard striking jets should be avoided.

The spread of the fire to neighbouring buildings should be dealt with as for Class 1 explosives fires.

(j) *Class 7 explosives fires.* If detected early, or if breakage or leakage occurs, the fire alarm should be sounded immediately and the package removed if possible. The prompt application of available fire-fighting first-aid measures may prevent the development of fire.

To deal with burning packages in a fire that has taken a strong hold, it is essential to wear breathing apparatus and full protective clothing. The fire should be fought in the early stages using hoses with spray nozzles.

Once the fire has taken a firm hold beyond the scope of first-aid fire-fighting measures, it will be necessary to continue from behind suitable cover such as an earth mound or substantial building, using copious supplies of water.

(k) *Class* 8 *explosives fires.* If detected early, the prompt application of fire-fighting first-aid measures will prevent the development of a serious fire. The fire alarm should be operated immediately and all non-essential personnel evacuated from the vicinity. A general alarm should be given to put the police and civil population on the alert and enable them to proceed in accordance with the pre-arranged scheme. A minimum exclusion distance of 457 m should be observed for non-essential evacuated personnel, up-wind if possible.

Once the explosives have become involved, the fire-fighting services should only continue to operate from behind substantial cover such as that afforded by intervening traversed buildings or suitable natural features. The spread of fire to neighbouring buildings should be dealt with as for Class 1 explosive fires.

Fire-fighting personnel should wear full protective clothing including respirators fitted with suitable filters or self contained breathing apparatus. They are to be checked for cuts, abrasions and weeping type dermatitis before being permitted to remain in the vicinity. Personnel who suffer cuts or wounds in the course of these duties should be withdrawn for first-aid treatment to prevent contamination entering the body. Smoking, eating and drinking should be prohibited in areas where contamination exists or is likely to occur.

Success in fighting the fire is largely dependent on action being taken before the fire can develop and the availability of copious water supplies (see footnote, page 22). In general the best method of safeguarding Class 8 explosives involved in a fire is by application of the water in the form of a spray from a diffuser nozzle with the object of cooling the material to a temperature below its ignition point, the spray being kept in constant movement so that the whole surface is cooled. Vaporising liquid chemical extinguishers are NOT to be used. Foam has very little cooling effect on explosives but may be used on adjacent fires.

All personnel are to be monitored for contamination before leaving the excluded area and should undergo personal decontamination as necessary. Equipment and appliances used in fighting fires should be retained in the area until checked for contamination and cleared by qualified personnel for removal. Equipment and appliances found to be contaminated should be clearly marked and segregated pending decontamination or disposal.

Salvage operations should not be undertaken in the vicinity of the fire until guidance has been given by the appropriate authority that it is safe to do so.

(m) *Class MP fires.* If detected early, the prompt application of fire-fighting first-aid measures may prevent the development of the fire. The fire alarm should be operated immediately.

To deal with the burning powders, extinguishers such as powdered talc, powdered asbestos, asbestos graphite or dry sand should be used, care being taken to avoid agitating the powder. Dry chemical extinguishers are allowed subject to the powder being non-hygroscopic and not unduly toxic.

Water, foam or chemical extinguishers (carbon dioxide or vaporising liquid) should NOT be used.

A notice to the above effect should be displayed at each building where these powders are used or stored.

8. FIGHTING FIRES IN UNDERGROUND SITES

(a) The term 'underground sites' in the fire-fighting sense covers explosives storage sites where the whole of the storage space is below the natural ground level.

(b) Self-contained breathing apparatus is essential for underground fire-fighting.

(c) The fire-fighting action follows the general lines set out for above-ground sites with the following exceptions:

Class 2. If these explosives become involved, the flames are likely to sweep through the workings, the available oxygen will be quickly exhausted and the chances of escape or of fire-fighting arc vcry small.

Class 4. Explosives belonging to this class are not stored in underground sites because of the difficulties in fire-fighting which would arise from the flames or smoke given off by these substances when they are seriously involved in a fire.

Class 5. These weapons should only be held at underground sites in very exceptional circumstances.

Should they be involved in a fire at an underground site, the fire can be fought so long as it is clear that only the packages are involved. When there is a risk of the weapons bursting, through heat or explosion, fire-fighting should be suspended.

Full protective equipment should be worn in all operations of this nature. During fire-fighting operations personnel in close proximity to the fire should be sprayed with water to prevent their protective clothing becoming ignited.

The site can be entered at a reasonable interval, two or three days after the fire-ceases, and inspected by fully protected personnel.

Class 8. The conditions given in paragraph 7 (**k**) are applicable with the exception that fires involving explosives of Class 8 should not be fought once the fire has taken a hold and personnel should not be permitted to enter the underground site after the fire until it has been certified by qualified experts that it is safe to do so.

(d) No person, unless equipped with self-contained breathing apparatus, should enter an underground explosives storage site after an outbreak of fire until the area has been certified free from noxious gas.

9. FIRE-FIGHTING INVOLVING EXPLOSIVES IN TRANSIT

(a) By sea

The strict and detailed control of the carriage of explosives in ships and in harbour areas makes it certain that the fire officer can always ascertain the quantity and location in the ship of explosives which may be threatened in a ship fire. The fire classification of the explosives will also be available to him, and thus the behaviour which is to be expected of them in fire. The correct fire-fighting technique and safety measures to be adopted are therefore a question of adapting the instructions given in paragraph 7 above to the particular circumstances of the fire.

The Inter-Departmental Explosives Storage and Transport Committee have also issued a booklet 'Notes for guidance on fire-fighting in harbour areas when Government explosives are conveyed', which allocates the responsibilities of the various officers involved (including the appropriate fire brigade officer) and modifies the Code of Safety Conditions referred to in paragraph 5 above to suit their application to ships.

(b) By train

The transport of explosives by this means is also strictly controlled under railway transport legislation as well as the *Explosives Act*, 1875. Broadly speaking, explosives so carried must be clearly marked and specially packed in specified types of wagons; may only be carried in goods trains, with the exception of only small quantities of specified types; and are to be subjected to a number of safety precautions.

Here again, the procedure to be adopted will vary according to the fire classification of the explosives carried. When the type of explosive combined with an early arrival at the scene of the outbreak make it practicable, a quick and resolute attack on the fire is the best means of averting all danger: if this is impossible or unsuccessful, every effort must be devoted, with the co-operation of railway officials, to getting the affected train or the trucks involved clear of inhabited areas, as far as possible consistent with the safety of fire and railway personnel.

It has been agreed with the British Railway Executive that if, for any reason, it is known that a trainload of explosives may be held up for 24 hours or more in a built-up area, the person in authority at that place must advise the chief officer of the local fire brigade so that he may be prepared to deal especially with any fire which might occur in the vicinity of the explosives.

(c) By road

Similar restrictions and safety measures to those mentioned above, but appropriate to road transport, govern the carriage of explosives by lorry. Loads of more than 45 kg require that an attendant in addition to the driver be present on the vehicle, and both attendant and driver are required to know the fire classifications of the explosives carried and to inform police and fire brigade accordingly in the event of fire occurring. (If the explosives are in Class 5 they will also warn of the necessity for 'anti-gas equipment', which as applied to the Fire Service, should be interpreted as meaning 'breathing apparatus', and of the need to fight the fire from upwind of the vehicle). Standing instructions also include driving the vehicle clear of a built-up area if it is reasonably possible before the fire reaches the packages or weapons.

Match manufacture and storage

1. MATERIALS USED

Most of these are fire hazards, in differing degrees: they include potassium chlorate, sulphur, paraffin, phosphorus and, in small quantities, many other compounds and substances. Few of the materials used are themselves explosive, though explosive mixtures could be formed by two or more of them coming together. The raw materials used for the matchsticks, cartons, boxes and the like are mostly highly flammable.

2. PROCESSES

Many of these are similar to those employed in explosives manufacture and include grinding, mixing, sifting, melting and dipping. The same care is exercised to maintain the purity and correct constitution of mixtures, correct humidity and temperature, etc. More use is made of large-scale and complex machinery than in explosives manufacture, since segregation of different degrees of fire risks is possible and much of the manufacturing work involves comparatively little hazard.

3. CAUSES OF FIRE AND PROTECTIVE MEASURES

The incidence of serious fires in match manufacture is low. A number of fires occur, partly due to process, partly to the human element and to other causes: unlike the situation in explosives manufacture, match factories frequently contain flammable material in benches, partitions, floors, doors and structural woodwork, and thus the initial outbreak usually leads to a normal fire in somewhat flammable surroundings. Because of this and the nature of the materials used, it is characteristic of such fires that spread will be extremely rapid, perhaps explosively violent in the initial stages.

Safety for workpeople primarily relies on means of escape, though first-aid fire-fighting appliances and fixed installations such as sprinklers are widely provided.

4. FIRE-FIGHTING

In most parts of the factory the fire-fighting technique required will be normal to the character of the building. Breathing apparatus will be essential in most parts of the plant, and very violent burning with rapid spread may be expected. Since a degree of separation between buildings or parts of the building is usually provided, a first objective will normally be to confine the fire within the compartment involved and prevent the spread of fire and heat to other areas of high hazard. Care should be taken as far as possible to avoid the indiscriminate use of high-velocity jets so that the explosive materials are disturbed as little as possible. Some materials such as metallic powders, on which water should not be used, may be found here and there but usually in small quantities only: the fire officer should be guided by the management as to the presence of large quantities of these or other materials requiring special extinguishing agents. In the majority of the substances likely to be found in quantity, however, water is the best extinguishing medium. The special and dangerous properties of phosphorus (q.v.) should be borne in mind.

Part 6c, Chapter 45, Section 3

Fat and wax fires

NATURE AND PROPERTIES OF FATS AND WAXES

FAT FIRES IN KITCHENS

1. **Causes and types of fires**
2. **Fire-fighting technique:**
   (a) Small fires
   (b) Larger fires

FAT FIRES IN MANUFACTURE AND STORAGE

1. **Solid fat fires**
2. **Liquid fat fires, contained**
3. **Liquid fat fires, flowing**
4. **Wax fires**

INDUSTRIAL PROCESSES AND HAZARDS

1. **The production of fat**
   (a) General
   (b) Refining
   (c) Hydrogenation
   (d) General hazards
2. **Fats splitting**
   (a) Construction
   (b) Process
   (c) Principal hazards
3. **Soap manufacture**
4. **Margarine manufacture**

Nature and properties of fats and waxes

CHEMICALLY, the substances called fats are the glycerides of fatty acids. They embrace (1) edible vegetable fats such as cacao and other butters, (2) animal fats such as butter, lard, tallow and wool fat, and (3) synthetic substances such as margarine and glycerine, the latter being obtained by the hydrolysis of oils and fats. The dividing line separating vegetable oils and vegetable fats is an arbitrary one, since they are similar in nature and properties, but for purposes of this Section the animal and vegetable oils are treated separately in Section 8.

Waxes comprise animal waxes (beeswax, and spermaceti from the whale); vegetable waxes such as carnauba wax; and mineral wax or paraffin wax. They are used for polishes and electrical insulation among other purposes, and in all important respects they may be regarded as having the same properties as fats.

Fats and waxes are composed of carbon, hydrogen and oxygen. They are used in a wide variety of industries; the principal manufactures in which fats form the bulk of the raw materials are margarine, soap, polishes; these industries will be described in later paragraphs of this Section.

Fats present a hazard to the fireman more because of their behaviour when burning than of their flammability. At normal temperatures they can only be ignited by the prolonged application of a source of ignition; the flash points of most fats are from 240°C (464°F) upwards and the ignition temperatures considerably higher. Once melted, fats have most of the characteristics of flammable liquids in that they will run freely and thus cause a very rapid fire spread; they burn, moreover, with intense heat and give off large quantities of heavy black smoke. If water reaches them after burning is established at a high temperature, the reaction is violent and may be dangerous: in most circumstances foam is the most satisfactory extinguishing method, though small fat fires can be dealt with quickly and efficiently by other means. (Methods of extinguishment are dealt with separately in later paragraphs). All fats are lighter than, and are not miscible with water.

Fats also present a hazard additional to those associated with flammable liquid fires: there are risks of explosion when large quantities of fat are involved in fire at high temperatures. One reason for this is that, because the surface of a fire involving fats which are still partly solid is not level, as a liquid fire would be, the pattern of distribution of the flammable vapour is irregular, and may lead to pockets of explosive mixture being formed and eventually exploded. Another reason is that in some circumstances the fat may still retain incorporated in it some of the organic tissues which it contained in its raw state, and this in turn may contain moisture; when the moisture reaches the temperature at which it turns into steam, the effect is explosive, though in these circumstances only minor explosions are to be expected.

Although, as has been said, fats are not easily ignited at ordinary temperatures, they have some hazardous properties even when cool, since they are capable of producing spontaneous combustion in organic materials. The principle behind this is the property possessed by fats and oils of absorbing oxygen at a high rate when in contact with fibrous materials, the absorption of oxygen being a reaction which automatically produces heat. If the situation is such that the heat cannot escape (and particularly if additional heat is already present in the materials or their surroundings) a steady and sometimes rapid rise in temperature will occur. This, in time, will lead to the emission of flammable vapour from the fat, and, with the possible access of extra oxygen by some means, a fire may start; this fire, although only smouldering in its early stages, may eventually break out fully. This accounts for numerous outbreaks in which fatty, oily or greasy rags, greasy mop-heads, etc., have been the initiators by spontaneous combustion.

Waxes are in themselves a somewhat lower fire hazard than fats. They melt at temperatures between 125° and 160°F (51° and 71°C), giving off flammable vapours, but the vapours are only in dangerous quantities in the case of mineral waxes. However, it should be noted that flammable solvents almost always play a part in the treatment of waxes, so that the wax itself will be rendered more highly flammable because of mixture or association with the solvents.

Some wax polishes, etc., are produced by emulsification with water, and this has the reverse effect: water-based waxes are of considerably less fire hazard than natural waxes.

Fat fires in kitchens

1. CAUSES AND TYPES OF FIRES

A very large number of fires is reported every year as starting on domestic and commercial cooking stoves, about 1,400 of them occurred in fish-frying ranges in 1967 though these may be in oils equally as in fats. This number is not surprising, since cooking fat can quite easily be brought to above its ignition temperature by heating in an open pan; when this is so, it only requires the flash from a few drops of fat coming into contact with the heating flame or element to produce a fire in the pan itself. Alternatively, if the pan or container is left on a high heat for long enough without attention, the fat will eventually reach its ignition temperature and even without the aid of a flash-back from the heating source a fire may easily start.

The worst risks of such a fire, when occurring in a pan on a domestic cooker, are that it may light curtains or cloths, etc., near the stove, or, more commonly, that the cook will attempt to carry the pan across the room to the door or the window. This is a most dangerous practice, often leading to the spilling of burning fat on to the floor or even on to the clothes of the person concerned: the unexpected backward rush of heat and smoke resulting from a forward movement of the pan is frequently enough to startle the holder and lead to serious results.

In fish-frying kitchens and those of the larger catering establishments the use of cooking fat is usually confined to the frying range (Fig. 6). In this are provided anything from one to six rectangular

Fig. 6. An electric fish-frying range with four compartments and hood above.

sink-like recesses provided with separate lids; the lids are either hinged at the back or close with a swinging movement on pivots at each side. The pans themselves are most commonly heated either by gas or electricity, and in both cases there will be a local control on the appliance itself and a master cock or switch elsewhere. A dwindling number of fish-frying ranges are still heated by solid fuel. The only course available when it is desired to cut off the source of heat in these cases is to draw the fires. Over the range is a hood to catch heat, smoke and fumes, and this in turn is connected to a system of trunking leading to the open air. The draught in the trunking is either natural or, in the larger installations, fan-assisted. The most common cause of the numerous fires in frying ranges is overheating of the fat, allowing the ignition point to be reached and the gases to ignite. It is not the general rule for a thermostatic control to be fitted to the gas or electricity supply, and if a relatively small quantity of fat is left for a long time on full heat without much air movement to cause a loss of heat, the ignition temperature can eventually be reached. If this occurs when the kitchen is attended, the cook can usually turn off the supply of heat and close the lid at an early stage and the fire is then certain to be more or less starved except in unlucky circumstances. If there is no one present, the fire is very likely to get into the ducting system, and this presents problems and is likely to lead to extensive damage.

2. FIRE-FIGHTING TECHNIQUE

(a) Small fires

The special properties of burning fat, already referred to, as regards its reaction to water should be borne in mind, namely that drops of water which sink into the fat will instantly be converted into steam with explosive effect, and that the fat will float on the surface of any large quantities of water which are not immediately vaporised. This means that the use of water should not be considered for liquid fat fires in most circumstances, and, indeed, that it is in many cases dangerous as well as ineffective.

Two effects are desirable in any method used for the extinction of fat fires: first and most essential is the cutting off of contact between the surface of the hot fat and the air above, since without this contact burning cannot continue. If this can be done permanently by a physical barrier, no other effect is necessary, and extinguishment will be complete and permanent. If, however, the hot layer of mixed air and fat vapour above the surface is simply displaced by some less solid means, such as by the rush of inert gas from an extinguisher, for example, a second effect may be necessary in that some cooling of the heated fat must take place. Otherwise, the inert gas may rise through convection or be dispelled by draughts and there will be a danger of re-ignition of the vapours.

This means, in practice, that a small fat fire in a pan or on a cooking stove is best extinguished by a damp cloth laid without violence on the burning surface: the dampness is somewhat important, since most dry cloths will not form an air seal and a damp one will, and it also helps to ensure that the cloth is not ignited. An asbestos or woollen cloth, even if not damp, is likely to be equally effective. If, of course, the fire is in a pan and it is possible to cover this with its own or any airtight lid, or a pastry-board, etc., this will be completely effective also. (The various types of hand fire extinguisher which are suitable for use on fat fires would also unquestionably be effective, but it is not suggested that an appliance is necessary for such a small fire as a single pan of fat.) Extinguishment of the fire must be accompanied, of course, by cutting off the heat supply and removing the pan from the source of heat.

(b) Larger fires

In any instance, however, where more than one pan might be involved, or any substantial quantity of burning fat might be spilt, such as in a larger domestic kitchen or that of a school or institution, it is important that an extinguishing appliance be available, and it is essential that this should be of a suitable type. It has been almost universally accepted for many years that the most suitable type is the foam extinguisher, and to recommend this appliance would always be correct; the foam extinguisher is very well adapted for use on fat fires, which are the principal likelihood in any kitchen, and it is safe, reliable and easy for the inexperienced to use. The proper technique to aim at is not to discharge the stream of foam directly at or into the burning mass (see Book 3 of the *Manual,* Chapter 4, 'Foam extinguishers'), since any foam which penetrates below the burning surface tends to be melted ('broken down'), and may in any case splash the burning liquid about. Instead, a vertical surface just above the burning area should be chosen at which to direct the stream, so that it falls relatively gently on to the burning surface and spreads across it.

Foam is not the only extinguishing medium which can be successfully applied to a fat fire in a kitchen, though it is probable that for the non-professional fire fighter it is the safest and most reliable. Both carbon dioxide and dry powder extinguishers are capable of very rapid and efficient extinguishment of liquid fires, including fat. It is also claimed on their behalf that they make less mess than foam; while this is true in many circumstances it is not of great importance in cooking fat fires, since it is unlikely that the fat will remain fit for use whatever method of fire-fighting is used. There is some slight risk in the use of carbon dioxide or dry powder appliances by inexperienced people on this type of fire, partly because these appliances must be directed not too close nor yet too distant from the fire, and partly because of the risk of re-ignition of the fat or continued burning if extinguishment is not at first complete. If this occurs after the appliance has been completely discharged (and this may be with some types in the short time of four seconds) the operator is unable to continue his attack. A nine litre foam extinguisher, on the other hand, discharges its contents in up to 90 seconds.

Other methods of a less orthodox kind are sometimes used for attacking cooking fat fires, by fish friers especially; one of them consists of throwing soda crystals into the fat receptacle and closing the lid. There is a sound foundation for this practice, since both steam and carbon dioxide are liberated, and it would probably have good effects even without the help of the closed lid. But it is not practicable as a general recommendation in all circumstances, and such methods are best left to those who are well versed in their use.

Once a fat fire has made its way into the exhaust ducting over the cooker it is no longer as a fat fire that it presents difficulties but rather as a duct fire. The duct may, it is true, be loaded with deposits of solid fat, and these may spread the fire quickly and be difficult of access: there will be no great mass of heated fat, however, and cooling the burning material sufficiently to produce extinguishment will be quickly effected if the fire can be reached. A hose-reel jet or water spray is the best means to apply; foam is not well adapted for this sort of work, and carbon dioxide or dry powder, while they might be immediately effective, might equally fail through inability to cool the ducting sufficiently to prevent re-ignition. Fires in fume ducts are dealt with fully in Book 11 of the *Manual*, Chapter 5.

Fat fires in manufacture and storage

Details of the hazards to be expected in various industries using fats on a large scale are given in later paragraphs in this Section. Some general principles can be stated, however, on the technique of fire-fighting required when fats are encountered in factories, warehouses, shops, etc., as follows.

Fat fires larger than those found in kitchens will usually take one of three forms: the fat may be solid, such as tallow in bags, or it may be liquid, in vats for example, or it may be liquid and flowing.

1. SOLID FAT FIRES

In theory, of course, there can be no such thing as a solid fat fire, since before the fat can ignite it must be at an elevated temperature and at that temperature it will no longer be solid. In practice, however, it is frequently possible to encounter a mass of fat solids, such as slabs of margarine or bagged tallow, whose exterior surface is burning because hot enough to melt and give off flammable vapour, but whose main bulk yet remains solid. This gives all the characteristics of a solid fat fire, and it can usually be dealt with quickly and effectively in its early stages at least by the use of water sprays, or even jets. The quantity of fat which will have melted at this stage is relatively small, and the character of this type of fire is such that the use of water will have no very violent consequences, while the cooling will produce rapid extinction. If, as is often the case, the fats involved are only a part of a mixed goods storage or are packed in containers of flammable material, the use of water on the fats may well be operationally unavoidable in any case.

When a solid fat fire has been in progress for some time, however, the technique required may be somewhat different. Since the production of liquid fat from the solid by melting will be more rapid than its reduction by burning, there will be a steady increase in the amount of free liquid fat, and thereby a greater necessity for an attack by foam branches to be made. The use of high expansion foam could also be effective especially assisting in the reduction of smoke and heat.

All fat fires of substantial size are fierce, very hot and accompanied by very bad smoke conditions including the characteristic sooty lumps in the smoke. Breathing apparatus should be worn and frequent reliefs should be arranged.

2. LIQUID FAT FIRES, CONTAINED

A liquid fat fire in a vessel or container is a comparatively simple matter, resembling exactly a flammable liquid fire and effectively dealt with by the application of a foam blanket. Care should be taken to use the proper technique in the foam application but no specially high rate of application will be needed, nor need any special type of foam be used.

When a vat of liquid fat is encountered among miscellaneous storage of other substances for which the proper extinguishing agent is water, circumstances will dictate whether it is worth getting a foam branch to work or whether the water which is used for the surrounding storage can be used on the fat without risk or likelihood of failure. Generally speaking, where the vat or container of fat is not exceptionally large and has not been raised to a very high temperature by prolonged burning of the surroundings, water sprays will be successful and will not be accompanied by risk to fire-fighting personnel except at close quarters. Care should be taken, however, not to permit water to enter any container in sufficient quantity to cause it to overflow.

There may, equally, be circumstances in which the use of dry powder as an extinguishing agent would be effective on liquid burning fat, but the risk is stronger with this type of agent that re-ignition may occur, since the cooling effect is relatively slight.

If fat is involved in fire when packed in closed containers such as tins or jars, these may be expected to burst at some time after reaching temperatures of about 450°F (232°C). To prevent this, it is essential that such storages should be cooled with water sprays and the surrounding fire quickly reduced.

3. LIQUID FAT FIRES, FLOWING

Here again, the effect of flowing liquid fat alight is a similar problem to any other flammable liquid fire, the main difference being that liquid fat will solidify or at least become more viscous when substantial cooling has been effected, and will thus flow less readily. The treatment required is the application of a stiff foam covering or, in appropriate circumstances, high expansion foam, though, as before, water sprays or dry powder may be used if they are already in use on the surroundings and if there is no risk of re-ignition of the fat. At the same time, steps must be taken to estimate the direction of flow and ensure that the fat does not reach some point at which it could further feed the fire.

4. WAX FIRES

The remarks made in the paragraphs above as to fat fires may be taken as applying equally to wax fires, which require the same treatment and show the same characteristics and dangers.

Industrial processes and hazards

1. PRODUCTION OF FATS

(a) General

The methods used vary according to origin of the raw material. Edible vegetable fats, such as cacao butter and also butter itself, are produced by solidifying, either by emulsification or hydrolysis (see later) of vegetable oils, which have in turn been extracted from oil seeds by crushing and other methods (see Section 8, 'Fires in Oils'). Animal fats such as lard, tallow and wool fat are produced by direct extraction (rendering) or by solvents in combination with pressure. Finally, glycerine is produced by distillation and margarine by emulsification or hydrogenation.

In the extraction of animal fats, animal tissues are chopped and rendered, i.e. heated in open or closed pans with or without water until the fat is released and flows off. The residues (after the removal of oil) are mixed with offal fats of low grade material and autoclaved to give brown greases for soap. Among the hazards in fat extraction, tending to produce flammable vapours, are overheating and 'ebullition' (boiling) attended by fat decomposition. The solvents employed in fat extraction under pressure may include ether, benzole and carbon disulphide.

Ideally, separation of the raw material, the solvent and the fat should be effected as a single train of operations within the apparatus, none of which should be open to the air; in modern processes of fat extraction, the tendency is towards continuous-production machinery of the totally enclosed type. Somewhat high temperatures are employed in the conversion of fat into fatty acids and glycerine, (600°F or 315°C) in the distillation of fatty acids (palmitic acid, 662°F or 350°C) and in distilling glycerine (392°—590°F or 200°C—310°C). The addition of rosin to soap lyes is particularly dangerous due to the possibilities of violent reaction.

(b) Refining

The object of fat refining is to remove the following:—(1) suspended matter, (2) colloidally dispersed mucilage, (3) colouring matter, (4) free acid and (5) odoriferous members.

Suspended matter is removed by settling, centrifuging or filter pressing. (2) is removed by hydration, *i.e.*, by warming with water to cause the colloids to coagulate, or treating with superheated steam at high temperature. Linseed oil is sometimes treated with sulphuric acid. Removal of free acid is achieved by agitating the product with a weak caustic soda solution which neutralises fatty acids.

Modern development is towards a continuous or semi-continuous process: the sequence is—heater—agitator—neutralisation—centrifuging. In the United States the Clayton process employs a sodium carbonate solution.

(c) Hydrogenation

This is a process of converting liquid to solid fats. The oil is agitated with hydrogen at 302°—392°F (150°—200°C) in the presence of a nickel catalyst. Variations in temperature and pressure affect the process considerably. Hydrogenated whale oil is often used in the cheaper margarines.

(d) General hazards

In general, the hazards associated with the industrial processing of fats are as follows:

(i) Explosion or fire due to the ignition of flammable vapour from fat rendering, distillation, etc.
(ii) Explosion or fire due to the ignition of flammable solvents.
(iii) Explosion or fire due to incompatible storages of substances, either in the raw material or finished product stage.
(iv) Outbreaks of fire due to spontaneous combustion (see page 31).

2. FATS SPLITTING

At these works the fatty acids and glycerine are obtained from fats or vegetable oils.

(a) Construction

Large stills, erected on steelwork well above ground level, are the principal feature of those works.

(b) Process

Raw materials arrive at the works as animal fats (*e.g.* tallow) or vegetable oils (*e.g.* rapeseed oil). The contents are removed from the drums or barrels in a pit, and, if necessary, are converted into a liquid by steam. This liquid is then pumped from the pit to storage tanks.

Fat can be split in a number of ways, but the most usual process is as follows:

The fat passes from the tanks to steam heated hydrolysers where the material is treated with live steam at approximately 27·5 bar and is kept agitated by means of paddle gear. The fat is broken down into its constituents, chiefly glycerol, olein and stearin. The glycerol is held in solution in the water, and is drawn off for concentration and purification, after which it is run into barrels which may hold up to 500 kilograms. The oleic acid and the stearin go to separators which pass the two constituents and to filter pressers whence the stearin is removed in blocks. The stearin is again distilled and pressed, and is then placed in sacks for despatch. The heavier products are run into furnace-heated stills whence tar is drawn off. This tar resembles bitumen in appearance and properties, and is extensively used in the manufacture of roofing felts.

The oleic acid passes forward for purification and is then run into drums.

(c) Principal hazards

(i) The stills, which contain quantities of hot liquid.

(ii) The raw and finished materials, all of which are flammable, though none presents a high fire risk. (The glycerine is mixed with water).

3. SOAP MANUFACTURE

The manufacture of soap is often carried on in the same premises as margarine (*q.v.*), as the same oils may be used in both processes. Such commodities as brilliantine, lipstick and bath salts may also be made, in which case raw materials other than those mentioned below may be present. A diagrammatic representation of the process of soap manufacture by hydrolysis is shown in Plate 9.

(a) Construction

Most large soap works are of modern construction. Shed-type buildings are unusual.

(b) Raw materials

The basis of soap manufacture is the combination of a fatty acid with an alkali. Mineral oils lack the necessary properties, so that the oils used are animal and vegetable ones. The fatty acids obtained are stearic acid (stearin), palmitic acid (palmitin), and oleic acid (olein or red oil). The most commonly used animal oils are tallow or bone grease, whale oil or herring oil, the rendering being done on the premises or at a fat splitter's. The vegetable oils found to be most suitable are palm oil, palm kernel oil, coconut oil, olive oil (now largely replaced by groundnut oil), linseed oil, soya bean oil and cottonseed oil. At soap works situated near the sea or a canal it is often the practice to prepare these oils from the seeds, rather than buy ready-prepared oils. For the preparation process see SEED MILLS. Other raw materials commonly found are:

(i) Alkalis—caustic soda and caustic potash, usually manufactured on the premises from sodium or potassium carbonate, unless the soap works is near a chemical works, in which case it is brought in rail or road tank cars.

(ii) Tallow and rosin in drums and casks. Rosin, though not a fat but the residue from the distillation of turpentine from the pine tree, will form soap when treated with alkalis.

(iii) Chlorosulphonic acid (in some works), sulphuric acid (for glycerine treatment), and hydrochloric acid. (See Section 16).

(iv) Glycerine manufactured as a by-product, and stored in large tanks and in drums.

(v) Petroleum naphtha or other spirits are added to soap manufactured for laundries.

(c) Special hazards and fire risks

(i) The greatest hazard is the presence of large quantities of oils and derivatives, frequently at high temperatures, sometimes under pressure, in vats, tanks and kettles. Oils and melted fats, in addition to burning, spread fire by flowing from floor to floor and from one building to another.

(ii) The hydrogenation process used for hardening the fats. Hydrogen was made in older plants by the action of steam on red-hot iron, giving iron oxide and hydrogen, but in modern plants it is produced electrolytically (*see* Plate 6). In either case the hydrogen storage tanks are a hazard, and an explosion risk arises from possible leaks in hydrogen piping. Activated fuller's earth treated with sulphuric acid is used for bleaching. The spent earth can take fire, and is best dealt with by removal. The process is shown diagrammatically in Plate 8.

(iii) The manufacture of soap powders by grinding, as in certain circumstances explosive mixtures of soap and carbonate dust may be found.

(iv) Soap will burn at 350°—508°F. (177°—264°C.) approximately, but it is far less combustible than its constituents, due to a certain water content. It will of course melt at lower temperatures, and tends to block drains, etc.

(v) Timber flooring, where it exists, becomes oil-soaked in most parts of the building and burns freely.

(vi) Drying machines, especially if they are of timber manufacture.

(vii) Quantities of fibreboard, cartons and sacking.

(d) Fixed fire protection

Soap works are usually sprinklered.

4. MARGARINE MANUFACTURE

Margarine making is sometimes associated with the manufacture of other commodities such as soap, in which vegetable and fish oils are used; in other cases it may be carried on in a separate factory. The construction of the margarine factory is usually modern, and follows the general lines of soap works.

(a) Process

The principal raw materials used are vegetable and fish oils; hardened oils such as whale, groundnut or palm kernel oil, and also soft groundnut and cottonseed oil being most common. Hardening is carried out by hydrogenation, a process described under '*Soap Manufacture*'.

The oils are stored in special rooms in open tanks fitted with steam heating coils, and are stirred to prevent them settling. Thence they are led to mixing machines where various substances such as salt, milk, water, boracic acid powder and vitamins are added. The resultant mixture is cooled in a brine type refrigeration plant and is then taken to the packaging machines.

(b) Special risks

Generally speaking margarine factories do not present a high fire risk. The oil stores and the refrigeration plant are the most likely to cause difficulties in a fire.

(c) Fire-fighting

(i) Stocks of margarine will have to be re-processed if they are affected by water.

(ii) Smoke will seriously taint margarine. As much ventilation as possible should, therefore, be provided.

Part 6C, Chapter 45, Section 4

Fires in fibrous materials

CHARACTERISTICS OF DIFFERENT FIBRES

1. **Animal fibres:**
   (a) Wool
   (b) Silk
   (c) Hair

2. **Vegetable fibres:**
   (a) Cotton
   (b) Jute
   (c) Hemp
   (d) Flax
   (e) Kapok
   (f) Rayon

3. **Synthetic fibres**

INDUSTRIAL PROCESSES INVOLVING FIBRES

1. **Worsted and woollen manufacture and storage**
   (a) Construction of mills
   (b) Processes—worsted
   (c) Processes—woollens
   (d) Processes—shoddy and mungo
   (e) Storage
   (f) Causes of fire
   (g) General fire problems
   (h) Fire protection
   (i) Fire-fighting in mills
   (j) Fire-fighting in warehouses

2. **Cotton spinning, weaving and knitting**
   (a) Construction of mills
   (b) Processes
   (c) Causes of fire
   (d) General fire problems
   (e) Fixed fire protection
   (f) Fire-fighting:
      (i) Spinning mills
      (ii) Weaving sheds
      (iii) Cotton knitting mills
      (iv) Dyeing and finishing works
      (v) Warehouses, ships' holds & other storages

3. **Rayon manufacture**
   (a) Viscose rayon:
      (i) Process
      (ii) Special hazards
      (iii) Fire-fighting
   (b) Cupra-ammonium rayon: as above
   (c) Cellulose acetate rayon: as above

4. **Jute spinning and weaving**
   (a) Construction of mills and warehouses
   (b) Processes and risks
   (c) Fixed fire protection
   (d) Fire-fighting in jute warehouses

5. **Flax manufacture**
   (a) Construction of buildings
   (b) Processes
   (c) Causes of fire, fire protection and fire-fighting

6. **Lace manufacture**
   (a) Construction of factories
   (b) Processes and risks
   (c) Fire-fighting

INDUSTRIAL PROCESSES INVOLVING FIBRES—*continued*

THE fire hazards of the fibre-using industries arise much more from the peculiarities of the industry than those of the fibres. The majority of the materials used at any stage in these industries are what can be called 'ordinary carbonaceous materials': although they have certain characteristics when burning which are peculiar to themselves, they present few unforeseen dangers to the fireman and are easily extinguished, in theory at least, by the use of water. In this, they are in sharp contrast with many other industrial materials which need special extinguishing techniques or offer serious personal hazards, or both, when involved in fire.

Nevertheless, prior knowledge of the processes used in these industries is just as important to the fireman as in any other: without it, he will be severely handicapped and his chances of taking the right action in time to prevent serious loss are much reduced. Moreover, the amount of prior knowledge needed is considerable; the factories concerned are of many and varied types and each contains a big variety of processes. Generally speaking, the factors which contribute most to fire risk in the fibre-using industries are the age and style of the buildings, the amount and complexity of the machinery and the great extent to which the materials are exposed in their most flammable forms.

In few other industries is the construction of the buildings so intimately bound up with the processes. They are of a type of construction peculiarly their own, often incorporating much wooden construction, and their walls are usually riddled with openings, chutes, trap doors, shafts, ducts and the like, any or all of which are liable to make an unexpected contribution to the spread of fire. Their rooms are frequently very large and undivided, and filled with machinery to an extent which makes access and movement difficult, and the wooden floors on which the machinery stands are commonly oil-soaked and thus readily flammable. Finally, most of the processes used are such that large quantities of raw and woven fibrous materials are continually exposed, sometimes as woven fabrics, sometimes as threads, either wound on bobbins or stretched with air circulating freely around them, and sometimes as fibres, fly or fluff, or dust. In these latter cases, where there is free percolation of air through the mass of material, it is obvious that the possibilities of ignition and rapid spread of fire through the exposed fibre are high.

Common causes of fire in these industries, aside from the same evidences of human carelessness as are found in other industries, are friction and overheating in machinery bearings, foreign bodies drawn into the machinery, electrical faults, overheating of electrical and other appliances, and spontaneous heating in raw materials. In many cases any of these causes can give rise to a severe fire within a very short time, which, if not detected and tackled quickly, can have disastrous results. The fire at the Eastwood Mills, Keighley (*see* Plate 19), in February, 1956, occurred in daylight during working hours, as a result, it is thought, of the use of a blow-lamp in plumbing work; although the fire was detected immediately it started, the spread was so rapid that, in combination with other circumstances, it resulted in the loss of eight lives.

One major change which is taking place in these industries, and which closely affects the fire risks, is the progressive introduction of synthetic fibres in addition to, and in substitution for, the natural fibres traditionally used. Since all truly synthetic fibres are to all intents and purposes non-combustible, their introduction will have the effect of reducing the fire risks of virtually all industries in which fibres are used. Because the situation is changing rapidly, it is not possible in the following paragraphs to make mention of each instance in which synthetic fibres are being substituted for natural ones but the reader may take it for granted that the process is widespread and steadily increasing. Nevertheless, the time has not yet arrived when any general reduction in fire hazards in the industries concerned is perceptible.

Characteristics of different fibres

Fibres provide the materials for a wide range of industries, such as textiles, brush and rope manufacture, and may be animal, vegetable or non-organic in origin.

1. ANIMAL FIBRES

Animal fibres such as wool and silk are solid, without internal air spaces, and will thus burst into open flame only when considerable heat has been generated. They burn slowly, but the subsequent loss of heat is very gradual; they also tend to smoulder.

(a) Wool—*burning characteristics*

(i) Wool is difficult to ignite and tends to smoulder and char rather than to burn freely, unless it is subjected to considerable external heat. It will, however, contribute towards a fierce fire.

(ii) Charring wool forms a sticky, black, tar-like substance.

(iii) Burning wool gives off dense, greyish-brown smoke.

(iv) Spontaneous heating and ignition can occur, usually where compressed wool contains oxidising oil. Moisture alone is believed not to cause heating.

(v) Wool absorbs a large amount of water.

(vi) Unprocessed wool is impregnated with a natural grease. This has a sufficiently high flash point not to present a fire risk in itself, but it saturates the storage buildings and contributes to the flammability of the premises.

(b) Silk—*burning characteristics*

(i) Silk is the least dangerous fibre. It is difficult to ignite, and burns sluggishly, combustion usually being supported by an external source of heat.

(ii) Burning silk produces a large amount of spongy charcoal intermixed with ash, which will continue to glow or burn only in a strong draught.

(iii) Once alight, silk retains heat longer than any other fibre.

(iv) Burning silk emits quantities of thin grey smoke, somewhat acrid in character.

(v) Silk attracts both liquid and solid substances. This tendency is exploited in the practice of loading the fibre with metallic and other substances which give the fabric its well-known sound ('*scroop*').

The capacity of silk to absorb water is very great, and that of loaded silk even greater.

(vi) Spontaneous heating and combustion are possible in wetted silk, especially when it is loaded.

(vii) No external evidence that a bale of silk has ignited may appear for some time, until the fire burns through to the outside.

(viii) A great deal of fly settles in rooms used for the preparatory silk processes, but this is not likely to flash, although it will burn.

(c) Hair (horsehair, animal hair)—*burning characteristics*

(i) Animal hair is largely composed of keratin, a hard protein of which sulphur is a constituent. Hair has properties similar to those of other animal fibres, but in the mass will normally be found to burn quickly and readily, since it is not easily packed into dense masses and tends to have more air spaces between the fibres than, for instance, wool or silk.

(ii) On burning, hair produces whitish, sulphurous smoke and leaves a sticky, black deposit.

(iii) Water easily penetrates a compressed mass of hair and is not absorbed by the fibres themselves.

(iv) Due to the maintenance of air spaces and to the ventilating effect this gives, spontaneous heating is not known.

2. VEGETABLE FIBRES

Vegetable fibres consist largely of cellulose. They include cotton, jute, hemp, flax and kapok, all of which have a cellular structure with internal air spaces and a greatly increased surface area. Surface absorption of oxygen may take place.

Rayon, classed as vegetable because of its cellulose origin, is made artificially and differs in being solid, without internal air spaces.

All partially burnt vegetable fibres are liable to present a fire risk (even after they have been extinguished), since:

(i) before ignition temperature is reached the fibres are liable to decompose, liberating carbon which may be of a pyrophoric nature, and

(ii) should the fibres have been in contact with a drying oil, or other oxidisable material, spontaneous heating may result. Such half-burnt fibres should therefore always be removed from the building.

Vegetable fibres, and to a lesser extent animal ones, suffer risk from contact with sulphuric acid (which causes charring) and nitric acid (which produces nitro-compounds) both processes causing a rise in temperature which may easily reach a dangerous degree.

BLEACHING PROCESS IN SOAP AND MARGARINE MANUFACTURE

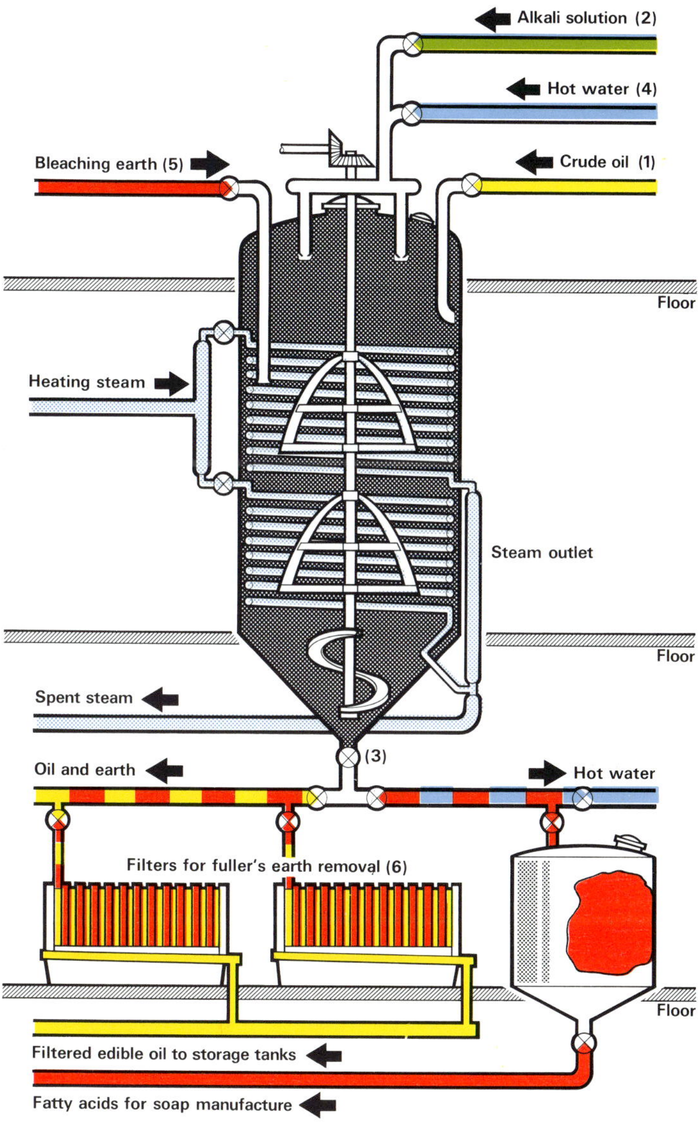

Plate 8. The crude oil (1) is heated by steam. It is neutralised by being sprayed with an alkali caustic solution (2). The fatty acid solution (3) settles to the bottom and is drawn off. The residual oil is washed with hot water (4) and the water is run off. Fuller's earth (5) is mixed with the oil to bleach it. The mixture is drawn off through filters (6), which retain the spent fuller's earth and pass the clear neutral oil. (Section 3)

CONTINUOUS PRODUCTION OF SOAP AND GLYCERINE BY HYDROLYSIS

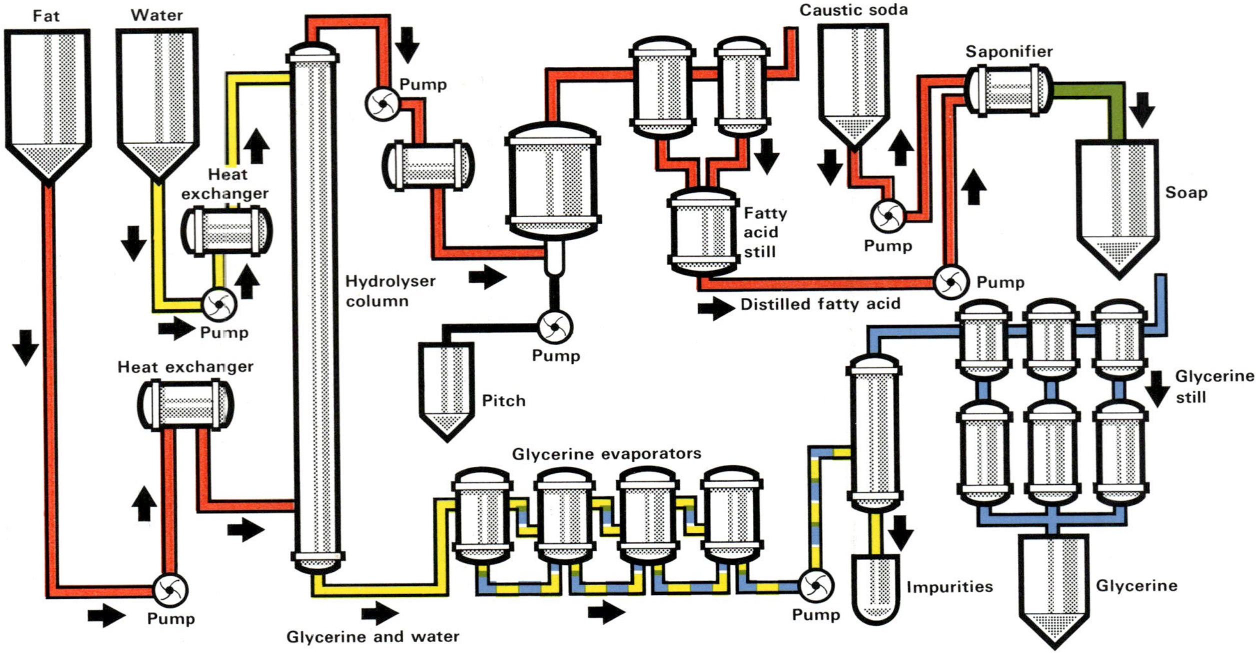

Plate 9. Heated fat is treated with live steam in a hydrolyser, causing the fat to break up into its constituents, chiefly glycerol, olein and stearin. The glycerol and water solution is drawn off at the bottom and the glycerol is purified. The olein and stearin pass to separators and filters which remove the heavy pitch and pass the fatty acids. These are purified and treated with a caustic soda solution to produce soap. (Section 3)

PROCESSES IN A SEED-CRUSHING MILL

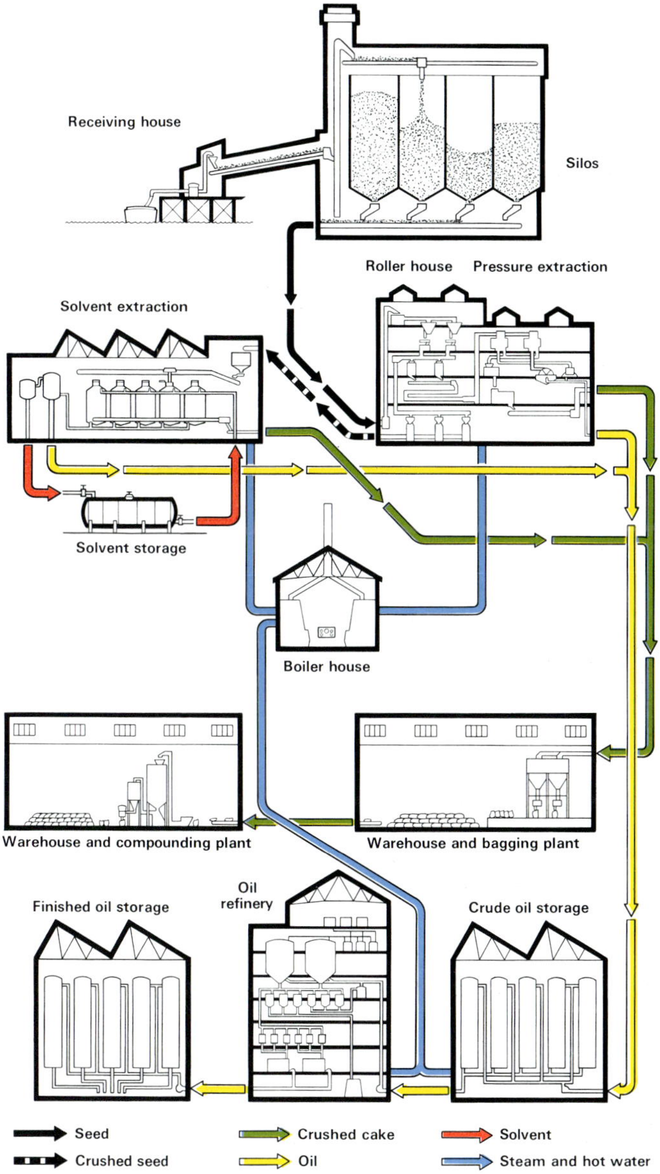

Plate 10. After removal of the linters (not shown) the cottonseed is crushed under pressure ("expressing") in the roller house and oil is driven out. Some of the residue is further treated in the solvent extraction plant where more oil is obtained. Crude oil is piped to storage tanks before passing to the refinery. The cottonseed residue is conveyed to a warehouse for bagging, then passes to the compounding plant where meal and vitamins are added to produce cattle cake. (Section 8)

PRODUCTION OF SUGAR FROM SUGAR BEET

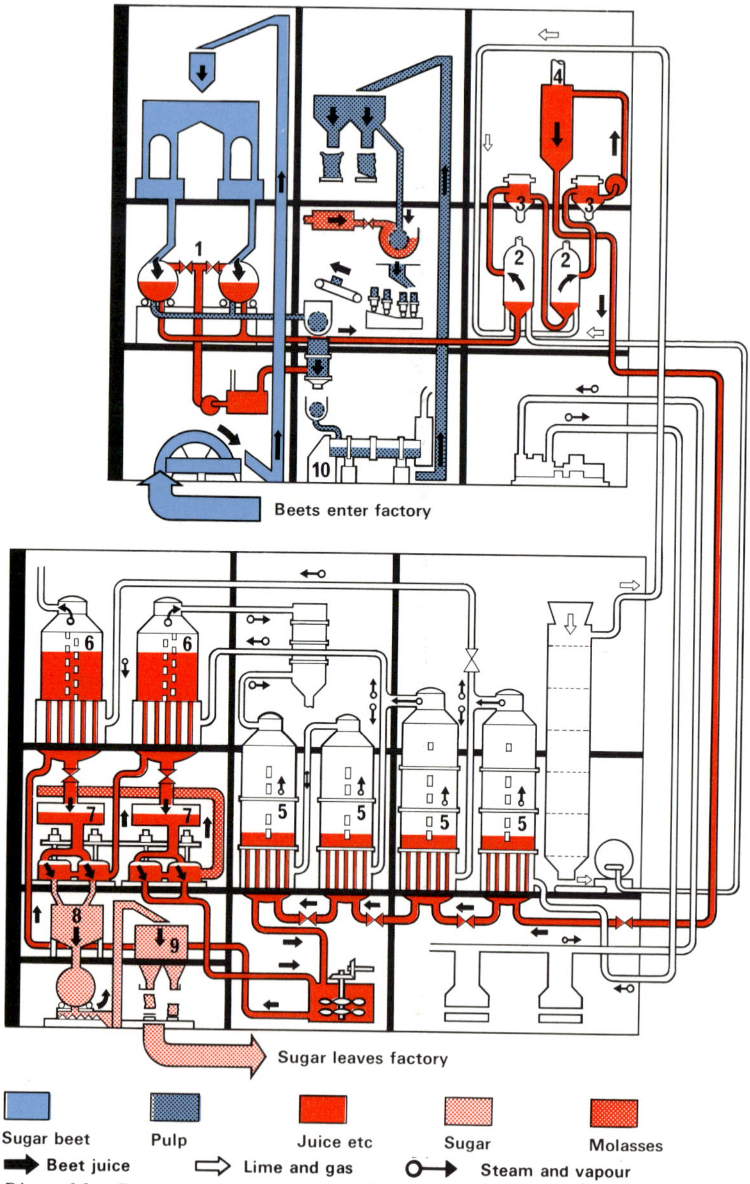

Plate 11. Raw juice is extracted from the washed and sliced beet in diffusers (1), then mixed with quicklime. CO_2 is passed through it in carbonators (2) and impurities are filtered off (3). Treatment with sulphur dioxide (4) follows and after more filtering the thin juice passes to evaporators (5), the vapour from one vessel being used to heat the next. Further boiling and concentration under vacuum (6) produces crystals. Centrifugal machines (7) spin off the liquid syrup, leaving the crystals. The granulator (8) dries and cools the sugar ready for packeting (9). Spent slices of beet, after pressing and drying (10) are made up in block form for animal feeding. (Section 15)

(a) Cotton

Cotton is the fibre which covers the seeds of the cotton plant, from which it is removed before it is shipped to this country. From the seeds are subsequently obtained the short fibres known as cotton linters. (For the use of linters *see* 'Rayon' (page 51) and Section 10, 'Plastics'; also Section 8, 'Seed-crushing Mills').

Burning characteristics

(i) Cotton consists of almost pure cellulose and ignites and burns easily and quickly. The lighter its form the greater the danger; this is especially the case when it is found as fluff or fly in cotton mills, for then it is capable of spreading fire at high speed (flashing).

(ii) Burning cotton gives off dense smoke.

(iii) Spontaneous heating will not be caused by moisture, but the risks affecting vegetable fibres in general apply (see above).

(b) Jute

Jute is obtained from a tall reedy plant (GENUS CORCHORUS) grown principally in Pakistan and India. It is prepared and shipped in machine-pressed bales about 1200×450×400mm weighing about 180 kilograms. Jute is used for the manufacture of hessian which is used for bags, wrapping cloths, and the backs of linoleum and carpets; for cords and ropes; and for paper. Dundee is the principal manufacturing centre in the United Kingdom.

Burning characteristics

(i) Jute ignites and burns easily, especially in the form of fluff and dust.

(ii) Burning jute gives off large amounts of dense acrid smoke.

(iii) Jute is liable to spontaneous heating but this will not normally result in combustion unless it has been contaminated by an unsaturated vegetable oil.

(iv) Jute absorbs water readily and swells: a bale will take in about its own weight of water and in doing so will assume about 1½ times its original girth.

(c) Hemp

Hemp is an imported fibre. The main varieties are sisal (nearly white) and manila (various shades of brown), the latter having the longer fibre. It is used for making ropes, twine, sackcloth, sailcloth, types of canvas, and by upholsterers for stuffing. Fine grades of manila hemp may also be used in making muslin and cigarette papers.

Sisal is usually found in steel-banded bales weighing about 250 kg, manila comes in 125 kilogram bales, wrapped in matting or banded with cane.

Burning characteristics

(i) In most ways its behaviour is similar to that of jute, including its capacity to absorb water and so to increase the loading of the building.

(ii) Hemp burns very fiercely and with great heat. Fire will at first flash along the lint on the outside of the bales (*see* Fig. 7), and then

Fig. 7. Sketch showing how fire will flash along the lint on the outside of stacked hemp bales.

burn inwards at fair speed. Great care should be taken not to ventilate a building until it is certain that a fire in baled hemp is under control.

Hemp dust lying under machines and on beams is liable to flash very rapidly.

(iii) The smoke of hemp has a narcotic effect, and breathing apparatus should be used at a hemp fire in a confined space.

(iv) Self-heating can occur rapidly should hemp be wetted, especially when pressure is being applied. This seldom reaches the point of ignition, but it may lead to the destruction of the fibre. Hemp which has been affected by water in the course of a fire should, therefore, be removed and exposed to the air as soon as possible.

(d) Flax

Flax is a plant which grows some 1·2m high, the fibre being obtained from the stem.

Burning characteristics

(i) The raw material burns fiercely. Processed flax burns easily, but not so violently as raw flax since it does not contain seeds.

(ii) Burning flax gives off heavy smoke.

(iii) Wet flax tends to heat and rot.

(iv) Flax absorbs large quantities of water, but although it does swell, it does not do so to the same extent as jute.

(e) Kapok

Kapok is a soft, light downy, oily fibre, short and silky, obtained from the silk-cotton tree. In appearance it is softer and creamier than cotton. It is used for filling quilts and pillows, good-class upholstery, and lifebelts.

Burning characteristics

(i) Being light and fluffy, kapok ignites easily and fire spreads through it with great speed.

(ii) Kapok gives off dense, acrid smoke when burning.

(iii) It is not considered liable to spontaneous heating.

(iv) It is capable of absorbing 36 times its own weight in water.

(v) Kapok being very light, a jet will scatter it widely.

(vi) It easily retains sparks. When turning over articles stuffed with kapok, or kapok found in bulk, great care should be taken to ensure that all sparks have been extinguished.

(f) Rayon

Rayon (originally called artificial silk) is reconstituted from cellulosic material (normally wood pulp or cotton linters). Its uses are now very wide.

The three principal types of rayon are viscose, cellulose acetate and cupra-ammonium rayon. Their burning characteristics vary enormously.

(i) *Viscose rayon*

(1) The thread burns as readily as cotton. When spun as staple fibre it is even more flammable because of fluff on the thread.

(2) It burns with less smoke than cotton.

(3) Wetted viscose rayon will heat spontaneously in three to seven days, but ignition resulting from this heating is not a likely possibility.

(ii) *Cupra-ammonium rayon*

(1) This rayon is flammable, but does not burn quite so readily as viscose rayon since it does not fluff to the same extent.

(2) It burns with more smoke than viscose rayon and with some fumes.

(3) Authorities do not quote this material as a spontaneous heating or ignition hazard.

(iii) *Cellulose acetate rayon*

(1) Acetate yarn is not easily set alight and requires a continuous source of external heat to keep it alight.

(2) On heating, it decomposes into hard, glue-like lumps which can cause serious burns.

(3) Neither yarn, nor woven fabric is likely to heat spontaneously even if damp.

3. SYNTHETIC FIBRES

Certain minerals can be formed into threads and woven into fabric, e.g. asbestos. Synthetic fibres such as '*nylon*' and '*terylene*' are outside the scope of this section and are dealt with in Section 10, 'Fires involving Plastics'.

Industrial processes involving fibres

1. WORSTED AND WOOLLEN MANUFACTURE AND STORAGE

In worsted manufacture only the longer, pure, new wool fibres are used. They are laid parallel to produce a strong, firm, smooth, even yarn and hence cloth. Sometimes specially prepared rayon or synthetic fibre is blended with the wool but the product cannot then be correctly described as pure worsted.

In woollen manufacture, the bulk of the new wool used consists of the shorter fibres which have been rejected from worsted manufacture and are called '*noils*'. Except in the case of genuine native tweeds and the better quality overcoats and blankets the wool, is mixed with other fibres such as cotton, rayon or hair and with reclaimed wool fibre (*shoddy* and *mungo*). In woollen yarn the individual fibres are not laid parallel.

(a) Construction of mills

In both trades, mills for spinning and earlier processes are usually substantially built, averaging four to five storeys. Common features are stone slab roofs, timber floors in timber beams and solid, stone staircases with the addition, in woollen mills, of timber stairways between departments. Woollen mills tend to be older and to have more extensions and additions than worsted mills. In both cases certain machinery may be accommodated in separate single-storeyed sheds.

In worsted mills the upper floors are used for sorting and storage, while mechanical processes are carried out on the ground floor. In woollen mills, however, the machinery is on all floors and is very congested. Gangways between machines are narrow and there is little clearance below overhead belting and pulleys.

Weaving mills in both trades consist partly of two or three storeyed premises; partly of single-storeyed 'sheds' with roof lights. The weaving is normally carried on in the 'sheds' and not on upper floors because of the heavy vibration of the looms.

(b) Processes—worsted

The four main processes—combing, spinning, weaving and dyeing and finishing are generally carried out separately by independent firms.

When the wool has been cleaned, dried, carded (see *Cotton*) and straightened it is combed, the short fibres (the '*noils*') being rejected while the long fibres are laid into a sliver (the '*top*'). Vegetable oil or olein is added to the wool as a lubricant. (A by-product of wool-combing is the wool fat, which is separated from the washing liquor by sulphuric acid). In the spinning mill the worsted '*top*' is drawn out and twisted into a thread in a 'spinning frame', several threads being then twisted in a 'twisting' (or 'doubling) frame' to make the finished '*yarn*'.

The weaving mills make the yarn into cloth on looms.

The dyeing and finishing processes are carried out in premises known locally as 'dye works'. The dyed cloth is dried in a tentering machine and the surface is then '*raised*' (by minute wire pins on revolving rollers), '*cropped*' (by knifed rollers) and '*singed*'.

(c) Processes—woollens

(i) *Willeying*. In this the fibres, *e.g.* noils, shoddy or mungo, cotton, hair and rayon fibre, all sprinkled with process oil, are mixed and separated from dirt and waste matter in the willey. This is a large, steel-toothed roller which revolves in an enclosed drum in opposition to a number of smaller toothed rollers. The willeys normally occupy separate small stone-floored sheds.

(ii) *Scribbling and carding*. The blend is fed into the hopper of the scribbler whence it goes through a series of opening-out and mixing processes, being finally wound on to bobbins as coarse threads, Scribblers and cards resemble cotton carding machines (*see* 'Cotton') and are in the main mill building.

(iii) *Carbonising* is employed in various branches of the wool textile industry to free wool or pure silk from material of vegetable origin. The process employed can either be wet, involving immersion of the material in a sulphuric acid solution, or dry, involving exposure to hydrochloric acid gas. In either case a fire in the plant produces dangerous fumes.

(iv) *Mule spinning* is often carried out in the same room as the scribbling and carding processes. The mule has two frames, about 20m long, one fixed and the other moving on a carriage running on rails fixed to the floor (*see* Plate 15).

(v) *Weaving and finishing* are as in worsted.

(d) Processes—shoddy and mungo (The heavy woollen trade)

These are the basic materials used in the production of cheap woollen cloth, and shoddy is also used extensively throughout the woollen trade. Shoddy and mungo are the two qualities of wool reclaimed from rags. The rags are sorted, washed, oiled and passed through a rag-pulling, or rag-grinding machine (*see* Plate 13), known locally as a '*devil*' (similar to a *willey*). The fibre is thence blown by fan through a duct into a '*blowpipe*'. The latter is usually a fire-resisting compartment within the same enclosure as the machine. Each combined machine and blowhole is usually in a separate compartment within a stone-floored shed.

'*Garnetting*' is another process for reducing certain types of rags and waste. Unlike the rag-pulling machine, the garnetter is fully open to the room.

(e) Storage

The warehouses, substantially constructed with load-bearing walls, tend to have five or more floors and may be built on two levels. Usual features are stone-slabbed or blue slated roofs with match-board interior linings, timber floors on timber beams, or unprotected steel girders supported by unprotected iron columns, numerous trapdoors and on each floor timber two-piece loopholes. (These loopholes cannot normally be used for access, being secured by a stout timber crossbaulk). Large timber bins for holding loose wool are on various floors.

Imported wool bales weigh from 45 – 135 kilograms and are covered with jute and banded with iron.

(f) Causes of fire

The fire risks are much greater in woollen mills and in the heavy woollen trade than in worsted mills for two reasons:

(i) Foreign bodies which may strike sparks in the machines are often present in the reclaimed fibres used in the woollen processes and

(ii) the inclusion of cellulosic fibres with the wool is much more common in these processes.

Fires are most likely to arise in the following processes:

Worsted—carding. A spark from the machine may ignite fly on the roof fittings and round the machines. The risk is not very high unless rayon fibre is mixed with the wool.

Spinning. Friction in the machinery can cause a fire which may be spread by the oil impregnated floor and by wicker yarn skeps and wooden bobbins.

Woollens—willeying. Foreign bodies in the reclaimed fibre may cause a fire, which can be serious if cellulosic fibres are present, but this seldom spreads to the whole mill as the willeys are normally in a separate part.

Scribbling and carding. Fire is more likely to start at the hopper or first roller of the scribbler than at any other place in the mill. It can spread rapidly through the large accumulations of fly, part of which is cellulosic fibre, and which is found on all raised objects such as lighting pendants, shafting supports, roof members, etc. The oil-soaked floors increase the risk of a severe fire. The fire risk is most serious where blends containing cellulosic materials are handled in old buildings.

Mule spinning. Friction can cause fire in the fly accumulated at the base of the spindles. Flame is liable to travel along the length of the machine inside the long timber case in which the spindle bases are sunk.

Worsted and woollens—Drying the cloth. There is some risk in all driers such as the tentering machine.

Raising and cropping. Friction on the warm cloth generates static electricity and a spark may ignite the fine fibrous waste. This is less likely if the product is pure wool.

Singeing. A breakdown in the machinery can cause fire.

Shoddy and mungo (the heavy woollen trade)—Rag pulling presents the greatest fire risk of all wool textile operations, due to sparks from buttons, fasteners and other metal objects. The usual practice is to keep the machine running so that the burning fibre is blown into the blow hole, where it can be dealt with by hand appliances. Fire may become serious if it spreads beyond the machine or if it breaks out after processing has finished, due to accumulated heat in the rollers.

Garnetting rags. Since these machines are open, the quantity of fly presents a risk, especially as cellulosic fibres are often handled.

(g) General fire problems in worsted and woollen mills and in storage warehouses

(i) *Floors* in all mills and warehouses are saturated with grease.

(ii) *Smoke* is likely to be dense in all wool fires. It irritates the respiratory organs and eyes. Slow-burning fires, such as those in bales, produce most smoke.

(iii) *Spontaneous ignition* is particularly likely in baled, oiled fibres which may be left for some time awaiting processing. The risk is greater in the woollen than in the worsted trade.

(iv) *Fluff* in quantity collects in both worsted and woollen mills and is more dangerous in the latter.

(v) *Heavy machinery* in the mills sets up vibration. This is worst in woollen mills, where the machinery may be on all floors and very congested.

(vi) *Sulphuric acid and hydrochloric acid gas.* These are used in carbonising plants in the wool textile industry, and sulphuric acid is used at an early stage of the worsted processes for separating wool grease from the wool washing liquor. In the event of fire there is a danger from fumes, and also from the acid and gas if these escape.

(h) Fire protection in mills and warehouses

Fire-resisting doors are far from general, and each main mill building can usually be rated as one risk. Most woollen mills are sprinklered and have wet risers and landing hydrants and also soda-acid extinguishers. The fire brigade advocates the installation of hose reels in rooms where willeying, rag pulling, garnetting or scribbling are carried on. Bulk supplies of static water for fire-fighting are readily to hand in the majority of mills, either in the mill dam or from nearby waterways. The standard of fire precautions in different mills varies enormously. Many maintain works fire brigades.

Town warehouses usually have open timber staircases and no

fixed fire-fighting installations. Warehouses attached to mills usually have substantial enclosed stone staircases, and are often protected by sprinklers and wet risers with landing hydrants.

(j) Fire-fighting in woollen and worsted mills

Fires in the machines tend to spread and linger. Fluff accumulates, and the space beneath the machines and the floor, beams, etc. should always be extensively searched for smouldering remains.

The machines are often separated by only very narrow gangways. In the initial stages of a fire the management of the mill may be reluctant to stop the machinery, but it is most dangerous for firemen to enter and search a smoke-filled room while the belt-driven machinery is still working, and the fire brigade officer should insist on the machines being stopped before fire-fighting commences.

Dense smoke is encountered, but, as the material is more finely divided, the fire burns more fiercely and with far more flame in the initial stages than a fire in a warehouse. Fire in bales of wool or rags, however, burns more slowly. Once such a fire has been extinguished, the affected bales should be taken outside the building and spread out where they can do no harm, as it is essential that no half-burnt fibrous matter should be left on the premises to smoulder.

Fire may spread from room to room or floor to floor by means of the vertical and horizontal shafting holes and rope races (*see* Fig. 8). Separating walls are often not extended above the roof. Fire-resisting doors are sometimes left open to facilitate movement from one room to another, and it should therefore not necessarily be assumed that they are closed. Attached to the mill, but in most cases separately constructed, is the tower staircase. This has stone steps and doors giving access to each floor.

The staircase is probably fitted with a wet rising main. The tower is usually a safe position from which to work. It is fairly remote from heat and smoke, and many cases are on record of floors and walls collapsing while the tower has remained intact.

The floors of the older mills are of wood, and in many cases are double floors with a cavity between them. Cutting away is, therefore, very difficult, particularly as so much machinery is bolted to the floor.

The hose reel jet is sufficient to deal with the majority of fires in worsted and woollen mills. For the remainder, the diffuser or hand-controlled branch is normally used.

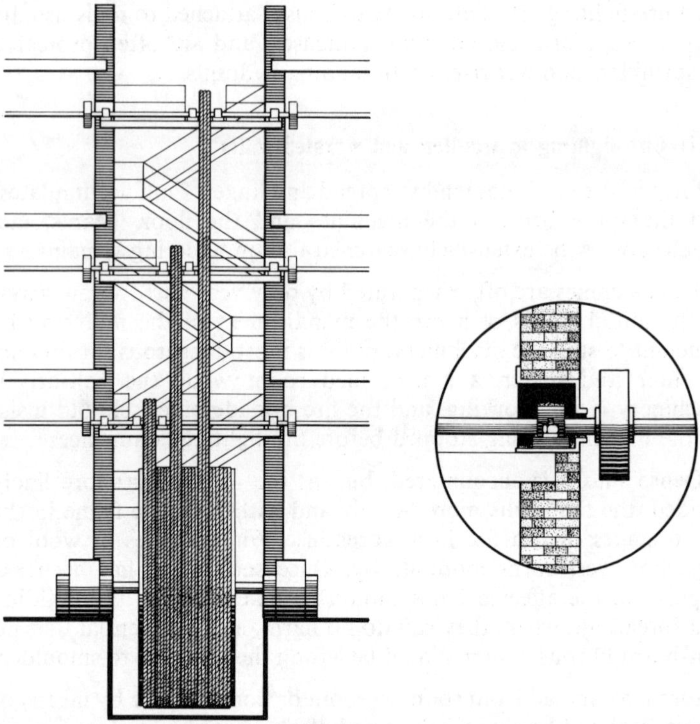

Fig. 8. Diagram of a rope race showing the driven shafts on each floor. The power is supplied by a steam engine (not shown). INSET: *Fire can spread rapidly through the holes carrying the shafting.*

Salvage plays an important part in fire-fighting operations at worsted or woollen mills, as a fire which occurs on the third floor, for example, may give rise to severe damage to machinery and materials on the lower floors. In particular, the carding, garnetting and wool combing machines should be sheeted to minimise damage by water, and water used in fire-fighting should, if necessary, be removed to prevent it over-loading the floor.

(k) Fire-fighting in warehouses

The dense smoke will necessitate breathing apparatus. Fire spreads quickly, flashing along the jute coverings of bales. It penetrates the bales themselves slowly. The building is liable to burn fiercely because of grease-impregnated timber.

Smoke pouring from a window or other opening may indicate the whereabouts of the fire. It should be remembered, however, that smoke from a fire on one of the lower floors or in the basement may rise to the floors above by means of the staircases or any trap doors (over the wool chutes in the floor) which may have been left open.

It may be necessary to break into the premises by forcing doors or windows to fight a fire when no staff are on the premises. Loading doors, however, are generally secured by a wooden or iron bar from inside and it is useless to attempt to force these doors from the outside.

Powerful lights are normally required to assist in the search inside the building. Firemen should watch for open trap-doors in the floors and unprotected stairways; each step should be tested when moving around.

A hose line or lines with hand-controlled branches should be taken into the building. When the fire is located, water should be turned on and the fire got under control before any attempt is made to ventilate the premises.

It is easier to locate the fire in a sprinklered building than in an unprotected building, as the sprinkler heads can be heard operating and the spray from them can be felt in the vicinity of the fire. Flame may not at first be visible as combustion may be taking place inside the stacks of bales; the latter may have to be pulled apart and removed to get to the seat of the fire.

Care should be exercised in the use of water as wool will absorb large quantities and the increase of weight may cause the floor to collapse.

Wool warehouses often adjoin each other in a continuous row, in which case they are generally divided by separating walls which do not extend above the roof. Precautions must be taken to ensure that a fire on the top floor does not spread by means of the roof timbers. A considerable salvage value is attached to baled wool, which can often be recovered from the bottom of a warehouse which has collapsed as the result of a fire.

2. COTTON SPINNING, WEAVING AND KNITTING

(a) Construction of mills

Spinning mills are normally of several storeys and with load-bearing walls. Features of modern mills (built since about 1900) are double-boarded wooden floors, laid on reinforced concrete and supported on unprotected cast iron columns; a flat reinforced concrete roof covered by a water seal between 150 and 300 millimetres deep; two or more enclosed stone staircases; and hoists

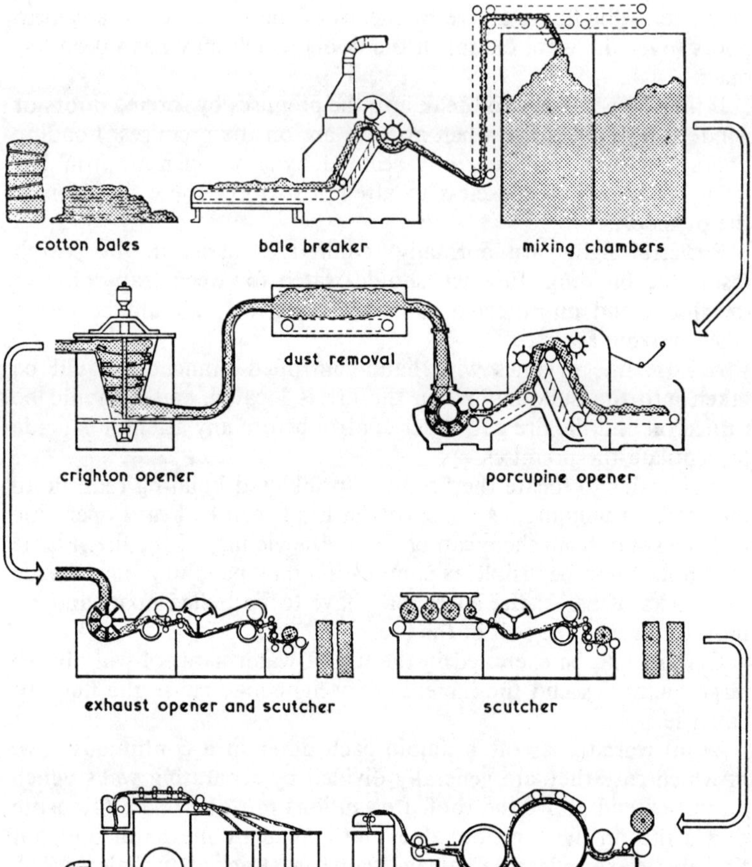

Fig. 9. The various stages in the process of cotton spinning. The final process may be ring spinning, as shown, or mule spinning (Fig. 11.)

communicating with the various floors by self-closing or sliding doors.

Older mills may have wooden floors laid on brick arches, supported by girders and unprotected cast iron columns; pitched slated roofs or roofs slated on close boarding; very large roof voids with a sawdust filling in the spaces between the joists; and lath and plaster ceilings (especially on the top floor).

Power is (i) from a steam engine, the rope race (Fig. 8) being at the centre or one end of the mill, and the machines being driven by belts off shafts; (an old mill may have an internal vertical shaft with bevel gearing on each floor, involving floor openings which can spread fire), or (ii) electricity, usual in ring spinning mills, where a central transformer and switch-gear room and a separate motor to each machine will normally be found.

(b) Processes

(i) *Spinning*. Cotton is fed through the bale-breaking machine (*see* Plate 12) into wood-enclosed bins called mixings. Thence it is passed by the hopper feeder into openers, which repeat the cleaning and opening process, finishing in the scutching machine. Fig. 9 shows, in diagrammatic form, the complete process.

Dust is extracted in these processes, and it passes through ducts to a dust chamber, a large room at the centre or end of the lowest floor. From this a dust tower rises the full height of the building or higher. It may have a wooden roof, but is more usually open, its floor being covered with a layer of mud in wet weather. The floor and walls of the dust chamber and tower are coated with fluff (*see* Fig. 10 and also Plate 16).

Among subsequent processes are carding, in which the cotton becomes a thin sliver, and spinning, either on ring frames (for coarser counts) or on mule frames (for finer counts). Carding machines (each $1 \cdot 85 \times 2 \cdot 75$m) are of great delicacy and value. They are close together and there may be 200 in a room. Spinning mules, in pairs facing each other, are 39·6m long with a central headstock or driving mechanism. The wheeled carriage containing 2,000 revolving spindles, travels backwards and forwards 1·6 metres on rails. It is of wood, or of metal with wood doors. At newer mills a dust extraction system may be fitted in the ring spinning room. An illustration of ring spinning is shown in Plate 14, and spinning mules in Plate 15.

Generally, the basement is used for the storage of raw cotton, waste and finished yarn, the ground and first floors for bale breaking, scutching and carding and the upper floors for intermediate processes and spinning.

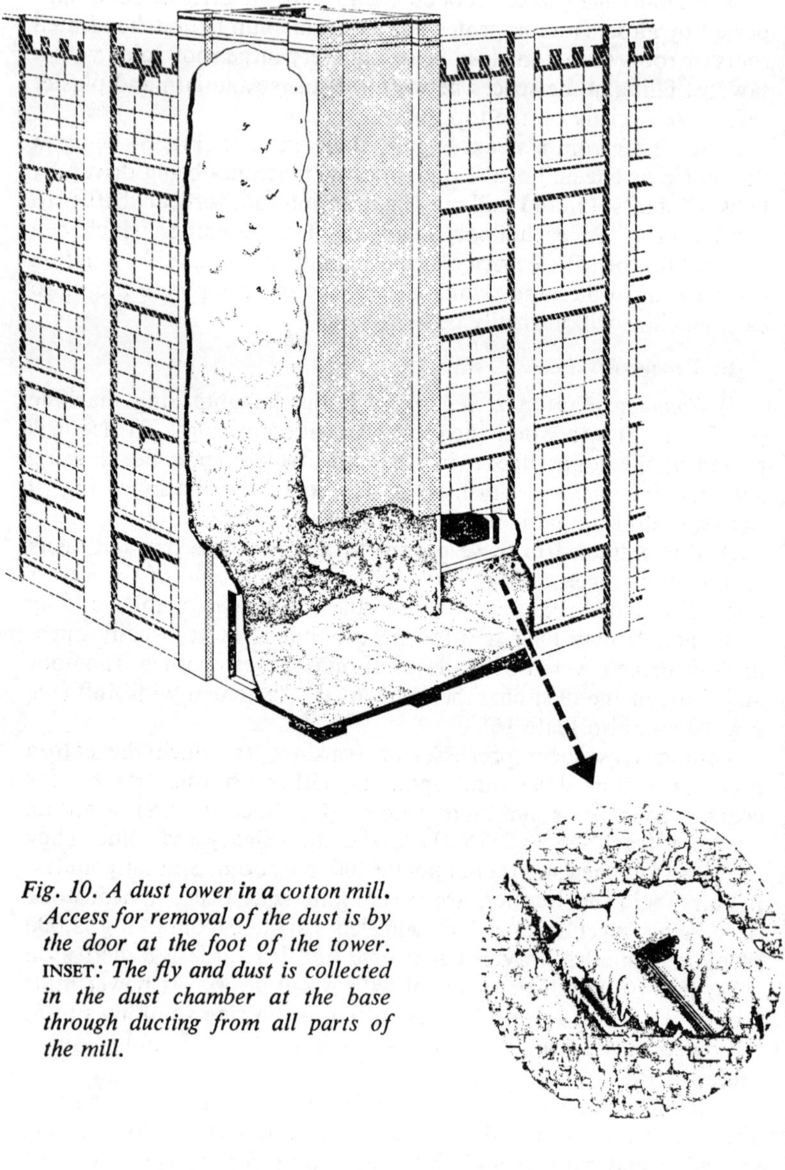

Fig. 10. A dust tower in a cotton mill. Access for removal of the dust is by the door at the foot of the tower. INSET: *The fly and dust is collected in the dust chamber at the base through ducting from all parts of the mill.*

(ii) *Weaving*. Some preliminary weaving processes may be carried out either in spinning or weaving mills.

Those of interest to the fireman are '*taping*' (or sizing), in which the thread is finished with size in large steam-heated machines, above which is usually a timber roof fitted with louvres; and '*drawing-in*' (*i.e.* drawing the ends of the warp threads through the healds). The healds, made of varnished threads on frames, are highly flammable. Those not in use are hung in rows from the ceiling or kept on racks.

Pitch-covered reeds used for pushing tight the weft threads on the loom are also highly flammable. (Reeds and healds are usually made in separate small factories).

(iii) *Cotton knitting mills*. In these mills the thread is knitted on knitting machines instead of being woven.

(iv) *Cotton finishing processes*. Razing, cord cutting, brushing and shearing machines are used. Some of these produce a great deal of fluff.

(v) *Storage*. Cotton is imported in highly compressed bales of approximately 180–320 kilograms weight, covered with jute sacking and banded with steel. Baled cotton is found in ships' holds, warehouses, open stores and cotton mill receiving rooms.

(c) Causes of fire

Fires in the machinery are very frequent in cotton mills. They may, for example, be caused in the earlier processes by foreign bodies or in the later by overheated bearings. The great majority of these outbreaks are effectively extinguished by the operatives, who deal with them by means of hose reel jets in some processes, or buckets and brushes in others. Cotton burns so readily, however, that a fire can get out of hand and spread at great speed.

Machines in which fires are particularly common are:

(i) bale breakers and waste recovery machines;

(ii) the hopper feeder, from which it may spread to the mixing and thence into the scutching or blowing room.

(iii) the carding machines, where spread is rapid, and

(iv) the spinning mules, where it is likely to be carried the length of the machine by the threads, or inside the carriage by the strings.

Fires may be carried from the machines to the dust chambers by the ducts.

(d) General fire problems in cotton mills are:

(i) the oil-soaked floors,

(ii) the vast area of the rooms, unbroken by any form of fire division wall,

(iii) the congestion of the machinery, and

(iv) the large wall area taken up by wooden-framed windows, which facilitate the spread of fire from floor to floor.

(c) Fixed fire protection

Most mills have:

(i) Sprinklers, except in the basement and dust chambers. Sometimes there are none in the carding room due to possible water damage. The sprinklers are supplied from the town's main or from an elevated tank in the main tower. Sprinklers are seldom fitted in weaving mills because of the relatively low fire risk and the great damage likely to be done by water.

(ii) A dry or wet riser on the principal staircase, fitted with internal hydrants and sometimes supplied by a fire pump driven by steam from the main mill boiler, and thus available only when the mill is working. The pump may be fitted with deliveries to which hose lines can be connected. In the majority of cases, however, the riser is fed from the mains.

(iii) Hose reels, especially in the bale-breaking, blowing and mule rooms.

(iv) Self-closing doors, though these are seldom fire-resisting.

(v) A static water supply of 365,000 to 9,000,000 litres in the '*lodge*', the supply normally used for the steam engine.

(f) Fire-fighting

(i) *Spinning mills.* Cotton mill fires must be fought from inside the building with speed and persistence. The standard branches to use are the hand-controlled and the diffuser. Quite apart from its effect on the machines, water always damages cotton (although only to a limited extent when the cotton is in bale form). The nearer the cotton is to completion of processing the greater will be the damage by staining and shrinking.

Should a fire have occurred in a bale-breaking or similar machine, a search should be made for sparks or fire in the next machine in the process and in the dust chamber and ducts, which are fitted at intervals with inspection covers. The ducts can be cleared by starting the fans while a lookout is kept in the dust chamber. If the fire is in a bale-breaking machine and has spread to the mixings, efforts should first be made to separate one from the other using a hand-controlled branch, and only then to deal with the fire in the affected machine.

The amount of 'fluff' and 'fly' present will depend upon the type of wall and ceiling and upon the good housekeeping practised in the mill, but some quantity is inevitable and its capacity to spread fire by flashing should never be forgotten. Burning fluff on the ceiling is liable to fall upon flammable goods at a distance from the seat of the fire. When there is a cavity between floor and ceiling, fire can readily spread inside the cavity. Fluff settles in cavities of damaged ceilings. A further danger arises where the spaces between roof joists have been filled with sawdust. If this catches fire it can fall through the ceiling and spread the fire.

When a fire occurs in a spinning room, the smoke will usually be extremely thick and the seat difficult to locate. Breathing apparatus should normally be worn and the position of the fire must be found *before water is applied*. Where mule frames (of which there may be 12 or 15 pairs in one room) are involved, the position in which the frame has stopped should be ascertained, because the interior can, as a rule, be reached only from the side on which the operative works, and when two mules have stopped at the end of their travel close to one another there is inadequate room for the fireman to work (*see* Fig. 11). In such a case it is necessary to have the mule moved either by power or, if necessary, by hand, so that access can be gained to the gangway.

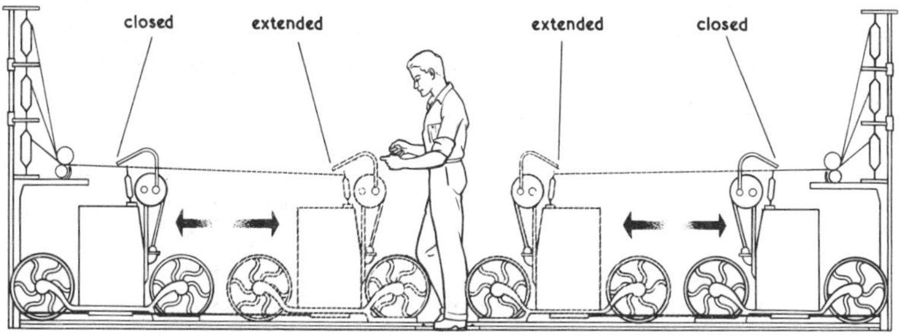

Fig. 11. Diagram showing a pair of mule spinning frames in the closed position, and the position of the carriages at the end of their travel. (See also Plate 15).

Fire usually starts near the centre of the mule and can spread in both directions in the interior so that it is necessary to operate from each end to prevent it spreading. The wooden doors must, of course, be removed or, if necessary, broken open to enable water to reach the interior. Care should be taken not to trip over the rails or fall over a truck carrying bobbins, as a fireman may be seriously injured by falling on to the mule frame.

The cotton mill must NOT be ventilated before the fire has been put out. If this is done, there is a serious risk of uncontrollable fires spreading to other storeys. The method of handling a fire in a cotton spinning mill should not be confused with that in a warehouse in which cotton is stored in bales.

Men should always be sent to the floor above the one on fire, as in some cases great heat will be generated there and in some cases will require a charged line for soaking the hot spot. In this event, it may be wise to project spray from a diffuser branch into the air for a short time to cool the atmosphere and to prevent windows cracking, but if this is done the greatest discretion should be used to avoid unnecessary damage. Immediate arrangements must also be made for carrying out salvage operations on the floors below. Particular care should be taken to protect carding machines and machines using metal rollers by means of sheeting, as they may be seriously damaged by water or steam if this is not done. If water has, in fact, affected the pins, the prompt application of French chalk may prevent damage. The management will advise on the placing of sheets, and often themselves keep stocks of sheets in bins. Machines which it is not possible to sheet should be wiped and oiled with a light grade oil as quickly as possible.

Fires in the rope race may occur through the ignition of waste at the bottom of the race or through hot bearings on one of the floors throwing a spark on to the cotton ropes. The engine can be stopped in approximately $2\frac{1}{2}$ minutes. During that time, however, the draught may cause the fire to spread, and the oily walls may possibly be affected.

Burning rope ends may be dipped in a bucket of water. Other parts of ropes which may be affected may be wetted with a wet rag or swabbed while the engine is slowly turned over. The sides of the race can be reached by the iron ladder running from top to bottom and also by doors communicating with the mill on some, if not on all, floors. In the case of a fire in the rope race, someone should be sent to the other side of the wall on each floor to make sure that the fire does not spread to the mill as a result of oil on the bearings. The top of the race is sprinklered.

A fire in the dust chamber is not necessarily dangerous. The draught should be shut off by stopping the machines. The fire should then be extinguished with diffuser branches through the access doors at the base of the chamber. Fire on the sides of the dust tower can be dealt with from the top of the tower by means of a diffuser branch or with a small jet. Crevices and loose bricks where fluff is likely to lodge should be sought out. After a fire in the dust tower has been extinguished, the roof and roof void should be searched for sparks that may have penetrated under the eaves or other likely places.

In most mills there is fitted at intervals on each floor a glass fronted box containing an emergency push button which, when pressed, stops the main engine. This is for use in case of accident or other emergency and can be used when necessary by a fire brigade officer. Where an electric drive is employed each machine can, of course, be stopped individually.

When the fire has been extinguished, the premises should be thoroughly searched for sparks, and a fireman should be left at the mill until machines are running again. If the fire has occurred outside working hours, a guard should be maintained throughout the night.

All cotton which has been involved in fire should be taken out of the mill in metal bins and thoroughly turned over.

(ii) *Weaving sheds.* Water will not greatly damage the looms, which, if wet, can be cleaned, but smoke and water will seriously affect the fabric being woven, and the material upon the beam. Extensive salvage operations should, therefore, be promptly put in hand.

The principal smoke and water damage can occur in the warehouse (always on the ground floor because of the loading bay) where stocks of finished goods are kept. Salvage here is of the greatest importance. It is often more satisfactory to remove goods from the warehouse than to cover them with sheets.

Fluff is usually concentrated under each loom, but though likely to flash it is not generally the cause of a serious fire.

Where there is a three-storied processing block (where fire is most likely to start), the wall facing the weaving shed is likely to collapse owing to the expansion of the steel joists as a result of the heat. For this reason men should not be stationed on the roof of the weaving shed nearer than the height of the processing block above the weaving shed.

A danger in weaving sheds is the extensive roof glazing which may damage the looms and fabric if it falls in.

The particular dangers arising in the case of the jacquard loom, which is occasionally used in cotton weaving, are dealt with under the heading '*Lace*'.

(iii) *Cotton knitting mills.* The machines themselves are, like carding machines, easily damaged by water and should be protected by sheets. The remarks with regard to the protection of finished stocks at weaving mills apply also in the case of cotton knitting mills.

(iv) *Dyeing and finishing works.* Dyeing processes are wet, and are not, therefore, hazardous in themselves. Fires do, however, sometimes occur in dyeing machines. Stocks of materials should be protected. Sprinklers are usually installed in dyeing and finishing works. Bleaching powder and various dye stuffs will be found on the premises. Many of the finishing machines, especially razing, cord cutting and shearing machines, are very easily damaged by water, and should be covered with salvage sheets if there is any possibility of water reaching them.

(v) *Warehouses, ships' holds and other storages.* Cotton bales are not liable to spontaneous heating as a result of moisture, though if damp is present some heat is generated. Water applied externally will not penetrate more than 75 – 100 millimetres into the interior of a bale, except after considerable time. Swelling of the bale due to absorption of water is therefore slight.

Fire in a cotton bale may start with a spark and may smoulder for some hours or even days with little external signs of burning. The smouldering very often works slowly into the interior of a highly compressed bale and travels between the layers of cotton. Much of the air needed to support combustion is available in the interstices of the fibres, and cotton will continue to burn even when the oxygen content of the air falls below 9 per cent. Once the fire obtains a firm hold it gives off great heat and dense clouds of light yellowish grey smoke. When this stage is reached, the fire is likely to spread rapidly throughout the bales as the sacking and loose cotton burns off, after which the fire will tend to settle back at the point of origin, while smouldering continues in all the bales affected by the spread. On the other hand, depending on the source of ignition, a rapid 'flash' fire may be spread by the surface fibres of the jute and of any cotton which may be exposed. Such a fire can run with great speed throughout a stack of bales and, though it may occasionally burn itself out, is very likely at a number of isolated points to catch a bale and in due course penetrate to the interior. A bale will not burn completely unless it is subjected to very great heat or the bands are off. Surface fire must be extinguished and all affected bales must be removed.

In the case of ships' holds into which the bales are tightly wedged when the cargo is loaded, the swelling between the bands caused by the water makes the removal of the bale difficult, and it may be necessary to break up a bale in order to create a space in which to work. If so, one not affected by fire should be chosen.

Once the bales are outside the building or vessel, the hose reel nozzle or a diffuser branch should be used to soak their sides and ends as they are removed, or handfuls of wet cotton should be pressed into those parts where burning can be found (*see* Fig. 12). Each bale should then be left with a space all round it sufficient to give access to all four sides. Bales isolated in this way should remain in the care of the mill or other officials, but the fire brigade officer is

Fig. 12. Dealing with fire in cotton bales.

wise to leave a man on duty for a day to watch the bales and to arrange for frequent return visits for several further days. Normally a bale should not be broken open, for the loosened cotton will burn much more rapidly and, because of its lightness, may be carried on the wind and spread the fire; the possibility of the successful salvage of the cotton will also be considerably reduced. If the breaking of a bale cannot be avoided, however, buckets of water should be placed close at hand and the affected cotton dipped in the water as it is removed from the bale.

Owing to the heavy smoke and fumes which are generated, it is usually necessary to wear breathing apparatus at a fire involving bales of cotton stowed in a confined space.

3. RAYON MANUFACTURE

(a) Viscose rayon

In addition to manufacturing rayon thread, the viscose rayon industry produces transparent wrapping film of the cellophane type, and information given under this heading can be taken broadly as applying also to such film.

(i) *Process.* The following are the more important stages in the manufacture of rayon by the viscose process:

Wood pulp is steeped in a caustic soda solution in iron tanks. The product is alkali cellulose. The latter is ground in a modified dough mixer, during which process heat is developed. The ground material ('*crumbs*') is then packed in boxes and kept in store at a constant temperature for ripening.

The crumbs are mixed with carbon disulphide in rotating churns. The product of this is termed 'cellulose xanthate,' and the latter is then dissolved in dilute caustic soda to form viscose. This is transferred into large tanks in a temperature-controlled room for ripening.

Wrapping film is made by extruding the viscose through a long narrow orifice into a bath of acid (7 per cent sulphuric acid) after which it is washed in water and treated with bleaching agents in a succession of tanks. It is immersed in a softener such as glycerine and dried on rollers. Finally, it is coated with a thick layer of nitrocellulose lacquer and the solvents are dried off by hot air during which sulphurous gases are evolved.

Rayon yarn is made in the same way as wrapping film, but a fine die is used through which the viscose is extruded into the acid bath. After staying in gentle heat it is wound on to bobbins or left as staple fibre.

(ii) *Special hazards.* The following are the hazards most generally associated with the viscose process:

A. Stocks of wood pulp.

B. Alkali cellulose stores. The material burns rather more readily than pulp.

C. Caustic soda and acids. External sulphuric acid tanks will usually be found above ground.

D. Carbon disulphide. Underground external tanks of carbon disulphide connected by piping to the works are found. If necessary it is possible to drain the solvent in the piping back into the tank. If the pipework or the tanks are damaged, there is a risk of carbon disulphide entering the drains together with hot water, in which case an explosion may occur.

E. Sulphurous fumes (sulphur and sulphur dioxide).

F. Nitrocellulose lacquer and accompanying solvents. (*See* Section 9, *Paints*).

G. The drying room, especially if the electric current driving the exhaust fans should fail. The wood is like tinder, as in the drying rooms of most factories.

(iii) *Fire-fighting.* Viscose thread burns as readily as cotton, but with less smoke, and fire will burn only slowly into a bale of thread. When spun as staple fibre, its flammability is increased because of the presence of a certain amount of fluff on the thread. Fluff is also found on the spinning floors. Water will seriously damage the thread (or fabrics made from it) by staining; heat will damage it at temperatures over 170°F (77°C), while smoke produces marks which make it necessary to rebleach the yarn. Should a box of viscose thread become wetted at a fire it is liable to heat spontaneously in from three to seven days. The occupiers should therefore be advised to remove stock affected by water and to arrange for it to be segregated and dried. The machines which carry out the lacquer coating process are often protected by water spray, carbon dioxide or steam fixed installations.

Because of the possible presence of dangerous fumes, breathing apparatus should be used inside buildings even if danger is not immediately apparent.

(b) Cupra-ammonium rayon

Rayon yarn is produced in this process by dissolving cellulose obtained from cotton linters in a solution of copper sulphate and ammonia, and reconstituting it as a continuous cellulose filament. The processes are wet. The chief likelihood of fire lies, therefore, at the beginning of the process in the cotton linter storage, and at the end during the drying or storage of the yarn.

(i) *Process.* The raw material consists of cotton linters which are subjected to the following stages in the manufacture of the yarn:

The linters are steeped in caustic soda solution, boiled and bleached with sodium hypochlorite. Cellulose is mixed with a solution of ammonia and copper sulphate, and the resulting solution is extruded to form the yarn. The latter is run through a bath of weak (8 per cent) sulphuric acid to remove the copper, which is subsequently recovered. The yarn is wound on frames, washed to remove the acid, then dried and wound on to bobbins.

(ii) *Special hazards*

A. Cotton linters may be found in bales weighing from

180 – 320 kg or in smaller paper-covered packages containing linters already cleaned elsewhere. Bales are covered with jute and metal banded.

B. Large storage tanks of ammonia.

C. Stores of sulphuric acid and caustic soda, both of which are corrosive, and copper sulphate: this has no fire risk, though it is poisonous.

D. Cylinders of chlorine gas, from which sodium hypochlorite is made by treatment with caustic soda.

E. Drying machines. A hank of yarn may become detached from the frame while being dried and remain in the machine, becoming fastened round an axle, where it eventually catches fire because of the heat generated by friction.

(iii) *Fire-fighting*. Bales of cotton linters will burn in a similar manner to cotton bales, and a fire in them should be handled as already described for baled cotton. Men can be injured by the metal bands bursting through the expansion of the contents. Packages of processed linters are much more flammable, since they are not so highly compressed. It may be possible to use an overhead crane to move bales or packages for salvage or in order to find the fire.

Cupra-ammonium yarn does not fluff to the same extent as viscose, but burns with more smoke and some fumes; the yarn is discoloured by smoke and has its properties altered by much heat, whilst water will cause staining. The machinery of the spinning room, which includes hundreds of glass containers in the machines, is liable to great damage from a jet and, if it is necessary to get to work, a spray should be used. The process carried on is so wet, however, that fire is unlikely to break out; the only risk is that it will be involved in fire originating in another part of the works.

There will normally be no danger from ammonia unless the store is involved, but the smell of ammonia pervades the works and small quantities may be released. The possibility of large scale escape when the fire is near the store must be borne in mind. Should water contaminated with copper sulphate (from the settling tanks) be used for fire-fighting, pumps and all gear should be thoroughly washed out at the end of operations in order to avoid corrosion.

(c) Cellulose acetate rayon

In this case rayon yarn is produced from a liquid made by dissolving cotton linters in acetic acid and acetic anhydride.

(i) *Process*. The linters are steeped in caustic soda solution, boiled and bleached with sodium hypochlorite. After washing and drying, they are acetylated or mixed with acetic acid, acetic anhydride and sulphuric acid and dried. The product is then ground to form cellulose acetate powder, used not only for rayon but also for *plastics*. In the case of rayon, the powder is then dissolved in acetone to form a dope, which is passed through a filter press.

The dope is passed through spinning machines and spun to air (no acid being used at this stage as in the other processes), the acetone vapours from spinning are collected and recovered by means of charcoal, and a certain amount of groundnut oil is added. There are no drying rooms, the yarn being wound direct on to bobbins. Where weaving is carried on, snippings and spoiled yarn are returned to an oil recovery plant to obtain the groundnut oil and to a mixer to break down the yarn to be used.

(ii) *Special hazards*

A. For linters see cupra-ammonium rayon.

B. Stores of acetic acid, acetic anhydride, sulphuric acid, sulphuryl chloride, chlorine, acetone and methylated spirits in acetic acid manufacture, and benzene in oil extraction, and the presence of these liquids and vapours throughout the plant.

C. Corroded metal resulting from contact with these chemicals. The metal structure of the building, and especially of parts like the acetic acid recovery plant, may have been weakened.

D. Once the spinning process has been completed, this type of rayon is liable to build up charges of static electricity.

(iii) *Fire-fighting*. (For linters *see* cupra-ammonium rayon, and for acetic acid and acetone *see* Section 16 of this Chapter).

The grinding of cellulose powder involves a risk of a dust explosion and the usual precautions against such explosions should be taken by the fireman. The dust is flammable as well as explosive.

Acetate yarn is not easily set alight and requires a continuous source of external heat to keep it alight. On heating, it decomposes into hard, glue-like lumps and serious burns can be caused if burning yarn falls on the skin.

The made-up material may be made unsaleable by damage by water. If water has to be used in fire-fighting operations, the damage—particularly to the yarn—may be less if the water is clean. The yarn or woven fabric is not likely to heat spontaneously even if it is damp.

This type of fire may well lend itself to an attack by high expansion foam.

4. JUTE SPINNING AND WEAVING

(a) Construction of mills and warehouses

Mills for the spinning and weaving of jute resemble cotton mills, except that single storey buildings are more common. Concealed roof spaces and wall linings are often found. Jute warehouses are also usually of single storey construction, sometimes adjoining one another with separating walls.

(b) Processes and risks

The risks of jute spinning and weaving are similar to those of other trades in which flammable fibre is processed. There is usually a good deal of fluff and dust about the machines, rafters, etc., and the least spark is sufficient to ignite this. Burning fluff will convey fire at great speed along rafters and other holding surfaces, and fiery particles are likely to drop to various points, starting fires in places remote from the original outbreak. Tramp metal in the machines or overheating of bearings are common causes of outbreaks.

(c) Fixed fire protection

Sprinklers are commonly installed in spinning mills but sometimes factories (weaving flats) or warehouses are found unprotected.

(d) Fire-fighting in jute warehouses

When an outbreak occurs in a jute warehouse fire will spread over the exposed hairy surfaces of the bales with great rapidity, at the same time working its way between the bales until the whole mass is involved.

Fire in a mass of jute bales cannot be finally extinguished except by the removal of the bales. Water in the form of spray should be used to control the burning of the material during removal, as sparingly as possible so as to avoid swelling of the bales and to preserve the salvage value of the jute (*see* Fig. 13). Because of the danger of re-ignition the damaged jute should be taken to a safe open space and distributed in small heaps. In certain circumstances high expansion foam and bulk supplies of carbon dioxide may offer an effective means of extinguishing jute fires. These particular systems require specialised control of ventilation.

During operations in a warehouse there is the danger that bales at the top of the stacks may fall, due to swelling, into the alleys where men are working. Smoke sometimes makes it difficult to keep a lookout to guard against this.

Burning jute evolves great quantities of dense, acrid smoke, and this necessitates the wearing of breathing apparatus. Since warehouses sometimes lack roof apertures, it may also necessitate breaking through the roof at strategic points to secure ventilation.

In stowage, a space should be allowed between the bales and the walls of the buildings to allow for the expansion of the bales when wet. Where this has not been done, damage to the building is likely, since no conventional structure can withstand the pressure

Fig. 13. Fire travels in the air spaces between stored jute bales. Water applied from above penetrates into the interior of the stack. (The stack has been cut in half to show the effect).

of the bales as they swell. It has been known that a building with 610 mm stone walls has been virtually levelled to the ground within four hours of an outbreak.

5. FLAX MANUFACTURE

The preliminary processes are carried out in flax factories, and later ones in spinning mills and weaving mills or sheds.

(a) Construction of buildings

(i) *Flax factories* in this country are normally of modern construction, being single-storey brick and galvanised iron buildings with the flax store adjoining. The store can hold 100 tonne or more, and there may be Dutch barns containing up to 400 tonne and ricks near to the factory.

(ii) *Spinning and weaving mills.* There are only a few in England and these are often of single-storey construction, with lined roofs and walls. Most are in Scotland and Northern Ireland, and these are normally multi-storey buildings.

(b) Processes

(i) *In factories*

Retting. The de-seeded flax is soaked in warm water in retting tanks.

Drying. In modern factories mechanical driers, heated by hot flue gases and fitted with spark arresters are used.

Scutching. The fibre is extracted from the hard outer stalk with rollers and beaters. In old factories scutching may be by hand, using a wheel with wooden blades. Hand wheels are also used in many mills for re-cleaning flax, and normally they are in a room by themselves because of the high fire risk.

Processed fibre is stored in 100 kilogram hessian-covered bales.

(ii) *In spinning and weaving mills*

Processes are roughly similar to those used in the jute and cotton industries. One worth mentioning is:

Hackling, in which fibre is separated from tow, which falls into boxes beneath the machine.

(c) Causes of fire, fire prevention and fire-fighting

(i) *Ricks and barns.* Many of these are near factories, so that protection of the factory and other stores, to prevent spread of the fire, is the first essential. Burning fragments of flax carry a long way.

Fire in a flax rick is very difficult to extinguish. The ricks are loose in construction, so that fire and air penetrate them easily. The fire is fiercer and hotter than a straw fire because of the oil in the flax seeds. Possibly the best method is to use one or more 19 mm jets to knock out the fire in the upper portion of the stack, while employing diffuser branches near its base.

(ii) *Stores*. In these, sprinklers are not usually fitted. Foreign bales weigh up to 300 kilograms and are compressed; English ones weigh up to 100 kilograms and are loosely packed. Fire travels rapidly over the bales. Flax absorbs a large quantity of water and increases in weight and bulk, but does not expand as much as jute.

Processed flax burns easily, but not as fiercely as raw flax since it does not contain the seeds.

Damage from water is slight to unprocessed flax, considerable to processed flax and most serious to seed stores. Wet seed will germinate or become like jelly.

(iii) *Factories*. Sprinklers are not usually fitted (except inside the driers). Points where fires may start are:

A. mechanical driers, which may become overheated or in which a spark may pass the arresters. There are hand-operated sprinklers inside the driers; and

B. the scutching room, where flammable dust and chaff collect on the floor, roof beams and machines. However, the dust does not flash and only a low explosion risk is attached to it.

(iv) *Spinning and weaving mills*. Flax spinning mills are usually sprinklered, but the weaving sheds seldom.

In the case of fire on the hackling machine, it is essential to clear the tow boxes and loose material lying around the machine. Both tow and fibre are stored in loose form in partitioned rooms. There is thus more loose material than in other textile mills. A great deal of dust and fluff are also present throughout the mills, especially when unretted flax is being worked. Various machines have dust extraction ducts.

Smoke at flax fires is heavy, and the use of breathing apparatus is normally necessary. The general procedure in handling a mill fire is similar to that in dealing with other textile fires. When flax is left wet it tends to heat and rot: material affected by water should, therefore, be removed and spread out to dry.

6. LACE MANUFACTURE

The fire risk in lace manufacture is high, especially since most factories are old.

(a) Construction of factories

These may be lofty single-storey shed-type factories, but most are substantial four-, or occasionally five- or six-storey buildings. Usual features are brick load-bearing walls; slated roofs; wooden floors supported at intermediate points by cast-iron columns; extensive window area—sometimes 50 per cent of the wall area; and lantern lights, lantern attics and mansard roofs in older factories. Where the building is used by more than one firm the floors may be extensively subdivided by wooden or iron partitions.

The macnines are often driven by belting from line shafting, powered from a mill engine through drive ropes in a rope race or by internal shaft and bevel gearing. In modern factories each machine is driven by its own electric motor.

(b) Processes and risks

The lace is made on huge jacquard looms. Important points are:

(i) In older factories part of the loom mechanism may be held in wooden boxes built out from the window frames, often on upper floors. These boxes increase the exposure risk and collect dust and waste.

(ii) The looms are so high that in some older factories a portion of the floor above has been removed to give the looms headroom. Only occasionally is the opening enclosed.

(iii) Hundreds of varnished strings, which become greasy, are used in the harness above the jacquard loom. They are highly flammable and, if burning, fall into the loom among bobbins and threads.

(iv) Cards, used in forming patterns, and made of layers of thick paper pasted together, lie on the floor or in racks or hang from metal or wood frames or from the ceiling (*see* Plate 18). These cards become very dry; if suspended cards catch fire they can set fire to the roof timbers, and burning fragments can fall among stock and machines. Cards can impede the sprinkler system. If card storage of any kind is involved in a fire a careful search for hidden sparks is afterwards necessary.

The jacquard loom may also be found in the cotton, woollen, worsted, flax (linen), hemp and jute industries.

(c) Fire-fighting

(i) The great weight of the machines makes the floors especially liable to give way when affected by fire, and should this happen the lower floor, too, may fail if the upper floor collapses on to it.

(ii) Water and steam can cause very considerable damage to the machines and early steps must be taken to sheet up any looms or stock which might be affected.

(iii) The great size of the jacquard looms makes it difficult to move about the floor and get an uninterrupted jet upon the fire.

(iv) Net factories, though using similar machines, have an additional risk. Whereas the treated thread used for lace making does not throw off any appreciable fluff, the untreated thread employed for net making deposits enough fluff to assist materially in the spread of fire.

(v) The use throughout the industry of graphite (blacklead) as a lubricant causes all floors and stairways to become very slippery and care should be taken to avoid falls.

7. TEXTILE WAREHOUSES

(a) Construction

Textile warehouses generally present a high fire risk in themselves. Furthermore, in certain large towns they are situated so close to one another as to present a danger of widespread conflagration.

In some cases, textile warehouses are as much as seven stories high, and, as they usually have large window openings facing other buildings, they often present a severe exposure hazard. The rapid spread of fire is often facilitated by skylights, inadequate separating walls, wooden floors and roofs, partitions, linings to walls and ceilings, racks for storage, open hoists and chutes, and common areas or light wells. In some northern towns warehouses are built on the side of a hill, one elevation being two or even three stories higher than the other. On the other hand, in some cases, such as jute, single storey construction is usual, the walls being of stone, the roof wood-lined and slated, and the floor of earth with gangways of stone setts.

A fire which occurs in a large warehouse outside working hours may remain undetected until it has acquired a firm hold, because of dirty or boarded-up windows. It may be difficult to put out the fire because of the height of the buildings and lack of road space for fire brigade appliances, whilst the large cubical capacity and the absence of internal fire-resisting division walls provides favourable conditions for a very large fire to develop.

(b) Fire-fighting

The following fire-fighting suggestions have much practical support, but must not, of course, be taken as universally applicable.

(i) In the case of a fast-spreading fire (one which travels over the outside of jute or cotton bales, for example), the procedure might be as follows:

A. Check the superficial spread of fire with the smallest essential number of hose lines.

B. When the fire is under control, knock off the branches and shut down the sprinkler installation (if there is one) in order to reduce the amount of smoke generated. Cover the fire with carefully sited branches.

C. Thoroughly ventilate the premises. If fire shows itself, make a concentrated attack. If not, examine the smouldering bales or stocks and determine the extent of fire.

D. Isolate and extinguish the fire by the removal of bales and the careful use of small quantities of water.

(ii) The capacity of some bales (especially jute and hemp) to absorb water, should be remembered. This fact is also true of rags and of loose textiles such as wholesale clothing stocks. The additional weight of water, as well as the sideways pressure on walls, may lead to the collapse of the building.

(iii) Fibrous material is liable to choke the building drains and lead to further damage by water.

(iv) Stocks of finished material of a delicate or expensive kind are easily made valueless by water, heat and smoke. The damage can, however, be minimised by efficient salvage operations.

8. CLOTHING FACTORIES

In these factories cotton, wool, silk and artificial silk fabrics are made up into clothes. The fire risks are usually connected with the buildings rather than the processes and materials used. The following risks are characteristic of the majority, but will also be found in modified form in buildings of newer design or construction.

(a) Construction of buildings

(i) A common defect is the excessive amount of wall space occupied by windows. The glass is usually set in wood frames.

(ii) The height of the buildings varies from three to seven stories, the higher ones generally being found in the more congested areas.

(iii) The quantity of woodwork is considerable. Wood is used for floors, wall and ceiling linings, encasing beams, partitions, work benches, storage racks, and chutes from floor to floor. Wall linings sometimes have cavities which collect dust and fluff and allow

Plate 12. Bales of cotton being opened up prior to being fed into the breaking or opening machines. (Section 4).

Plate 13. A rag-pulling machine, used in the production of cheap woollen cloth. Metal objects in the rags can cause friction sparks, presenting the greatest fire risk of all wool textile operations. (Section 4).

Plate 14. Composite processes room in a cotton mill showing ring spinning and doubling. Note the restricted access between machines and overhead belt drives. (Section 4).

Plate 15. Spinning mules in a cotton mill, showing the lack of space between the mules when the carriages reach the end of the metal runways. Fire can spread with great rapidity along the mules. (Section 4).

PHOTO: *C. E. Willis, Ramsbottom*

Plate 16. The base of a dust chamber in a cotton mill, where cotton dust and fly is collected by means of pneumatic trunking from the various machines. Fires often spread to the dust tower through the trunking. (Section 4).

Plate 17. Looms in a weaving shed for fancy silk or cotton. Again, there is great congestion of machinery and lack of wide gangways. (Section 4).

PHOTO: *C. E. Willis, Ramsbottom*

Plate 18. Jacquard cards, used in lace making, stored suspended from the ceiling; they can add greatly to the danger of fire spread. (Section 4).

Plate 19. The burning shell of the Eastwood Mill, Keighley, Yorkshire, which was destroyed by fire in February 1956 with the loss of eight lives. This fire has had a profound effect upon factory legislation as regards means of escape. (Section 4).

PHOTO: *Keighley News*

fire to spread behind the linings. As a constant temperature (usually about 60°F. (15.6°C.)) is maintained in the building the woodwork is invariably dry. Fire may, therefore, spread rapidly.

(b) Processes and risks

(i) Throughout the building a quantity of dust and cuttings from the fabrics accumulates on the floors and round the electric motors which are usually employed to drive cutting, sewing and other machines.

(ii) There are large quantities of paper patterns, wrapping paper and cardboard boxes.

(iii) Garments and fabrics are left on work benches every night, as are gas or electric irons.

(iv) Small quantities of petrol, benzine, and carbon disulphide will be found in finishing rooms and warehouses for removing grease and stains.

(v) Where waterproofing is done, stocks of rubber and naphtha will also be found.

(vi) Buildings are often occupied by a number of different firms.

(c) Fixed fire-fighting installations

The installation both of sprinklers and of risers is unusual in these factories.

(d) Fire-fighting

(i) Damage by water and smoke may occur to fabrics from the time they enter the factory as rolls of material to the time they leave as finished garments, but the further advanced the stage of working the greater the damage is likely to be. Garments once wetted lose nearly half their value at once owing to the drying and special treatment they must have (*see* Plate 23); salvage by removal or by sheeting is, therefore essential. The months from November to March are those in which the greatest stocks of finished clothing will be found in the factory.

(ii) The machines used are not of great value but efforts should be made to protect them from the effects of steam and water by sheeting.

9. CORDAGE WORKS

(a) Construction

Cordage works, where the manufacture of ropes, lines and yarns of all kinds is carried on, are mainly of single-storey construction.

The principal feature is the ropewalk (*see* Plate 24), a building of 300 metres or more in length and often about 15 m in width.

(b) Processes and risks

The raw material used is principally hemp, sisal or manila, and occasionally coir or cotton. Cord may also be made from jute, and there is a growing use of synthetic fibres throughout the industry. The hemp is put through several hackling machines for combing and, at one point, for application of rot-proof solution; drawing frames, which pull the fibres together into yarn; spinning frames, where the yarn is twisted, and the ropewalk where a setting machine twists the yarns into rope, the ends of the yarns being attached to a moving carriage which is driven down the ropewalk as the rope is made. Similarly, ropes are twisted into cables. Short-length fibres, known as '*tow*', are carded, teased and made into coarse string.

Fire normally originates from hot bearings. There is risk of rapid spread because of piles of hemp lying around the machines, and also of hemp fluff and dust beneath the machines and on beams and girders. The dust flashes like cotton fly.

(c) Fixed fire protection

Sprinklers will generally be found in the machine rooms of cordage works.

(d) Fire-fighting

The behaviour of hemp and jute is dealt with under those headings, Generally speaking, one must expect a fast-spreading fire, which will give rise to a large quantity of smoke.

10. HAT MANUFACTURE

The following paragraphs deal only with the manufacture of felt and straw hats for men and women. A large proportion of the hat trade is concentrated at Luton: other centres exist at Manchester, London and elsewhere.

(a) Construction of buildings

Most hat factories are comparatively small. The true factory building is uncommon; many factories are in converted private houses containing a large number of small rooms on several floors, with modern extensions to the rear of the older property. Other factories may be in two- or three-storey brick extensions at the rear of private houses, and often accessible only through the latter. This type of building is particularly common at Luton.

Hat factories are generally brick-built and slated. Inside the building the fire risks are increased by matchboard ceilings, partitions and walls; wooden staircases; and the large window area proportionate to the total wall area. Many of the rooms may be sub-divided into smaller compartments. Housekeeping is generally poor.

(b) Raw materials

For high-quality felt hats, the raw material is animal fur; for those of the usual grade, it is wool. Straw hats are made from imported straw plaits, which arrive in bales packed either in plaited material or three-ply wood.

Other materials likely to be found in a hat factory are cotton wadding, ribbons and trimmings, methylated spirit and similar solvents, amyl acetate (in straw hat making), dyestuffs, and in some cases stearin, glycerine or neat's foot oil for softening the felt.

(c) Processes (felt)

(i) The fur or wool is opened out, cleaned, rolled, spread on formers, shaped (by compressed air) and steamed. (Many hat factories buy ready-made rough felt hoods).

(ii) The felt is hardened with flammable solvents, usually shellac dissolved in methylated spirit.

(iii) The felt is smoothed on grinding machines, and dust is extracted by means of fans and vents.

(iv) Linings, ribbons, etc. are added, and are protected by bands of paper or cardboard.

(d) Risks (felt)

(i) Fluff, especially in the early processes, on floors and beams. Fur fluff flashes.

(ii) The flammable solvents.

(iii) Stores of materials. Baled wool and fur are kept in the cellar. Paper parcels of fur may lie on wood racks on the ground floor. There may be stores of rough felt hoods, linings, ribbons, etc., and packing rooms full of paper, cardboard and wooden boxes.

(iv) There is a woodworking shop for making blocks and a boiler to produce steam. In older buildings the boiler room is often in the basement.

(e) Process and risks (plait)

The process here is simpler, and consists of stitching or weaving the plaits together by means of various types of sewing machines to form a shaped hat. Afterwards gelatine dissolved in amyl acetate is used as a stiffening and proofing agent. This is applied in spray booths, one or more of which will be found in most factories. The hats are then dried in moderately heated rooms, or hung from the ceilings on racks. Finally the fabric and other ornaments are added.

(f) Auxiliary industries

A number of auxiliary industries, such as printing, dyeing, box-making, straw-bleaching, and so on are sometimes carried on at the larger hat factories.

(g) Fixed fire protection

Sprinklers are seldom installed in the smaller hat factories, but they are usually to be found in the large ones.

(h) Fire-fighting

(i) Fur in the loose state is flammable, as also is plaited straw and wool.

(ii) In congested areas, many of the factories have no access road suitable for an escape. In such case there is a considerable risk to life should a fire break out.

(iii) The factories are often in converted houses, organised in small rooms. The location of a fire is, therefore, often difficult, and there is a risk that staff or firemen may be trapped.

(iv) The passages as well as work- and stock-rooms are often congested with hat-boxes and hats. The ground floor, especially, of most factories will be in this state in the late afternoons, when boxes are awaiting transport.

(v) A fire at a hat factory is likely to be very smoky. Care should be taken, however, not to ventilate the premises too soon, in view of the large amount of wood in the construction of the building and the high proportion of windows to total wall space.

(vi) Hat factories are particularly susceptible to damage by water.

(vii) Broadly speaking, the further a straw hat is towards completion, the more flammable it is. In the manufacture of felt hats, it is the raw material and the solvents used during hardening that provide the main risk: on the other hand, the further advanced the process, the more damage can be done to the felt by smoke and water.

11. BEDDING MANUFACTURE

(a) Construction of buildings

In earlier days this industry was not often found in large units:

in only a few cases single firms made mattresses, pillows and quilts under one roof, since each of these was usually made separately by firms in a very small way of business. The modern trend, however, is towards a concentration in larger units, though it is still more common for a firm to occupy one floor of an industrial building than the whole building. Most of the premises are of old construction, involving a high degree of risk. Where a whole building is concerned in bedding manufacture, hoists, chutes and staircases are commonly found.

(b) Processes and risks

The filling materials for mattresses are usually hair, shoddy and cotton flock; pillows are filled with kapok, feathers or cotton waste. Storage of all these is in sacks, normally on the lower floors of the building; the cotton fabric for covers is kept in bolts on shelves. Synthetic fibres and textiles are penetrating the industry.

The cotton fabric is made up into covers in the sewing machine shop or shops. Mattresses are filled either by machine or by hand in the filling shops: large quantities of stuffing material will be found on the floors and tables. A carding machine teases out the shoddy or flock, similar to the rag grinding machine used in sections of the woollen trade, but not the carding engine found in the cotton and auxiliary industries. Kapok and waste for pillows will usually be put through a willeying machine. Feathers are cleaned by heating; large quantities of them may be found in cubicles near the heaters.

In mattress repair, quantities of loose flock and shoddy will generally be found on the floor and tables in rooms where mattresses are being unpicked. In quilt manufacture, down is usually inserted into the covers by compressed air gun.

Completed mattresses, pillows and cushions (if the latter are made in the same factory) will be found in quantity in the warehouse awaiting despatch.

(c) Fixed fire protection

Sprinklers are usually fitted in large bedding factories. Small factories are, however, seldom protected in this way.

(d) Fire-fighting

There is little in the processes themselves likely to cause a fire. although sparks may occasionally occur in the carding machine as a result of the teeth striking metal. The raw material and the finished articles are, however, flammable, and if a fire does occur, it is likely to be smoky and to spread rapidly.

(i) Breathing apparatus will usually have to be used when fighting a fire involving shoddy and flock.

(ii) There is always a quantity of filling material on the floors around the work benches. The spread of fire is, therefore, likely to be rapid.

(iii) Water is likely to cause great damage unless the fire is found before a jet is used. Comprehensive salvage work should be done as soon as possible, especially in the protection of sewing machines and of the finished articles (this particularly applies where quilts are concerned).

(iv) Burning mattresses are difficult to extinguish, and sparks tend to linger in them. They should if possible be removed to an open yard and be carefully examined to make sure that no sparks remain.

(v) It sometimes happens that bed manufacture is combined with mattress manufacture. In such a case, a cellulose spraying plant is likely to be installed.

12. UPHOLSTERY

Certain firms combine the making of the wooden frames and the springs for furniture with upholstery work. It is usual, however, for upholstering to be carried on in small workplaces, including shops and private houses, by firms which buy the frames and springs. Where all three processes are combined under one roof, a sawmill, timber stacks, woodworking shop, and spring-making machines will be found on the premises.

(a) Raw material

The principal raw materials used for coverings are various types of cotton and wool cloth, leather and leather cloth. The usual fillings are cotton felt coir (coconut fibre), jute, hair, and kapok for cushions. The felt is usually obtained by putting cotton waste through a garnetting machine (*see* 'Rag Grinding' (page 54)).

The web of cotton that comes off the garnetting machine is folded back and forth upon itself until it forms a felt of sufficient thickness.

(b) Workrooms

The filling and covering of chairs and sofas leads to the material being scattered about the comparatively large workrooms in a manner that facilitates the spread of fire.

(c) Fixed fire protection

Generally speaking, the premises of only the large firms will be sprinklered.

(d) Fire-fighting

The type of premises and the materials found on them, give rise to a smoky fire which is likely to spread rapidly. It may be necessary to use breathing apparatus.

(i) It is essential to find the seat of the fire.

(ii) It is often more satisfactory to remove upholstered furniture which has been on fire to a yard before making sure every spark is out.

(iii) Fibre, rags and other similar debris are likely to choke the drains.

13. BRUSH-MAKING (INCLUDING BASKET-MAKING)

This industry includes the manufacture of all types of brush and broom.

The materials used are usually animal hair or vegetable fibres, set into wooden handles or heads. Other materials may, however, be found. Toothbrushes, for example, may have plastic handles with nylon bristles.

In this section only brushes with wooden backs and hair or vegetable fibre bristles are dealt with; notes on basket-making are also included.

For plastic materials *see* Section 10, 'Fires involving Plastics.'

(a) Construction

Brush-making is carried on in many different kinds of building. Although some factories are large, the greater number are small, and are often to be found in converted houses. Both the brush-making and the basket-making industry employ many blind people. Some factories may, therefore, be found in large institutional buildings, where other articles such as mattresses, pillows, fibre mats, and furniture are made under the same roof.

(b) Raw material

Hair, usually that of the horse, pig, squirrel or bear, is used for the better-quality brushes, and vegetable fibres and grasses of various types for those of cheaper quality. The hair arrives in short bundles, the fibre in long rolls. Both are kept in a store room in bins or on shelves. In the case of basket-making, cane arrives in hanks weighing about 150 kilograms each.

(c) Hazards

(i) *Store room.* It is here that the hair and vegetable fibre are kept until required for processing. Hair burns easily and brightly; vegetable fibres usually burn smudgily and with much smoke.

(ii) *Workrooms.* In the workrooms the bundles of hair or fibre are inserted in holes in the wooden handles or heads and are fastened with pitch, gum or resinous cements. Pitch is usually stored in 12·5 kg tins, and heated in open gas-fired containers. Each bundle of hair, etc. is then dipped into it before being placed in position in the brush. If the pitch boils over or is upset it can start a fire involving the whole room. Sand is, therefore, usually kept available for extinguishing such fires. In other cases, flammable gums or resins are stored and used. Fume vents from the pitch heaters are usually provided.

(iii) *Woodworking shop.* In this part of the premises shaping machines cut out handles and broom heads.

(iv) *Stock room.* In the stock room for finished articles, brushes will often be found not only in racks round the walls, but also hung in bundles from the ceiling.

(v) *Packing room.* Boxes and paper will be found here as well as stocks of the finished article.

(vi) *Basket making rooms.* In the case of basket-making by the blind, a common arrangement is to divide a large room by low partitions. A blind worker, surrounded by his raw material, generally sits in each division. The risk of a fire spreading is great.

(d) Fixed fire protection

Sprinklers are not commonly found at brush-making factories.

(e) Fire-fighting

(i) Small works present a very high fire risk, due both to the flammability of the materials used, and to the poor type of building which is frequently found.

(ii) The possibility that blind workers may be present should be borne in mind, as many firms, other than institutions exclusively for the blind, employ blind persons.

Part 6c, Chapter 45, Section 5

Fires in fuels

THE fuels treated in this Section are solid fuels, heavy oils, vehicle fuels and liquefied petroleum gases. (Wood as a fuel is not regarded as a special risk). Fires in oil installations have already been described in some detail in Chapter 44, as have those in town and other gases in Chapter 41; details of the chemical composition and behaviour of coal and coke, semi-heavy oils and petrol have been given in Chapter 1. The special hazards of dusts have also been dealt with, in Section 1 of this Chapter. The purpose of this Section is to describe a number of special types of fires either occurring in the use and storage of fuels or arising out of their peculiar properties.

Fuels are by their nature flammable, some very readily, some not. In all cases special apparatus is employed in which to burn the fuel, varying from a Lancashire boiler to the internal combustion engine of a tractor, for example, and it is generally true to say that the majority of accidental fires in fuels occur in association with the storage or handling of the fuel, and that relatively few arise out of the burning of the fuel in the apparatus in which this is intended to take place. This is no doubt because the burning apparatus is specially designed to provide safe conditions, while storage and handling of the fuel are in many cases entirely at the choice of the individual and may or may not be done under satisfactory safeguards.

The fact, then, that fuel fires largely occur in storage and handling produces a wide variety of hazards, and in solid fuels the commonest dangers are those of spontaneous combustion and of dust. In the case of the former, perhaps 'dangers' is too strong a word, since the heating and ignition of a coal stack is not normally attended by dangerous conditions; it can, however, present serious difficulties in extinction. Fires in oil fuels arise from a variety of causes, most of those in the heavy fuels being due to incorrect operation or unsuitable design and layout of equipment, and usually confined to the older installations. In the lighter fuels, including petrol, the commonest cause of danger is probably the ignorance and carelessness with which many users of these liquids approach them: petrol particularly, and kerosene to some extent also, have special qualities which render them a serious fire hazard wherever they may be and entitle them to the utmost respect in handling and use. This applies particularly to the rapidly increasing use of kerosene in small portable heating appliances.

Mention was made in Chapter 41 of liquefied petroleum gases, and this Section contains further details on this subject. There has been in recent years a very considerable increase in the use of these 'bottled gases,' and this has led to a rise in the number of incidents from this cause, though not, fortunately, an increase proportionately to the rise in the rate of use. The commonest cause is leakage of gas from the apparatus, whether due to lack of proper maintenance of the component parts or to careless handling. Regrettably, the type of occupancy in which this fuel is commonly found—cottages, caravans and boats—is apt, by its nature, to permit an outbreak of fire to reach serious proportions in a short time and all too often to lead to casualties. No particular problems in fire-fighting arise in connection with bottled gas except the hazard to firemen of cylinders exploding through being involved in a fire.

Coal and coke

1. NATURE AND PROPERTIES

Coal is created by the decay of cellulose below the surface of the ground. It varies in nature according to its constituents and to the pressure and other conditions to which it has been subjected over the ages: common types are bituminous coals, steam coals, brown coal or lignite, and anthracite. An average coking coal (*i.e.*, one suitable for the production of coke) may consist of about 88 per cent carbon, 6 per cent hydrogen and 6 per cent oxygen and have an ignition temperature of approximately 390°F. (199°C.) Anthracite, however, has an ignition temperature of 750°F. (398°C.). (Many details of several types of coal, with notes on combustion processes, are given in Chapter 1 in Part 1.) Most varieties of coal burn with large quantities of smoke, though anthracite is smokeless.

Coke is the solid residue left after coal has been carbonised and the volatile constituents distilled off. It, also, may be of different types, according to the process used in production; typical classes are gas coke, low-temperature coke and semi-coke. In general, coke burns without smoke and has an ignition temperature of 660°F. (350°C.) and upwards according to type.

2. FIRES IN COAL STACKS

Whenever coal is stored, whether in a small layer or a large stack, the phenomenon of oxidation (*i.e.*, the combination of oxygen from the air with the coal) is certain to take place. (See Chapter 1 in Part 1). Thus, there is in all coals a tendency to self-heating, which will be assisted by the presence of foreign matter such as straw. This tendency is affected by both the size and type of coal. The greater the area of the coal's exposed surface compared with its bulk (*i.e.*, specific surface area), the greater is the tendency to self-heating; in a stack, therefore, self-heating generally begins in the fine coal and dust. Similarly, the softer the coal the greater the liability to oxidation: at the one extreme, anthracites are free from self-heating; at the other, soft bituminous coal and lignites are most susceptible to it.

Coal should either be stacked under water or in some other way designed to keep out air, or sufficient ventilation should be provided to carry off the heat as fast as it is generated. With hard coals, anthracite duff (fine anthracite) is sometimes used to seal off a stack and so to exclude air; but in most cases reliance is put on natural ventilation as effective sealing is most difficult, especially when a dump includes different sizes of coal.

Coal is often stacked in conical heaps, in stacks which are too large or in which new coal has been laid upon old; such methods are the principal cause of stack fires (see also Part 7, Chap. 48). A well-built coal stack is of oblong shape with a 3 metres wide roadway all round it to facilitate access. The height for coals over 150 mm cube should not exceed 4·5 metres; for coals 75 to 150 mm cube, 3 metres; and for sizes less than 75 mm cube, 2·15 m. The quantity in each stack should not exceed 1,000 tonne; it is not usual to expect spontaneous ignition in stacks of less than 200 tonne.

Steam rising from the stack does not necessarily mean that the stack is in danger of catching fire, as—particularly in winter—this may be due to permissible heat. Thermometer tubes are inserted to ground level in large stacks, so that the temperature inside the stack can be tested. A maximum thermometer (Fig. 14) should be attached to a cord and dropped down the tube. If the cord is knotted every 900 mm the knots form a useful indication of the depth of the thermometer in the stack and so of the hottest portion; the temperature may vary greatly from the top of the stack to ground level. Where a tube has to be driven, one of good design has a 25 mm internal diameter with a closed pointed end, the top being screwed internally for 25 mm and having a screwed metal plug to keep out the rain. For more exact readings a thermocouple is used (*see* Fig. 15).

Fig. 14. A coal stack thermometer, designed by the Fuel Research Station, shown in diagrammatic form.

A figure for dangerous heating is difficult to give: a safe heat for one type of coal may be a dangerous one for another. It is normally wise to start operations in the case of a large stack when the temperature of the stack reaches 90°F. (32°C.). Smaller stacks need not be tackled until the temperature reaches, say, 100° F. (38°C.), as they take less time to deal with.

When the thermometer indicates severe heating, the extent of the heated area should be ascertained. Taking the tube where the reading has been obtained as the centre of a circle, metal rods should be inserted round the circumference and left for a while. Should they feel hot on withdrawal they should be put in again farther from the central tube until they come out cooled. The hot area can then be roughly defined. In small stacks, iron bars can be used to find the extent of the heat without a temperature reading being taken. This preliminary work will be found well worth the time and trouble.

When the temperature of a stack exceeds

the figures given above, the stack must be cooled without delay, as the rise in temperature is self-accelerative, increasing in speed until, at 400°—600°F., (204°—316°C), combustion takes place. Water is not in general a satisfactory cooling agent: when applied to red hot coal it turns to steam which, in combination with coal, produces

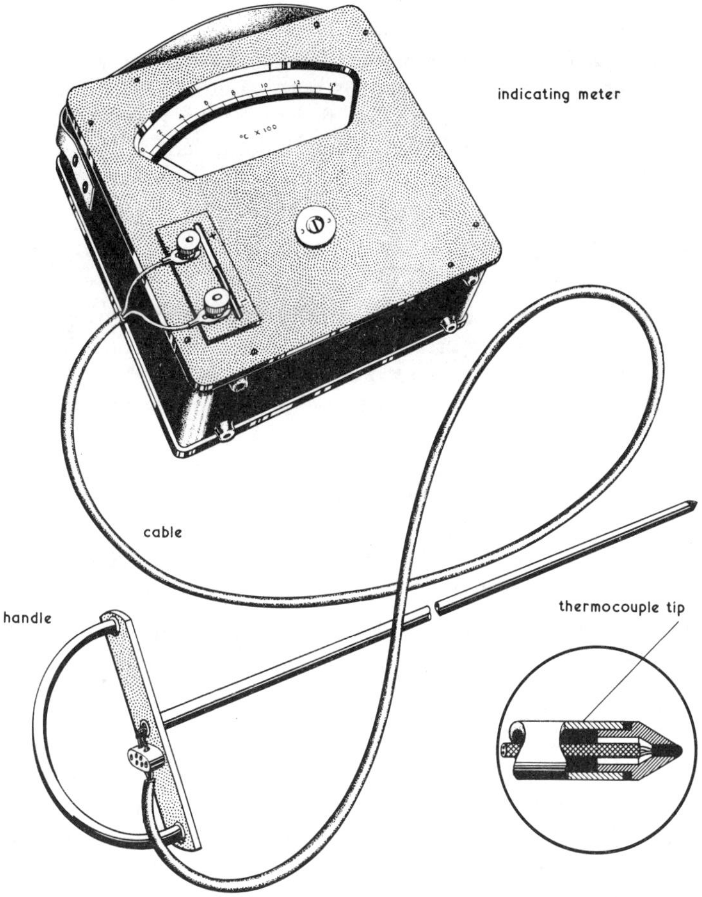

Fig. 15. Diagram of a thermo-couple type of temperature indicator for use in coal stacks. (Electroflo Meters Co. Ltd.)

water gas (*see* Chapter 41). This a mixture of hydrogen and carbon monoxide which is explosive when mixed with air. Water can only, therefore, be considered as an auxiliary, and the main work must be done by ventilation with or without removal of the coal. The management at the site should be consulted about mechanical equipment for the removal.

Fire in a stack of coal usually starts at ground level, where the coal becomes hot and dry. Coal in this state is much more dangerous than coal which, though hot, is in a moist and steaming condition.

The following information may be of value to the fire brigade officer:

(a) It should be remembered that though a stack may be warm throughout, only a small amount is burning at any given moment. Where burning can be seen, the fire probably does not extend for more than a few cubic feet. There is little danger of a general conflagration, for air cannot get into the stack in sufficient quantities to support it.

(b) The effect of wind should not be forgotten. The centre of the fire may shift with the wind.

(c) A stack may sometimes be cooled, *but only in the very early stages of heating*, by inserting a crowbar into the top of the heap and moving it round so as to produce a funnel-shaped hole. Fifty or sixty holes are quickly made, and may prevent the outbreak of a fire.

(d) In a flat stack, say 2·45 m high, a trench should be dug lengthwise along the top of the heap over the hot portion, the hot coal being cast down the sides of the heap. The trench may be about 900 mm wide by 900 mm to 1500 mm deep, according to circumstances. This may be sufficient to cool the area.

(e) A trench may be cut (preferably by mechanical excavator) right through the stack on each side of the hot area in order to isolate it (*see* Fig. 16). The practice of cutting *into* a heap in order to reach the heated spot is not usually successful, as the heat tends to travel through the stack in advance of the trench.

(f) *Use of water*. Generally speaking, the surface of the coal should not be watered, as this accelerates oxidation. The application of insufficient quantities of water may result in explosions of water gas, while the steam generated will create a draught that will increase the rate of oxidation in the interior of the stack. Small quantities of water at high pressure may, however, be used to keep the surface cool near men working on the stack.

Water can be used on the flanks of the heated portion so long as sufficient quantities are available. This can be done by scooping out holes in the sides of stacks and allowing water to percolate down through open-ended hose lines, or by building a dam round the base of the stack and then flooding it, making sure water does not escape to cause damage elsewhere. If an excavator is not available and manpower is short, one or more open-ended hose lines working in a line across the stack can usually apply enough water to make a firebreak.

Fig. 16. Bulldozers being used to isolate a fire in a coal stack.

Where the stack as a whole is not involved but a hot spot exists, an 'extended' branch (Book 2, Chap. 5) can be used to apply water to the exact place where heat is occurring. Water applied from above cannot always be trusted to percolate down, since it may tend to follow one channel rather than to distribute itself, while it is often also prevented from being effective by the formation of a layer of coke above the heated spot.

(g) *Use of* CO_2. Attempts to apply CO_2 down the thermometer tubes in coal stacks are often frustrated by an up-draught. CO_2 has, however, been successfully used to put out fires in small stores of coal, but as it has practically no cooling effect on the coal a watch must be kept to make sure that another fire does not break out. It should be noted that the passing of CO_2 over red hot coke or anthracite produces carbon monoxide.

(h) *Care of hose*. Hose can be considerably damaged if it is dragged over stacks of coal. As far as possible, therefore, it should be lifted from one position to another.

3. FIRES IN COKE STACKS

Coke is normally unlikely to fire by spontaneous combustion. Fires in coke stacks are usually the result either of insufficient quenching of the red hot coke, particularly at times when there are high winds blowing, or of the presence in the coke of a small proportion of insufficiently carbonised coal. In the case of piles made by tipping from steam trains, for example, there is also the risk of hot ashes from the locomotives dropping on to the coke and igniting it.

The first indication of a coke fire is usually the appearance at night of an almost colourless flame at some high portion of the stack. There is no smoke and very little smell.

Water must be used with care upon a coke fire which has attained any appreciable size, as its use results in the generation of water gas. This gas will permeate through the heap and spread the fire.

The usual method of dealing with coke stack fires is to dig the coke out while keeping the hot face of the stack cooled with water. It should be remembered, however, that burning coke alone, or red-hot coke that has had water on it, is liable to give off carbon monoxide as well as sulphurous fumes.

4. FIRES IN SLAG HEAPS

Slag heaps consisting of coal dust mixed with shale do not heat to the same extent as coal and may, therefore, be tackled with water, preferably through open-ended hose lines; if necessary, spray can be used to cool the surface sufficiently to get the open lines into position. Only in the case of an especially difficult slag heap fire is it necessary to remove the material. Slag heaps of this kind should not be attacked with a jet, as the latter makes a hole in the side of the heap and allows air to get to the burning material, risking a water gas explosion. It is not possible in all cases to extinguish a slag heap fire. In some districts there are burning heaps of such size that they are virtually not extinguishable, and combustion may proceed slowly for years.

5. FIRES IN MADE-UP GROUND

This ground consists of levelled-off areas in which is included refuse of a combustible nature such as partially burnt ashes, colliery shale or similar materials. In some cases buildings have been constructed upon such made-up ground.

Made-up ground will not normally ignite by itself, but, once it has been fired by an external source of heat, it will continue to burn until all the combustible material has been consumed. In one case, an excavation 15 metres deep had to be made for nearly a quarter of a mile to isolate a fire of this kind in order to protect industrial premises. The danger lay not only in the fire itself but in the subsidence of the ground surface as the combustible matter below it burned away. Water is useless in such cases and may even accelerate the fire, while the injection of CO^2 has not proved of practical value. Cement is sometimes used as a temporary barrier, but it is in any case always necessary to find the seat of the fire and remove the burning material or else to isolate the fire by digging round it.

6. FIRES IN PULVERISED COAL

Large quantities of pulverised (*i.e.*, crushed and finely powdered) coal are used as fuel. Coal in powdered form may also be found as a residue in and around crushing and grinding plant or remaining in bunkers. When pulverised coal is used as a fuel, it is usually carried from the mills through ducting by a stream of air and then burned with an additional supply of air in a furnace.

There is a risk of fire in plants using pulverised fuel at collieries, electrical generating stations, metallurgical furnaces, cement works, etc. Fires may also occur in the ducts of pulverised coal plants, or in grinding and screening processes, as a result of dust explosion—due, perhaps, to faulty ducting—or, in bunkers of spontaneous combustion, usually when the temperature exceeds 140°F. (60°C.).

In the case of a fire in a pulverised coal store, the fireman should remember that the danger is not great so long as the dust is not disturbed; but if air is allowed to get into the stack, or the dust is forcibly scattered, a fierce fire or possibly a dust explosion may occur. (*See* Section 1, 'Fires in Dusts').

If the coal is in a bunker, the application of water spray is almost useless, as the water forms a film over the coal dust and does not penetrate to the interior. Flooding, or the use of carbon dioxide is, however, effective. When the fire is in a pipe or duct any flow of air should be stopped, a joint should be broken or an inspection cover opened, the fire extinguished with water, and the fuel then raked out. Stacks in the open air should be treated as under par. 2 above.

7. COLLIERY SURFACE INSTALLATIONS

The technique of underground fire-fighting and rescue work in coal mines is outside the scope of this *Manual;* such work is seldom done by local authority fire brigades. Fires in surface installations at collieries occur, however, and present features that must be appreciated by the fireman if very serious consequences are not to follow.

(a) Main features of surface installations

The more important sections are:

(i) The headgear, erected above the shaft in which move the cages that carry men and coal to and from the surface. The headgear carries the wheels over which run the ropes carrying the cages. These ropes run to the winding engine in an engine house near the shaft. The headgear is usually of steel, or occasionally of wooden construction, and is always heavily loaded with a mixture of oil and coal dust.

(ii) Machine houses, containing machinery for ventilating the pit and for compressing air for the operation of coal cutters, etc. In some cases this machinery is driven by electricity, and in this event power may be either generated at the pithead or obtained from public sources.

(iii) Sawmills and timber yards where pit props are stacked.

(iv) The explosives store. The main store is usually some distance away from the pithead, but there is a small store (usually of nitro-glycerine) in the surface buildings.

(v) Plant for screening, breaking and washing coal, usually situated near the shaft. This plant is usually thickly covered with coal dust. In some cases the machinery is partly of wood.

(vi) The lamp room, containing large numbers of safety lamps, with stores of lamp oil and sulphuric acid.

(b) Fire-fighting

Many collieries maintain their own fire and rescue brigades, but the fire brigade officer may be called to assist them and should bear the following points in mind:

(i) In many pits there are men down the mine at all times of the day and night, and it can be estimated that it would take an hour to bring all of them to the surface. It is, therefore, essential for the fire brigade officer to discuss the position immediately with the manager of the mine or, if he is not available, the enginewright by day and the under-manager or underlooker by night.

(ii) The principal dangers are that smoke and sparks may be drawn into the downcast of the ventilating system and so into the mine, and that the telephone exchange on the surface or the wires running down the shaft may be affected by fire, in which case telephone communication with the mine will be cut off. The engine-house and the headgear must also be protected, or men may be cut off in the mine and be unable to reach the surface.

The ventilating system is always working, and power is always available to operate the cages. The upcast shaft presents some danger as it is usually encased in wood, and the draw-off to the ventilating fan house contains a fine coal dust deposit which, on catching fire, burns rapidly and fiercely.

(iii) It is always possible to stop the screening and washing machinery, and in case of a fire—usually caused by a spark igniting the coal dust—this should be done.

(iv) Spray or diffusers should be used whenever there is a danger of quantities of fine coal dust being disturbed.

Petrol and fuel oils

1. CHARACTERISTICS OF FUELS INVOLVED

The main types of fuel under this heading are as follows:

(a) 'Paraffin' hydrocarbons

(Petroleum derivatives):
Petrol
Paraffin or kerosene
Light fuel oil, gas oil or diesel oil
Heavy fuel oil
Extra heavy fuel oil.

(b) 'Aromatic' hydrocarbons

(Coal tar derivatives):
Creosote pitch.

Petroleum products as a whole are classified according to flash point: Class (A) products have flash points below 73°F. (23°C.), class (B) between 73°F. and 150°F. (66°C.) and class (C) above 150°F. This is independent of any classifications which may be used for individual petroleum products.

These fuels are used for very varied purposes, the main division being between those which are burned in internal-combustion power plants and those used in domestic and industrial heating plants of different types. In general, petrol and the lighter fractions of petroleum derivatives go into the engines of land vehicles, while the heavier paraffin hydrocarbons are used in ships' engines and large land-based power plants. For heating, almost the whole range of petroleum and coal tar products are used, with the exception of petrol. Table III, showing some properties of oil heating fuels is given opposite, and a diagrammatic sketch showing the main processes in a petroleum distillation unit is shown in Plate 22.

As will be seen from the table, the fuels vary principally in respect of viscosity, flash point and calorific value. The viscosity, i.e., the time taken for a measured quantity of oil at a given temperature to flow through an orifice of given size, is of fundamental importance to the type of apparatus required for efficient burning. Oils are measured in seconds to indicate viscosity. Flash point has been dealt with at length in Chapter 1; the flash point of petrol is usually quoted as below 0°F. (—17°C.), that of paraffin or kerosene as between 81°F. (27°C) and 150°F. (74°C.), and those of several other types of oil fuel as ranging from 168°F. (75°C.) for diesel oil to 230°F. (110°C.) for heavy fuel oil, and higher. Calorific value does not vary substantially with different types of oil fuel, and with most petroleum fuels will be found to be in the region of 42,000 to 46,500 kilojoules per kilogram while for coal tar fuels the figure is likely to be under 39,500 kilojoules per kilogram (For comparison, a good quality anthracite coal may be expected to have a calorific value of 35,000 kilojoules per kilogram.

All the oils, etc., mentioned in this section will give off sufficient vapour to burn at certain temperatures. In some cases, such as petrol, this process goes on at temperatures much below atmospheric; in others, heating is required to a given temperature before vaporisation commences. This temperature (which is indicated by the flash point) is important to the fireman, because, since it is the vapour that burns, the fire will be extinguished if, directly or indirectly, the liquid can be cooled to a point when it ceases to vaporise. It will be seen that substances with a low flash point, such as petrol, can never be extinguished purely by cooling, though a high-flash-point oil, *e.g.*, asphalt, can. In general the rule of fire-fighting is that smothering agents such as foam are used on low-flash-point oils and that as the flash point rises cooling becomes important as a secondary means of extinction, until with high-flash-point oils, water alone, in the form of a spray to give the maximum cooling effect, is usually sufficient.

The flash point also to a large extent is related to the quantity of vapour which an oil will give off at atmospheric temperatures. Some oils, like petrol, are highly volatile and therefore give off vapour freely; others, like kerosene give off hardly any. The variation is important since, in general, the vapours from all petroleum fuels become explosive when mixed with air in the proportions of from 1 to 8 per cent. When the oil vaporises freely, the atmosphere *near* the oil (especially if it is in a confined space)

TABLE III
Fuel Oil Characteristics

Property	Petroleum								Coal Tar	
	Domestic			Industrial						
	Paraffin (Kerosene)	Light		Heavy		Extra Heavy		Creosote	Creosote	Creosote-pitch
		Class A	Class B	Class E	Class F	Class G	Class H	C.T.F.50	C.T.F. 100	C.T.F. 200
Viscosity Redwood No. 1 @ 100 deg. F. (38°C) (seconds)	30	45 max.	250 max.	250 max.	1000 max.	3500 max.	7000 max.	60 max.	100 max.	1500 max. 1000 min.
Flash point (closed) (deg. F.)	Over 73 (normally 100—150) (38°—66°C)			150 min. (66°C)	150 min. (66°C)	150 min. (66°C)	150 min. (66°C)	150 min. (66°C)	150 min. (66°C)	150 min. (66°C)
Storage temperature (deg. F.)	Room	Room	45 (7°C)	45 (7°C)	75 (24°C)	100 (38°C)	130 (55°C)	50 (10°C)	90–100 (32°–38°C)	90–100 (32°–38°C)
Calorific value (gross) (Kilojoules per kilogramme)	46,520	44,194	43,031	43,031	41,868	41,868	41,868	38,379	38,379	38,379

may be too rich in fuel for an explosion to take place. For this reason, it is often necessary when a rich concentration of petrol vapour is present in a confined space, to avoid the addition of air within the space lest this make the vapour explosive. Vapours from oils which vaporise slowly, on the other hand, are almost always within the explosive limits.

The vapours of all oils are heavier than air. They tend to collect at ground level and to flow along in an invisible wave. It is possible for vapour to move many hundreds of metres in this way and to take fire at a considerable distance from its source, when it will flash back to the container. It should be remembered that vapour can, in certain circumstances, be forced out of a tank which is being filled, and flow away on the downwind side.

Petrol burns very fiercely, producing large sheets of swirling orange flame surmounted by thinnish clouds of dark smoke and giving off great heat. The heavier oils tend to produce less flame but more and thicker smoke, within which the flames are enveloped, and to emit large particles of carbon. The heavy oils also tend to burn more lazily than petrol, since they evaporate more slowly and thus create less violent convection currents.

Unburnt petroleum vapours are toxic and can, if sufficiently concentrated, lead to headaches, dizziness and often intoxication. They may sometimes induce a state of hysteria. It is possible to endure a concentration of 0·03 per cent without ill effects. Dizziness is caused by concentrations higher than 0·07 per cent. Concentrations higher than 2 per cent may lead to unconsciousness and death, and for this reason breathing apparatus and safety lines should be worn by the fireman when performing rescue work inside a tank, for instance. Persons affected by the fumes should be removed to the open air, and artificial respiration should be applied if necessary. The after-effects which they are likely to experience are a severe headache and nausea.

2. FIRES IN TAR DISTILLERIES

Both Tar Distilleries and Benzole Recovery Plants produce much the same type of product, but in the first case the raw material is tar obtained chiefly from gasworks, while in the second it is gas obtained from coke ovens.

(a) Processes

Tar is distilled by processes of evaporation and re-condensation into its component fractions, and the process is shown in diagrammatic form in Plate 21. The broad division is between heavy oil (*e.g.*, anthracene oil), middle oils (*e.g.*, cresylic acid) and light oils (*e.g.* benzole); in addition, solid naphthalene is usually found as well as the residue, pitch. Subsidiary activities such as the re-refining of waste lubricating oil or the manufacture of soft soap may also be carried on.

(b) Principal hazards

(i) Storage tanks for the various grades of products. These are seldom as large as on a tank farm, but they are often closer together. They are seldom bunded and seldom protected by a fixed foam installation. The tanks are often of unfamiliar shapes, being tall and of small diameter, or horizontal.

(ii) Stores of chemicals used in the process of distillation (especially concentrated sulphuric acid and caustic soda), and also of those produced in the distilling process, *e.g.*, phenol (carbolic acid).

(iii) Steam stills, where the raw material, after being washed with sulphuric acid, is worked up through fractionating columns to twin receivers, whence it is run off to the tanks and similar apparatus for obtaining other oils.

(iv) The tar itself—stored in underground tanks with wooden tops, in open tanks and in wooden casks—is not readily flammable but will burn when sufficiently heated. The casks and drums are liable to explode.

(v) Pitch in beds.

(vi) Naphthalene stores.

(c) Fire-fighting

In general, fires are most likely to occur when spirit leaks into the steam flues (in which case foam is most valuable) or in the naphthalene producing plant, due to gas pressure as the result of choking of pipes by naphthalene crystals.

In the case of a bad fire, blazing tar, pitch and oil may run out of the distillery, while drums and tanks may explode as they become heated or as released gases are ignited. In general, the flow should be stopped by damming, the fire should be put out by foam or water spray, whichever is appropriate, and exposures should be cooled.

Care should be taken not to walk into open tar or pitch stores in smoke or in the dark nor to fall into underground tanks.

3. FIRES IN BENZOLE RECOVERY PLANTS

(a) Process

The process, (which is shown in diagrammatic form in Plate 20), begins in coke ovens, where coke is produced by heating coal. The gas used for benzole recovery is given off in the process. This gas is taken through a cooling system which, lowering its temperature, extracts tar and whatever can be condensed (*e.g.*, water). The gas then goes to a filter plant, where by treatment with sulphuric acid, ammonia is extracted to give sulphate of ammonia.

The gas then goes to scrubbers, and creosote oil is added; in the benzole house steam is applied to liberate the benzole, which is cooled in a condenser and purified by distillation, carbon disulphide (among other products) being removed. The benzole is then washed with sulphuric acid and caustic soda and re-distilled, giving benzole, toluol and xylol. There is also subsidiary plant for reconditioning creosote oil.

(b) Fire-fighting

Nothing can be said that is not contained in the paragraphs on individual chemicals (see Section 16).

4. FIRES IN OIL-BURNING INSTALLATIONS

(a) Features of oil-fired apparatus

Heating apparatus for use with oil fuels may consist of industrial heating plants for a wide variety of purposes, such as ovens, furnaces and boilers; central heating and other types of water boiler for domestic and similar property; and space heaters for heating air in any type of property. The apparatus in all cases normally comprises the burner, the storage tank, and the pipework and controls, with the possible addition of a small service tank as well as a storage tank. *British Standard Specification* No. 799 specifies apparatus for oil-burning appliances.

It is usual for all except the smallest installations to be automatically or semi-automatically fired as opposed to hand fired. When automatic or semi-automatic firing is provided, this is usually worked by electricity. Burners may be of the atomising or the vaporising type: in the former, the fuel is broken up into a spray of very fine droplets, the oil being fed under pressure to the jet, and in the latter the oil is pre-heated to above its flash point before feeding it to the burner. The air supply may be by natural draught or by forced draught in the form of a fan. The oil supply pipework is usually of small diameter copper, buried where possible, and electricity supplies in connection with the burner will normally be in conduit. The controls associated with the burner may include an oil level regulating control, thermostats, flame failure devices, time switches and a fire valve. (See paragraph (c), page 108).

Storage tanks are preferably placed out of doors but may be indoors in a fire-resisting tank chamber. In older or smaller installations the feed may be by gravity directly from the main tank, but in many newer installations the better arrangement of a gravity supply from a small service tank, fed by a pump from the main tank, will be found. Storage tanks may contain anything from 1300 to several thousand litres: they may or may not be electrically heated to give correct viscosity to the oil, according to the type of oil used.

Certain types of small space heater are becoming more popular in which a metal casing contains the burner and controls, together with fans for the propulsion of heated air; the larger types of space heater are supplied with oil from a storage tank elsewhere in the same way as has been described above, but some of the smaller types contain their own fuel tank. Fig. 17 shows a type of oil-fired space heater suspended from the ceiling.

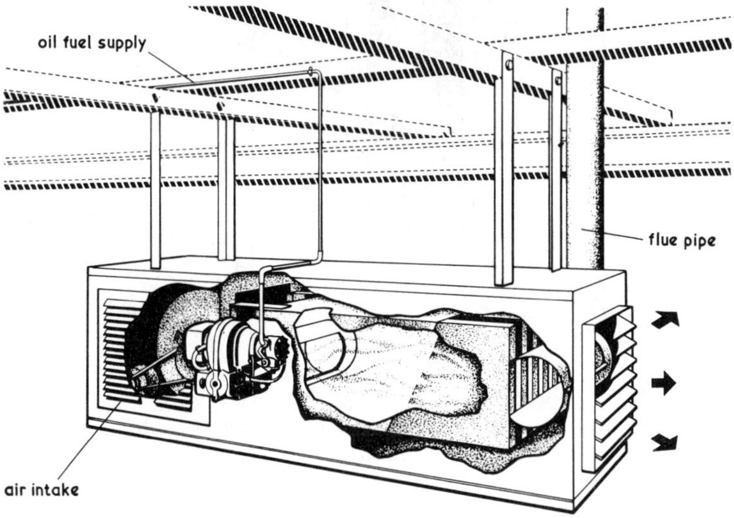

Fig. 17. An oil-fired air heater suitable for suspension from a ceiling. (Harris Engineering Co. Ltd.)

Little need be said here of the characteristics of the familiar portable paraffin (kerosene) heater: many of these employ a wick or wicks, and are referred to as 'blue flame' heaters, while others operate on the 'chicken fountain' principle, in which an inverted tank supplies the oil at a pre-determined level to an open burner trough with an asbestos 'kindler' immersed therein. The former are usually 'long-drum' burners (Fig. 18) and the latter 'short drum burners' (Fig. 19).

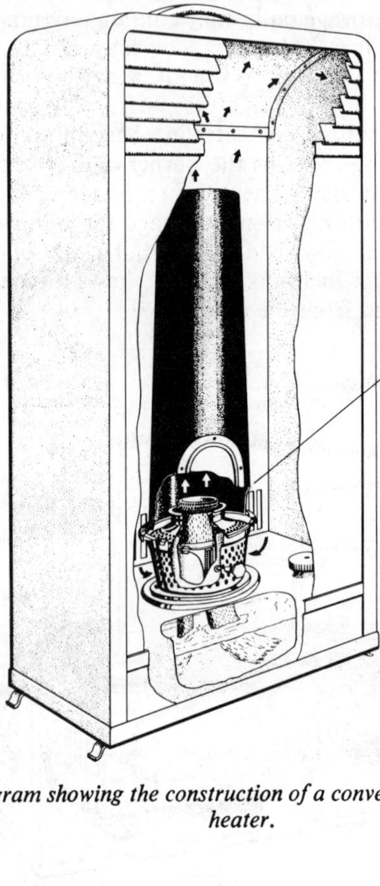

Fig. 18. Diagram showing the construction of a convector paraffin (kerosene) heater.

(b) Fire and explosion hazards

Among the causes of fire or explosion in oil-fired heating apparatus are the following:

(i) Faulty mechanism, *e.g.*, overheating due to failure of electrical or other controls; explosion of oil vapour in the combustion chamber; sooting up of thermostats in chimney flues; leaking pipe joints; (in portable heaters) flare-ups caused by draughts or by the heater being overturned.

(ii) Faulty operation *e.g.*, attempting to re-light the burner while still hot after flame failure; placing automatic safety controls out of commission; the use of incorrect fuel; allowing an accumulation of rubbish near the burner; etc.

Any of these, or other, causes may produce an explosion in the burner chamber sufficient to rupture the oil feed lines and start a fire in the vicinity of the burner. This may or may not lead to the

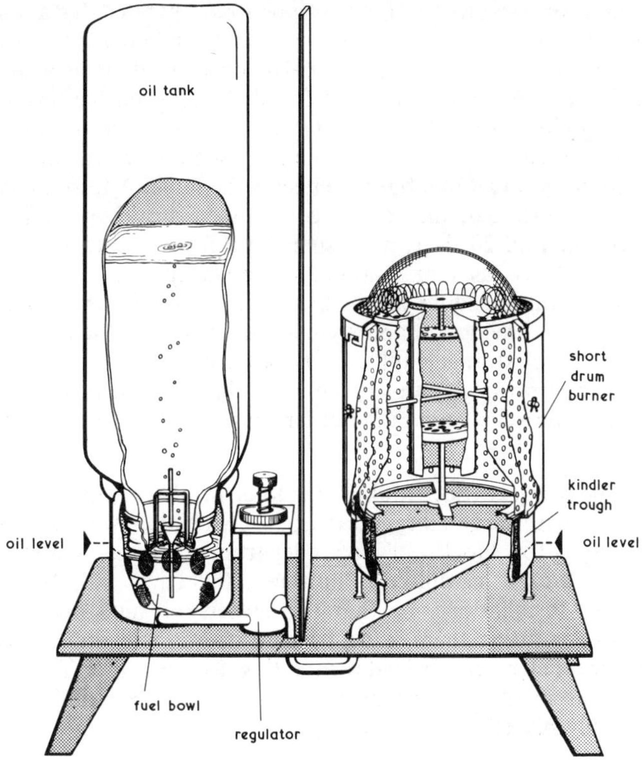

Fig. 19. Diagrammatic arrangement of a radiant type paraffin (kerosene) heater.

ignition of the oil in the storage tank and ultimately to the rupture of the tank itself according to circumstances. Some fires in storage tanks have resulted from an outbreak of fire nearby from other causes. It should be borne in mind that where a former hard-fuel burning apparatus is converted to the use of oil, there is an increased risk of fire around the flues because of the higher temperatures involved.

(c) Fixed fire protection

(i) *Fire valves.* Apart from the normal controlling devices associated with the burner the commonest fire protective device is the fire valve, which should be a part of every oil burning installation but is not always so found. It consists of a valve in the oil supply pipe, as near the storage tank as possible, operated by a rise in temperature at or over the boiler or burner chamber. One way of operation is by fusible link, the valve being held open against a spring by a wire connected to the fusible link; when the link parts, the spring closes the valve. A weighted valve may be used instead of a spring-loaded valve, or it is sometimes operated by the expansion of air or liquid in a tube, but the effect is the same. A third system is electrically operated, and all three methods are shown diagrammatically in Fig. 20. It is not usual for a fire valve to work on a rate-of-rise basis, as some automatic fire alarms do, but only on a given temperature being reached. If a fire valve is not fitted, the importance of having a manually-operated valve on the supply line from the main storage tank will be appreciated, especially if the supply is by gravity. However, the valve should be as near the tank as possible, or a fire round the burner chamber will prevent its being closed.

(ii) *Foam inlets* (See Book 9, Chapter 8). These may be found associated with large storage tanks, in connection with the larger installations for blocks of flats, department stores, government buildings, etc. It is unlikely, however, that they will be found in smaller installations, or that the owners of such installations will consider fitting them, except where operational difficulties are to be expected by fire brigades in extinguishing a fire. This might be in a basement or other instance where the tank is indoors and inaccessibly placed.

(d) Fire-fighting

In older installations of all kinds, especially those in which the layout has been poorly designed, or improvised, a fire originating in the burner apparatus may well involve the storage tank at an early stage. When this is so, the fire brigade attendance is faced with what is substantially an oil fire. In industrial plants this may also be the case, though here it is not possible to forecast the form which an outbreak may take, due to the variety of types of apparatus and of surroundings in which it may be fitted. However, in many

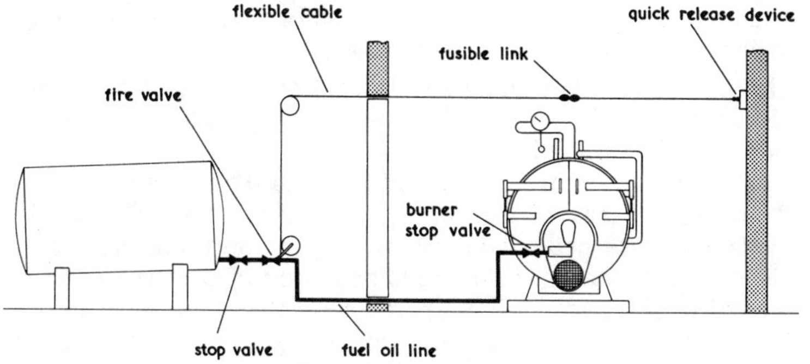

(*a*) *Mechanically by fusible link.*

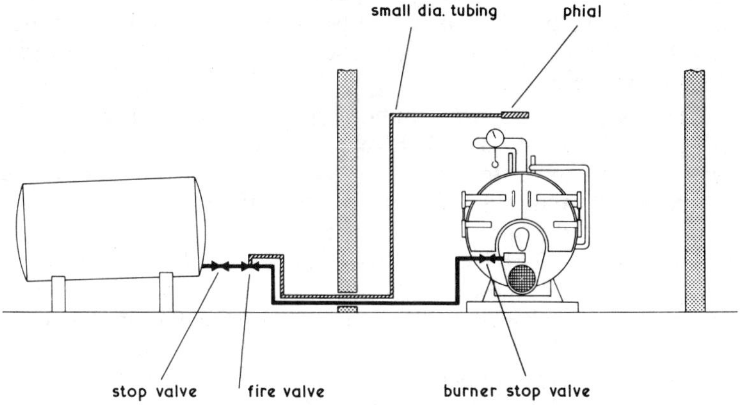

(*b*) *The pressure tubing system.*

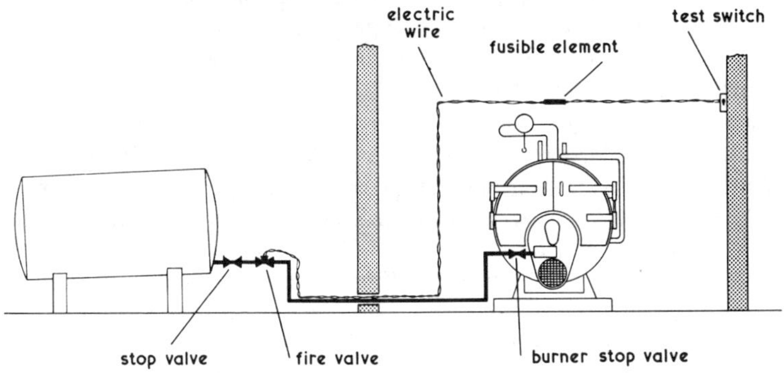

(c) *Electrically operated.*

Fig. 20. Three methods of operating the fire valve on an oil-fired boiler installation.

instances the pattern of events will conform in some, if not all, particulars with the outline given below.

In oil-fired central heating apparatus, from the position in which the heating installation is likely to be found, (*i.e.*, low down in the building and possibly in the basement), the fire is likely to be hard to fight since access may well be difficult and the heat and products of combustion may be trapped in confined quarters. Though this will in any case necessitate the wearing of breathing apparatus, it is also likely that air in the neighbourhood of the fire will be restricted so that smoke will be particularly thick, but burning may also be somewhat restricted, and the fire may be slow to develop. Foam will be an effective extinguishing agent, and, if the tank is involved, it will be necessary to get one or more foam branches to work. This type of fire may also be dealt with by means of water sprays, a delivery system giving a solid cone of reasonably fine spray being desirable.

When oil is burning in a confined space, it may be found useful to use a water spray to cool the burning vapour and the container itself, before using a smothering agent such as foam. Foam has little cooling effect, and it is necessary to allow time for the liquid, the container and the surroundings to cool down, otherwise there is always danger of flash back and re-ignition, either from hot surfaces or from deposits of red-hot carbon. The foam seal should, therefore, be maintained and increased for some time after the fire has been put out, especially when the oil is in a confined space. The more volatile the spirit, the thicker should be the foam seal used, as the vapour may push its way through and form an explosive mixture with the air above the foam. If this should happen, a flash-back may cause an explosion. Water spray can usefully be employed to cool heated surroundings. Nothing should be done, however, to interfere with the foam seal.

When using water spray upon the high-flash-point oils for which its use is suitable, a fan shaped spray will be found useful. A spray of this type is usually more effective than the spray produced by a diffuser branch. It provides a horizontal screen above the fire and cuts off its supply of air. The spray falls on the surface in a line that can be advanced as the fire is extinguished. If a fan spray branch is not available, the desired effect may be produced by holding the thumb or the blade of an axe over a jet.

In cases where the storage tank has been installed separately from the boiler or burner apparatus (the tank being very commonly out of doors in these cases), the fire originating in or near the burning apparatus is not likely to involve the tank itself but may involve oil flowing out the supply pipe to the burner; this is likely to occur when the fire or explosion has ruptured the supply pipe and no fire valve or other automatic cut-off valve is fitted—or, if fitted, is not in working order. The probable result here is an oil fire spreading over the floor of the boiler room, and it is essential that action be taken at once to stop the flow of fuel oil to the fire. Every installation is virtually certain to have a main cut-off valve at the point where the supply line leaves the storage tank, and this should be closed. The oil fire should then be dealt with by means of foam or water spray in the ordinary way. In more fortunate circumstances the ignition of accumulated fuel or vapour at the burner apparatus which ruptures the supply pipe will be followed after a short time by the cutting off of the oil supply, either manually as soon as the fire was discovered or by the fire valve. Since the operating temperature of the latter is usually 155°F. (68°C.) this will most probably have occurred before the arrival of the fire brigade and before any substantial amount of oil has escaped. This may be expected to leave an oil fire of small proportions in and around the burner chamber; such a fire will not have had the chance to build up great heat and should respond to lighter equipment such as a foam extinguisher, hose reel spray or even a carbon-dioxide or dry powder extinguisher.

The present tendency towards fitting oil-fired boilers in domestic property frequently results in the installation of the burner apparatus in the kitchen or scullery instead of a special boiler room with some degree of fire separation. Where this is the case, there is obviously a greater probability that a fire originating in the oil burning equipment will become predominantly an ordinary room fire and require to be dealt with accordingly.

5. FIRES IN ROAD AND RAIL TANK TRUCKS

Tanker trucks may be found either as road vehicles or as railway wagons. They commonly contain petrol or some type of oil; when, however, other commodities are involved (*e.g.*, carbon disulphide, liquefied petroleum gas, etc.), the properties of the commodity in question must be taken into account. (*See* Plates 25 and 26). The following fire-fighting points should be noted:

(a) Railway trucks

(i) Use a water screen or a good foam wall to isolate burning trucks from trucks not yet involved in the fire.

(ii) Watch for petrol, etc., escaping down culverts and into ditches, and either bank up the ground to prevent its escape or neutralise the liquid by dilution, if miscible with water.

(iii) Protect any exposures.

(iv) Foam should be used when petrol or oil is contained in a suitable form, but water is most useful upon relatively undamaged trucks.

(v) Salvage as much unburnt liquid as possible.

(vi) Remember the danger of a flash-back from heated metal after the fire is out. Foam and/or water should continue to be applied for some time after the fire has been extinguished.

(b) Road vehicles

(i) Watch where the liquid is flowing, and take steps to prevent it escaping into sewers. If it is impossible to avoid this, the authority responsible for sewers should be notified without delay.

(ii) Protect the surrounding area, extinguish fires and naked lights in buildings that may be affected by vapour, and stop by-standers smoking.

(iii) Work in towards the vehicle, making sure that no flash-back occurs.

(iv) Water spray will probably be found of greatest practical use unless the liquid can be confined.

(v) Salvage any unburnt liquid.

6. FIRES IN MOTOR VEHICLES

A fire in a motor vehicle generally involves petrol and is, therefore, likely to increase rapidly in intensity. Fires are generally caused by a back-fire through the carburettor, an electrical fault or, in the case of vehicles fitted with twin rear tyres, through the heat generated when only one tyre is punctured and the vehicle continues on its way.

The following points are important:

(a) Remove the vehicle if there is a danger of the fire spreading. Alternatively, remove or protect its surroundings if such a course is possible.

(b) When the engine itself is on fire and running, and a petrol tap is fitted, the tap should be turned off and the engine run until it empties the carburettor and stops. (The tap is usually clearly marked on those types of vehicle to which one is fitted). Private

RECOVERY OF BENZOLE FROM COKE OVEN GAS

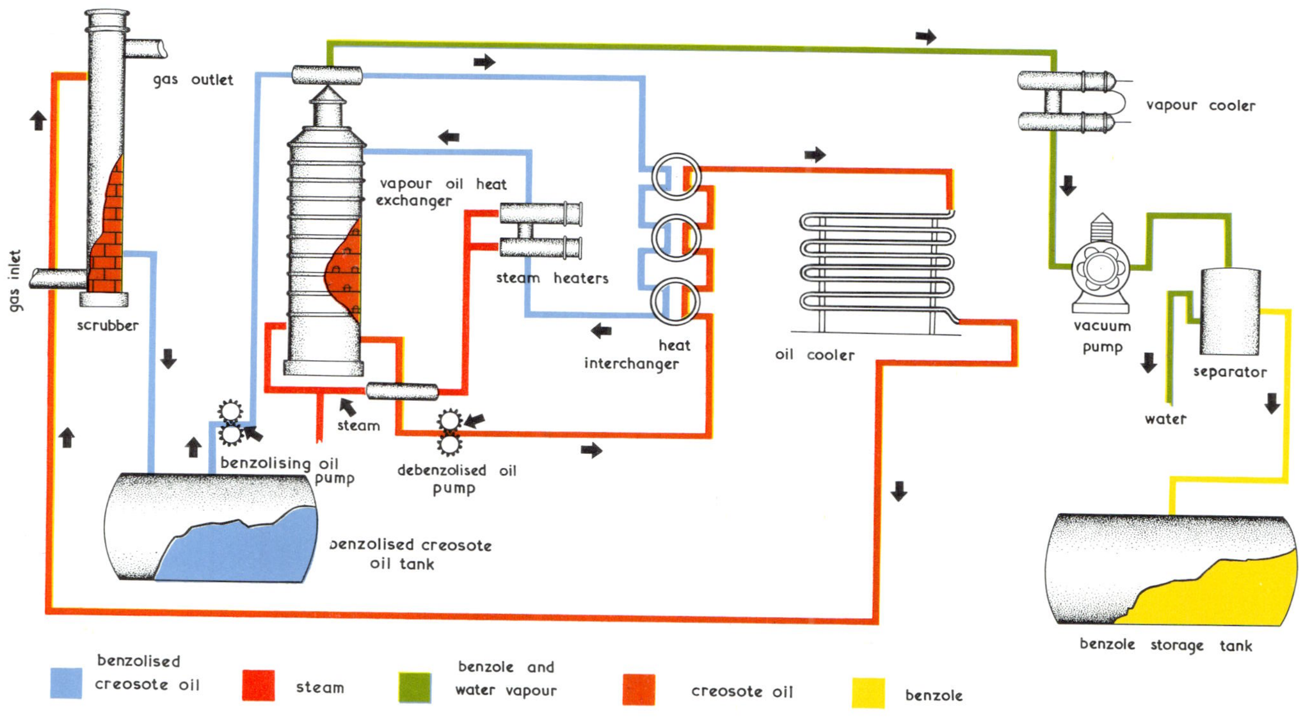

Plate 20. Benzole vapour is absorbed into creosote oil, which is then made to liberate the benzole vapour. This in turn is condensed into benzole. The creosote oil is recirculated and used again. (Section 5)

Plate 21.

MANUFACTURE OF COAL-TAR PRODUCTS

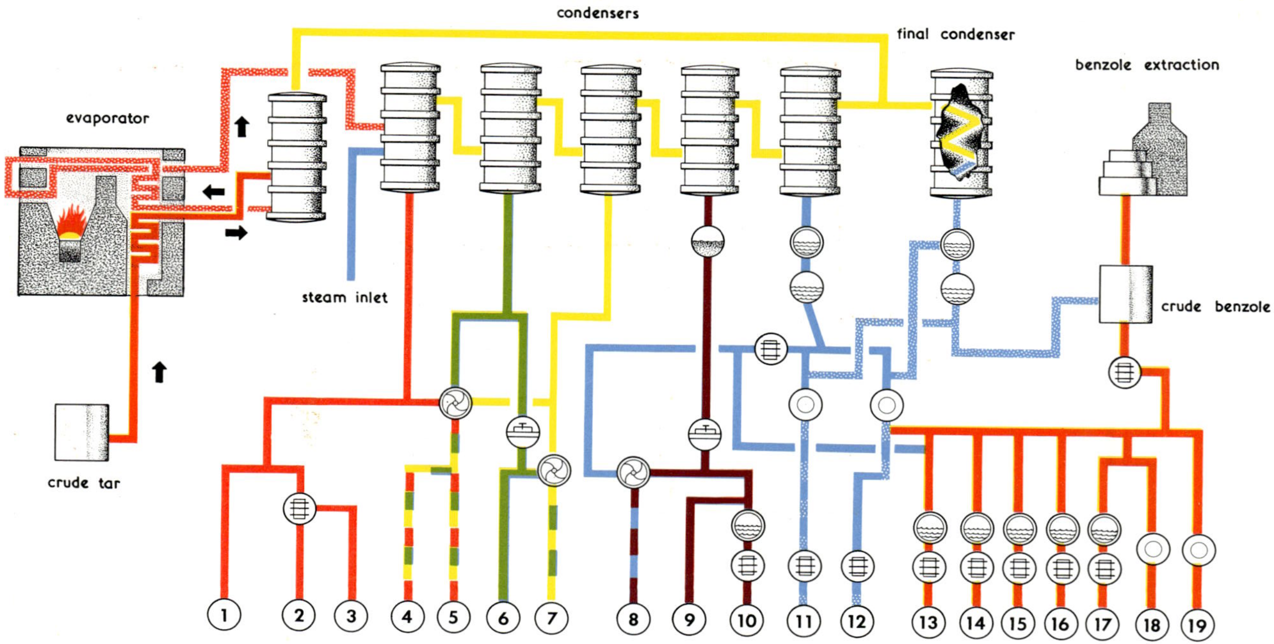

The process of tar distillation is effected by evaporation of the tar and re-condensation into its component fractions. A large number of products is obtained. (Section 5)

MANUFACTURE OF COAL-TAR PRODUCTS

KEY TO PROCESSES

- MIXING
- DISTILLATION
- PRESSING
- ACID WASHING
- SODA WASHING
- CRYSTALLISATION
- OTHER CHEMICAL TREATMENTS

	DERIVATIVE	USES OF DERIVATIVES
1	PITCH FOR BRIQUETTING	COAL BRIQUETTES PATENT FUELS.
2	HARD PITCH	PULVERISABLE FUEL, MOULDED ARTICLES, OPTICAL WORK, FOUNDRY WORK, ELECTRODES.
3	HEAVY OIL	LIQUID FURNACE FUELS.
4	COAL TAR FUELS	
5	REFINED TAR	ROAD SURFACE TAR MACADAM.
6	ANTHRACENE	DYES.
7	ORDINARY CREOSOTE	
8	CREOSOTE FOR HYDROGENATION	SPIRIT (MOTOR AND AVIATION.)
9	HOT PRESSED NAPHTHALENE	GLYPTAL PLASTICS, LACQUERS, SOIL INSECTICIDES.
10	PURE NAPHTHALENE	DYES, INSECT REPELLANTS, PRESERVATIVES, PLANT HORMONES.
11	PHENOL, CRESOLS, XYLENOLS, HIGH BOILING TAR ACIDS	BAKELITE PLASTICS, DISINFECTANTS, ASPIRIN, ORE FLOTATION, OIL REFINING.
12	PYRIDINE BASES	SULPHA DRUGS, VITAMINS, RUBBER ACCELERATORS, TEXTILE AGENTS.
13	HEAVY NAPHTHA	DISINFESTATION, INSECTICIDES, RESINS, FLOOR TILES.
14	SOLVENT NAPHTHA	WATER PROOF TEXTILES, PAINTS, RUBBER SOLVENT.
15	XYLENE	PRINTING INKS, PAINTS, PLASTICS, PERFUMES.
16	TOLUENE	OIL REFINING, EXPLOSIVES, DETERGENTS, SACCHARIN.
17	BENZENE	DYES, SULPHA DRUGS, INSECTICIDES, DETERGENTS, NYLON PLASTICS.
18	MOTOR BENZOLE	BENZOLE, PETROL MIXTURE.
19	CARBON DISULPHIDE	ARTIFICIAL SILK.

Key to the processes and products of tar distillation.

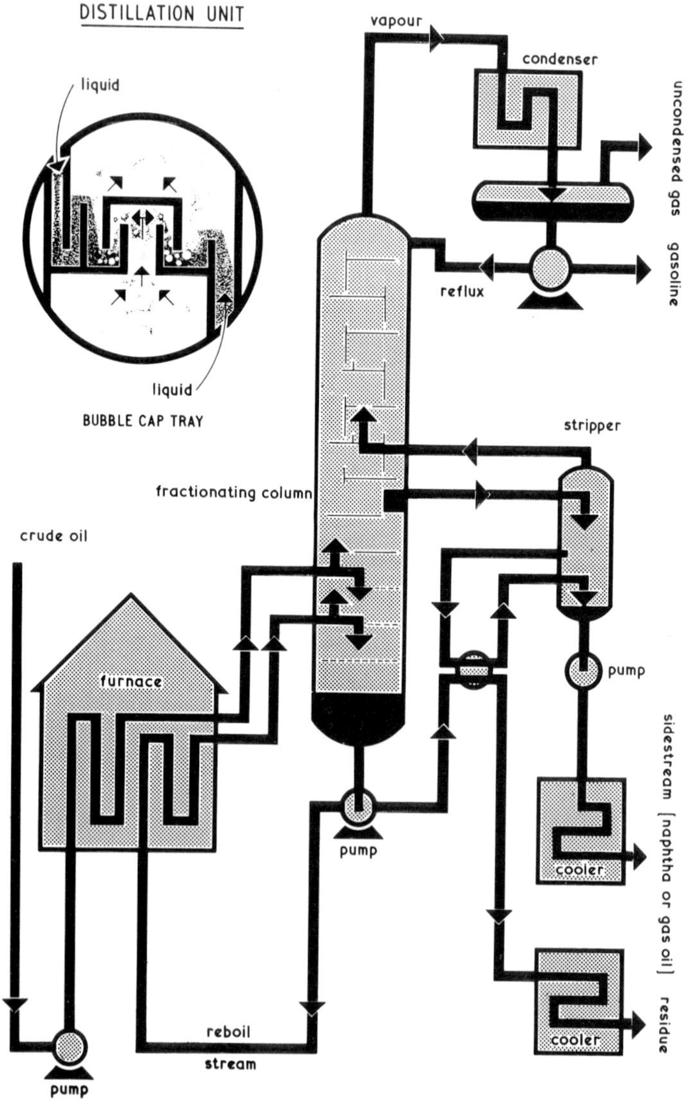

Plate 22. Crude petroleum is broken up into its various fractions in a distillation unit. The bubble cap trays (shown greatly magnified and with one cap only for simplicity) are the essential feature of the fractioning column. (Section 5)

cars are almost never fitted with taps, and in this case the engine should be switched off at once.

(c) When the fire is under the bonnet, the carburettor will generally be the seat of the fire, unless the fuel line is fractured. Insert the nozzle of a vaporising liquid extinguisher and discharge through any available aperture high up in the bonnet on the carburettor side but not through the front grille (*see* Fig. 21). If no aperture is available, lift the bonnet slightly and discharge into the side of the engine space containing the carburettor. If the fire is tackled early little damage will normally be done except perhaps to the ignition leads.

(d) If a vaporising liquid extinguisher is not available, a foam extinguisher may be used, or the fire can probably be knocked out or smothered with a small mat or rug. If nothing else is available, earth can be used.

Fig. 21. Method of tackling a fire under the bonnet of a motor car with a vaporising liquid extinguisher. The arrows indicate alternative positions.

(e) If the fire gains a hold, a small jet, or preferably a spray, may be necessary.

(f) When fires occur in other parts of a vehicle, *e.g.*, tyres, a jet or spray may be used unless the petrol tank bursts.

(g) For a fire involving electric wiring, vaporising liquid or other suitable extinguishers may be used. The battery, which is generally either under the bonnet or under one of the seats, should be disconnected as soon as possible.

(h) If the fire becomes extensive, the petrol tank should be cooled with a spray to prevent it bursting.

7. FIRES IN PETROL FILLING STATIONS AND GARAGES

(a) Fire and explosion risks

(i) *Petrol pumps and storage tanks.* It is very rare for a fire to break out in association with petrol pumps for storage tanks except when operations are in progress in connection with them; the operation may be that of filling the storage tank from a transport tanker, or filling the tank of a motor vehicle from the petrol pump. Very serious fires, sometimes following an explosion, have occurred from the escape of petrol vapour during either of these operations, the petrol vapour having travelled in some cases as much as 15 metres from its source and been ignited by some such agency as a slow combustion stove or an electric switch. On occasions, leaks have developed in the pipe or connections between a transport tanker and the storage tank, or in the filling pipe from a petrol pump, and, more commonly, a quantity of petrol has overflowed from the tanker or the petrol pump during the course of filling operations. If the vapour from this has become ignited, a fierce fire has resulted, and this has been distributed perhaps over the ground, on one or more vehicles and on the clothing of pump attendants, etc.

(ii) *Petrol, kerosene and solvents in working spaces.* A regrettable number of fires have been originated in garages through careless handling of relatively small quantities of highly flammable fluids in and around the working area of the garage. It is common to find unsuitable heating appliances in garage workshops, among them slow combustion stoves and open element electric fires. Open vessels containing petrol or other flammable fluids are frequently found on benches and on the floor, and only too often these fluids are applied by means of rags, etc. for cleaning purposes, the users' clothing, hands, etc., often becoming soaked with the liquid. A hazardous operation is the draining of vehicle petrol tanks, which not infrequently results in a flow of petrol vapour across the workshop floor, leading to its ignition by the nearest source of heat.

(iii) *Stocks of oil and grease.* Every garage carries considerable stocks of a wide variety of lubricating oils and greases, many of which are in closed containers but some of which may be in drums and even in open-topped vessels. The standard of housekeeping maintained in the garage will determine the degree of risk of a fire breaking out in association with these oils and greases, but in general the risk is not high. On the other hand, the likelihood of these flammable stocks becoming involved in a fire originating from some other cause is considerable. This will tend to produce a fierce running fire productive of much heat and smoke.

(iv) *Cellulose spraying booth or enclosure, stocks of cellulose and paint.* Many garages have a cellulose spraying installation whose layout is controlled by Regulations. Fires, particularly originating in and around the hood and ducting of a cellulose cabinet are not uncommon and tend rapidly to involve all cellulose deposits in and near the installation. Because, however, of the separation required between the cellulose spraying area and the remainder of the garage, the fire spread outside the area is not necessarily rapid. Stocks of cellulose and paint will intensify and spread a fire from other causes but are not a serious fire risk in a well maintained building.

(v) *Oxy-acetylene cutting gear and cylinders.* This type of equipment is very widely used in garage workshops and can very easily cause an outbreak of fire through incorrect or careless use. A fire in a garage should never be tackled without attention being paid to the possible presence of oxygen and acetylene cylinders.

(vi) *Petrol in tanks of vehicles.* A not uncommon cause of fire in filling stations and garages is an accidental outbreak in a motor vehicle itself. Such fires are relatively easily extinguished by the garage employees, or alternatively, it is frequently possible for an affected vehicle to be removed. Welding and cutting operations on vehicle petrol tanks have often been known to cause explosion and fire through being undertaken on tanks which have not been rendered gas and vapour-free.

(b) Fire-fighting

A fire in a garage may take several forms but whatever the form it is likely that one or more of the following points will need consideration:

(i) Although in some cases only an isolated quantity of petrol has been spilled and ignited, there are many other instances of fire involving petrol pumps, tanks, drums or vehicles in which the supply of fuel continues to flow to the fire. This must be halted at an early stage by turning off the appropriate tap or switch or by providing a barrier of sand or earth or by the removal of the vehicle providing the petrol supply. It is also of vital importance to ensure the cutting off of any possible vapour discharges near the fire which might allow a flash-back to the supply point.

(ii) Steps must also be taken to protect surrounding materials from the effect of the fire. This may best be done by cooling, as would be the case with an ordinary fire, but motor vehicles should be removed and stocks of flammable liquids should either be covered with a foam blanket (especially if their contents are exposed) or cooled with a water spray. Note, however, that no water spray should be allowed to reach a foam blanket and break it down.

(iii) In many instances where foam is used for extinction, cooling of the surroundings will have to be undertaken in addition to the actual extinction of the burning liquid. When a flammable liquid fire has reached an advanced stage of development it is likely that metal or other objects involved in it will have become greatly heated and these can cause severe danger of re-ignition of flammable vapours while they remain at high temperature with petrol vapour in the vicinity.

(iv) The extinction of burning petrol has been the subject of much experimental work, and there is no doubt that great improvements have been made in the technique of dry powders for this purpose in recent years. Used with skill, dry powder can produce extinction over a large area with spectacular speed, especially on an open or spill fire in its early stages which has not yet built up to great heat and does not involve metal or other material which could provide re-ignition due to the absence of cooling effect. Carbon dioxide shares to a lesser degree the properties of quick attack and extinction shown by dry powder. Foam, on the other hand, is less spectacular in action, but, if methodically applied, can provide very positive extinction with less fear of re-ignition and is equally suitable for fircs of all sizes and many different types. Dry powder extinguishers are not compatible with mechanical foams, they should not be used together because of a tendency for the foam to be broken down by the powder. Vaporising liquids are completely effective in the limited number of cases where a concentration of the vapour can be maintained within the space occupied by the fire; they are thus pre-eminently suitable for engine fires and are of much reduced value in the open air. Water spray is unsuitable for petrol because it is less effective on the lighter fractions of petroleum products than on heavy oils, it may also have the effect of spreading some types of fire (such as a spill fire on flat ground) with serious results.

(v) In a petrol fire it is essential, in addition to cutting off the supply of fuel, to contain the petrol flow at the earliest moment and prevent it from spreading the fire by its own travel. Its entry into flood-water and other street drains should be prevented, and the surface drainage channels of garages and filling stations are provided with traps for this purpose. The control of the flow of the burning petrol may best be done by means of a sand or earth barrier, by piles of wetted sacks or rags, or in some instances by a foam barrier.

(vi) When a closed tank is burning at the vent or filling orifice it is then urgent that the flame be extinguished as quickly as possible since there is a risk of the vapour in the free space in the top of the tank being exploded. Such an explosion might be of a devastating nature, and the risk increases as time goes on because the heating up of the metal of the tank may make a change in the rate of flame propagation or the proportions of the gas-and-air mixture inside the tank. However, it is dangerous to extinguish the flame without a positive assurance that there are no sources of ignition in the vicinity which could re-ignite the issuing vapour after the extinction of the flame. The extinction in these circumstances can sometimes be done by means of a stiff foam, though this may have to be reinforced by an earth or clay barrier to contain it or an improvised barrier of wet rags or sacking.

In these circumstances it is also very important to avoid taking any step which could cause the admission of air into the top of the tank; obviously, this might have the effect of producing a suitable gas-and-air mixture inside the tank for explosion. Thus when a petrol tank is flaming at a vent, petrol should never be exhausted or drained from the tank, as this will entrain air into the free space.

(vii) If oil, grease, paint, etc. are involved during the course of a garage fire, the choice is between the application of foam or of water spray. Much will depend on the possibilities of getting to close quarters with the fire, since a fine water spray cannot easily be applied at long range, and also on the degree of heat development which has taken place since the outbreak; it must be remembered that if the build-up of heat among metal objects such as drums, cans, cabinets, etc., has been considerable, much cooling effect has to be produced before extinction can be facilitated and completely effective.

Liquefied petroleum gases

1. CHARACTERISTICS

The term liquefied petroleum gas (L.P.G.) refers to varieties of hydrocarbons derived from crude petroleum processes or from natural gas, being gases at normal temperature and pressure but which become liquid with either a moderate increase in pressure or a moderate drop in temperature, or both. The term does not embrace such substances as methane which, though having certain of the same characteristics, are lighter than air and therefore require different treatment.

These hydrocarbons include propane, propylene, butane, isobutane and butylene. The more readily liquefiable gases of this group are commercial propane and commercial butane, each of which may contain in varying amount several of the other hydrocarbons mentioned. Butadiene also is a liquefied petroleum gas; it does not occur naturally, but is obtained by process. It is chiefly used in the rubber industry and not as a constituent of L.P.G. fuels. The properties of these three gases as used commercially are as follows:

	Commercial propane	*Commercial butane*	*Commercial butadiene*
Vapour density	1·4–1·56	1·9–2·1	1·87
Boiling point	–40°F (–40°C)	30°F (–1·1°C)	24°F (–4·4°C)
Explosive limits (percentage)	2·2–9·5	1·9–8·5	2·0–11·5
Cu.m. of free vapour per litre of liquid (m^3 per litre)	0·249	0·235	0·257
Approximate calorific value kilojoules/kilogramme	50,009	49,265	—
Gauge pressure at 60°F (15·6°C) bar	5·17	1·03	1·1

It should be noted that the figures quoted above are for the commercial liquefied petroleum gases and that they may differ slightly from the figures quoted in Section 16 for the pure gases.

The different names under which these products are marketed are *Calor gas*, *Bottogas*, *Scottish Rural gas* and others; they are produced by five main petroleum products manufacturers and by other large commercial concerns.

Propane is only found in industry, being largely confined to oxygen cutting apparatus ('oxy-propane'). Butane, on the other hand, is used for cooking, lighting, and heating in small houses, boats and caravans. L.P.G. is odourless and non-toxic but has anaesthetic properties when inhaled in quantities for a long period; inhalation in moderate concentrations produces nausea and headache. Heavy concentration in a pit or similar confined space may produce oxygen deficiency and cause asphyxiation; such concentrations would present a serious risk of fire which might originate with explosive violence.

To assist in detecting the presence of L.P.G., a stenching agent is added (before sale) except where the intended use requires a gas free from odour as, for example, in the manufacture of aerosols. The odour produced is sufficiently strong to ensure that leaks can be detected well before the gas concentration approaches the lower limit of flammability in air. A leak at a flange or valve may often be detected by the presence of hoar frost due to the evaporation of the L.P.G. causing a local reduction in temperature and the freezing of moisture in the surrounding air.

The density of the liquid is approximately half that of water, but the density of the gas is greater than that of air. Therefore, L.P.G. if released, will tend to accumulate at low levels, hugging the contours of the ground, filling valleys, ditches and other declivities. In a comparatively still atmosphere, it will not disperse easily but can travel for long distances from the point of release. In this respect it differs from the usual town gas which, being lighter than air, tends to rise and disperse readily.

L.P.G. becomes flammable when mixed with air in a concentration within the flammable range (see above) by volume. 4·5 litres of butane when vapourised will produce not less than 0·9 m^3 of gas at atmospheric pressure and 60°F. (15·6°C.); thus, at a 5 per cent concentration in air, 18·12 m^3 of flammable mixture would be formed. Leaks of L.P.G. are very likely to occur as vapour but if, at atmospheric temperature, there is a leakage from the liquid phase of a pressure container or from a pipe-line, an outflow of liquid butane or butadiene can occur and there may be some delay in evaporation to the gas phase. An out-flow of liquid propane is also possible but, owing to its lower boiling point, this would be more rapidly converted to gas. The pressure in a container is related directly to the composition of the particular variety of L.P.G. and the liquid temperature. In all cases, a moderate rise in temperature greatly increases the pressure; for example, at 100°F (37·8°C.) commercial butane exerts a pressure up to 4·85 bar gauge, and at the same temperature commercial propane will exert a pressure up to 15 bar gauge.

Liquid L.P.G. has a relatively high co-efficient of expansion and, therefore, a container is filled to a limit which permits liquid expansion due to a normal rise in temperature without danger of over-stressing the container by hydraulic pressure.

2. DOMESTIC USE OF L.P.G.

(a) Storage cylinders

Butane for domestic uses is handled in light gauge steel bottles of which the commonest type is about 460 mm high and 300 mm in diameter and holds 14·5 kilograms in weight of butane; it weighs in all about 27 kilograms when full. There are smaller sizes, and there is also a 38 kilograms capacity cylinder weighing, when filled, about 82 kilograms. Illustrations of various types of liquefied petroleum gas cylinders are shown in Fig. 22. It is usual for containers to have to pass a pressure test fixed by the manufacturers of the gas every few years but this pressure bears no relation to the ultimate bursting pressure of the cylinder. The cylinders are fitted with a closure valve worked by a hand wheel and some have a safety valve discharging at about 25 bar; most makers recommend that the cylinders should never be exposed to a temperature of more than 110°F. (43°C.) Propane cylinders are liable to be found anywhere in industrial premises, as can be deduced from the fact that they are used for cutting and welding. Butane cylinders may be found indoors in houses and cottages in rural areas, in houseboats and converted barges and in caravans, but *British Standard Code of Practice*, 339 recommends that they should preferably be stored in the open air. They may be kept inside providing adequate ventilation is afforded, and it is common to find them in lockers or cupboards to which access is from the exterior though the main valve can be manipulated from inside.

(b) Fire and explosion risks

About three-quarters of all accidents with L.P.G. resulting in explosion or fire are due to leaks, which may occur from valve assemblies, from the pipework or from the appliances connected to it. Common causes are mishandling of the apparatus, vibration (in the case of moving vehicles), the perishing of flexible connections (most types of rubber are slowly attacked by these gases), inadvertent extinction of burners by draughts or lack of ventilation, leaving burners unlit and turned on, and, in the case of propane, blow-backs. The manufacturers habitually warn users never to look for leaks with lighted matches but always to turn off the main valve when a leak is suspected and to put the installation out of use until examined by a competent service agent.

(c) Fire-fighting

(i) A small fire at a leak can usually be extinguished efficiently and quickly by means of a damp cloth. The fire brigade will almost never have arrived at the incident at this stage, and will more commonly encounter a normal fire in which the cylinder or piping may or may not be involved.

(ii) If the cylinder is not involved, it should be turned off at the main valve, disconnected and removed to a place of safety.

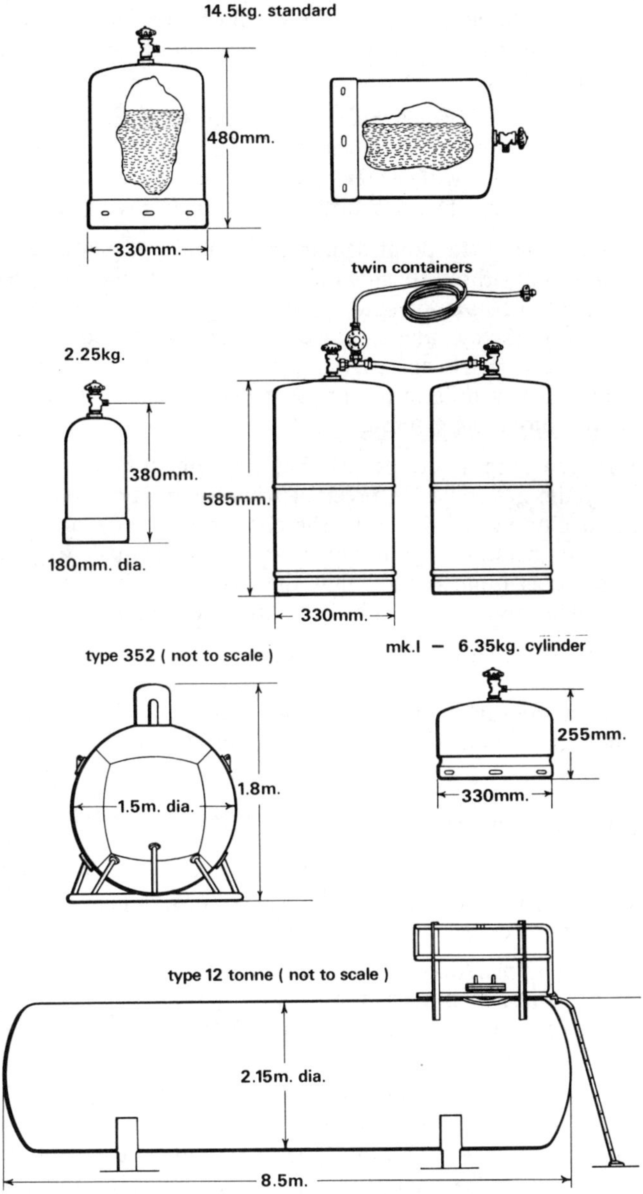

Fig. 22. Types of liquefied petroleum gas cylinders. Note that the liquid can emerge if a fairly full cylinder is turned on its side with the valve open.

(iii) If the cylinder is involved it should be cooled with a water spray until the main valve can be turned off. It should then be removed as above.

(iv) If the main valve has been damaged or for some other reason cannot be turned off, or if the gas is issuing from the cylinder itself, cooling the cylinder with water spray will reduce the gas flow sufficiently for the fire to be extinguished and the cylinder to be removed.

(v) The flame at the point of burning should not on any account be extinguished without turning off the main valve unless it is certain that there is no source of ignition in the vicinity. (If gas continues to escape from the cylinder it may collect and be ignited by a naked light or electric spark nearby.) If the leak can be assumed to be small, and local ventilation is good, it will be safe to extinguish a flame and remove the cylinder.

(vi) If there is reason to believe that the cylinder has been directly heated by the fire for an appreciable time there is a danger of the cylinder exploding. In this case, the cooling of the cylinder should be done from behind cover and no effort should be made to get close to it until it is quite certain that the explosion danger has been removed. This applies particularly to propane cylinders.

(vii) The main valve of a disconnected cylinder should never for any reason be turned on when the cylinder is not in a vertical position; in any other position, the valve may be below the level of the liquid, and liquid will emerge instead of gas (*see* Fig. 22).

3. BULK STORAGE OF L.P.G.

The storage and transportation of large quantities of L.P.G. in liquid form is both convenient and economical. Liquefaction may be achieved in two ways:

(i) by the application of pressure in excess of the equilibrium vapour pressure; at normal temperature (60°F., 15·6°C.), 1 bar gauge for commercial butane, 5 bar gauge for commercial propane and approximately 1·1 bar gauge for butadiene;

(ii) by cooling to temperatures below the boiling point. When liquefied in this way the refrigerated L.P.G. may be stored at pressure only slightly in excess of atmospheric provided it is continually refrigerated or the low temperature is maintained by the cooling effect of vapour boiling off and being continuously removed.

Refrigerated storage is only economical when large quantities of approximately 2,000 tonne or more are required to be stored at one location. The usual method of storage and transporation is as a liquid under pressure at ambient temperatures.

(a) Bulk storage sites

The storage of L.P.G. in large quantities occurs:

(i) at refineries and other major installations and distributing depots of the oil industry:

(ii) at producer and holder stations of the Gas Boards;

(iii) at the premises of some large industrial users.

It is normal to plan installations upon a site with due regard to fire exposures between containers (or groups of containers) and buildings, other surrounding risks, and boundaries. The contour of the ground, as it may affect the flow of heavier than air gases relative to roadways or other places where ignition might occur, is also a factor. Where several containers are required, alignment across rather than along the direction of the prevailing wind is usually adopted where site conditions allow. In cases of horizontal tanks, these may be positioned so that the tank ends, which are more susceptible to failure under excessive pressure, do not face towards buildings or plant.

(b) Containers

Containers may be cylindrical tanks, both vertical and horizontal, or spheres. The capacity of large containers is considerable; for example, spheres to accommodate quantities of the order of 3,500,000 litres are not unusual within the oil industry, and the capacity of a refrigerated tank may be many times greater. On most sites, however, storage requirements are met by smaller containers.

Safety fittings normally include excess flow valves at container outlets and pressure relief devices (including, in the case of refrigerated tanks, vacuum breakers) fitted in the crown of a container, the latter devices being designed to accommodate the pressure variation due to a normal rise in the ambient temperature.

Containers for refrigerated storage are constructed with an outer cladding to protect and contain the insulating material, which may be of the order of 900 mm in thickness; the interspace may also contain inert gas.

High standards of construction and maintenance make it unlikely that a container would fail under normal conditions. Any major spillage is more likely to result from damage to or failure of a valve or flanged joint on a product pipe-line.

(c) Catchments

In order that any liquid spillage will be directed away from containers and from ancillary equipment such as pipe-lines and valves the ground beneath containers is usually sloped or channelled to a catchment area. At the same time, in order not to interfere with the dispersal of gas by natural ventilation, catchments are kept as shallow as possible consistent with a suitable degree of containment and the superficial space available for this purpose.

Main valves are positioned where they are most likely to be accessible in an emergency and they may also be remotely controlled.

(d) Facilities for the cooling of tanks

In the event of fire, the exposure of any container to the heat will be accompanied by an increase in the internal pressure which, unless steps are taken to cool the container, may lead to the opening of the pressure relief devices, the involvement of gas escaping from these positions, and a general deterioration in the fire situation.

Various types of automatic water spray installations are in use which apply cooling water to a container either through a piping system with a large number of heads distributed over the whole of the container surface or, alternatively, over the upper parts of a container so that a film of water spreads downwards. With the latter arrangement it is possible that some of the lower parts of a container will be imperfectly covered but the upper parts, most often including the gas phase which is more readily susceptible to damage by heat, will normally be effectively protected. Any water spray systems may also be designed for manually-controlled hot-weather cooling to reduce loss of product due to an unduly high rate of vaporisation.

On some storage sites, fixed ground monitors are provided as an alternative or to supplement water spray installations on the containers.

The need for further water supplies, for fire brigade use, is mentioned later (in par. (h)).

(e) Accidents at bulk L.P.G. installations

The kind of accidental occurrence for which provision has to be made is the leakage of L.P.G. either as a spillage of the liquid—with vaporisation dependent upon the particular variety of L.P.G. and the ambient temperature—or as an escape of gas. In either event, there is the serious risk that fire or explosion will occur, not necessarily at the initial stage but inevitably if the escape is not arrested and the gas safely dispersed before it reaches a source of ignition. In all cases the first consideration is to stop the outflow of product, if possible by the closure of valves, and, if ignition has occurred, to afford immediate protection to the tanks and all surrounding risks.

Pre-planning between the fire brigade and the management concerned, instruction of the staff of the premises, fire brigade familiarity with the topography of the site, and exercises in which the site staff and the fire brigade train together are vitally important. These matters are dealt with in paragraph (h).

At the actual time of an occurrence, the appropriate technicians of the premises should, whenever possible, be on hand to advise fire brigade officers what emergency measures are possible or intended by use of the facilities of the plant itself; such technicians should also be fully consulted regarding action to be taken by the fire brigade. It is against this background of pre-planning and close co-operation that the following consideration of certain courses of action is set out.

(f) Leakage of L.P.G. with ignition

A small quantity of liquid butane or propane in depth and of small surface area will burn with a reasonably lazy flame and for a comparatively long time, not increasing the surrounding temperature sufficiently to increase greatly the rate of vaporisation. The larger the area of fire, the less readily can heat be dissipated and the more intense the fire becomes.

A fire associated with a leakage of L.P.G. should not be extinguished until the outflow has been stopped, or as part of a precisely timed operation to enable operatives to go forward and immediately stop the flow. If the fire were extinguished in other circumstances, the large volume of flammable atmosphere which would form (see par. 1, page 119), and the unpredictability of the movement of a gas cloud would be extremely hazardous.

A large area of burning liquid may, however, be reduced by the use of protein-based foam, gently applied from the edges in order to avoid disturbance of the liquid and consequent increase in the rate of burning. Care should be taken to leave sufficient points of burning to avoid the risk of a serious flash over.

Notwithstanding the early protection afforded by any automatic water spray installations, an incident involving a major spillage of L.P.G. or in which an L.P.G. leak cannot be immediately arrested, will almost certainly necessitate additional use of water for cooling operations as soon as fire-fighting personnel and mobile equipment reach the scene.

In the application of water spray from portable branches, particular attention should be given to any parts of a container surface or supporting structure not effectively covered by the fixed installations and especially any dry areas which may be above the level of liquid within a container. Any nearby product pipe-lines and installation structures should be included. In the case of refrigerated storage containers, cooling to preserve the structural integrity of external shells is equally important; external shells enclosing the legs of spheres may be particularly vulnerable.

The utmost discretion must be exercised in the use of water on a fire involving L.P.G. in liquid form since the effect is to increase the rate of vaporisation and intensify the fire. The application of a jet of water would raise a cloud of burning gas and liberate such heat as to be extremely dangerous for persons nearby.

Water should not normally be applied to a stop valve or a safety valve at which L.P.G. is leaking, nor to the associated piping nearby; to do so may render the valve inoperative due to freezing.

Where liquid is burning beneath a container, the use of water spray to cover both the fire and the underside of the container may prevent direct impingement of flame upon the vessel and assist in burning off the liquid.

If, due to continuing exposure to heat, the stage is reached at which pressure relief devices open, cooling operations should be massively increased in an endeavour to lower the pressure but if, nevertheless, the discharge of gas that is on fire becomes markedly more noisy, this must be taken to indicate a dangerous rise in pressure. It will then be imperative to withdraw all personnel from the vicinity without delay and, to this end, the early use of ground or trailer monitors and branch holders, will reduce the need for manpower in hazardous situations.

An extremely hazardous situation would arise if the contents of an L.P.G. container became exhausted during a fire. With the loss of internal pressure and the admission of air, an explosive mixture would occur within the container and if flame should enter a violent explosion would be likely to ensue. If the escape of L.P.G. is from outlets near the base of a container, this risk can be prevented where the installation includes facilities to enable the bottom part of a container to be charged with water.

When L.P.G. is burning only at the point of escape, the fire will go out as soon as the outflow is stopped and any small remainder of the product has been consumed. In the case of a large spillage, however, the quantity of liquid remaining after the flow has ceased may be sufficient to sustain burning for some time and, in these circumstances, the protection of any surrounding risks must continue while the L.P.G. is either allowed to burn off or, possibly, assisted to do so by judicious use of water spray.

(g) Leakage of L.P.G. without ignition

The essential difference which characterises a serious leakage of L.P.G. *without* ignition is the continuing process of vaporisation to form a gas cloud of unpredictable extent and behaviour which may be ignited by any source in its path.

The gas itself is not visible except in heavy concentrations but the vicinity of a liquid leak may be discerned because the fall in temperature will condense moisture in the air and be seen as a deposit or a white fog. Neither of these manifestations must be assumed to indicate the limits of a gas cloud which, in fact, is liable to be much more extensive. The most effective means for determing the approximate limits of risk is by means of an explosimeter. Such devices are normally available at major oil installations and production plants of the Gas Boards; arrangements for them to be brought into use in case of need will be a matter for inclusion in discussions between chief fire officers and the managements concerned.

In any escape of L.P.G. which has not ignited, emergency operations must be directed to shutting off the leak at source and, in the meantime, taking all possible steps to prevent ignition and to disperse the gas safely. At some major oil installations, quantities of gas oil are available and may be used to absorb a spillage of liquid and so reduce the rate of vaporisation and the likelihood of a flammable atmosphere reaching a source of ignition.

It may be that a sudden escape of L.P.G. could be intentionally ignited within moments of its occurring and so avoid the formation of a dangerous volume of flammable mixture but any such action would need to be taken at an extremely early stage and on the decision of the plant management. At any later stage the gas should not be ignited and, therefore, such a course is unlikely to arise in the presence of the fire brigade. A considerable gas cloud may have formed by the time of the brigade's arrival.

Fire brigade and other emergency personnel should approach from up wind, all vehicles being left outside the area. *Only in very exceptional circumstances should persons enter the gas cloud;* notwithstanding the taking of precautions, the risk of a chance ignition may be considerable and the heat due to ignition of a gas cloud is extremely severe.

The direction and force of the wind at the time will have a large influence upon the movement of the gas and the distance within which its concentration may be reduced below the lower level of flammability. Therefore, whilst all potential sources of ignition in the vicinity should be removed without delay, the most urgent consideration may need to be given to those which are downwind of the leakage. So long as any doubt exists as to the actual limits of the danger zone, precautionary measures should be applied over a more extensive area than is likely to be affected.

Within this area of precautions, all vehicles and other engines should be stopped, electrical equipment should be switched off, fires should be extinguished and the use of other kinds of heating appliances should cease. Telephones or radio equipment should not be used unless flame-proof. According to the proximity of other premises, roads and other transport routes, it may be necessary to bring movement to a standstill, prohibit smoking and evacuate the area. Steps must be taken to prevent persons from unwittingly entering the danger zone. Special consideration should be given to the possibility of gas accumulating at low levels, including basements where ventilation may be poor, or following the course of ditches and similar channels.

If any irremovable ignition source lies in the path of a gas cloud, it may be possible to divert the gas by interposing a dense curtain of water spray. The massive use of water sprays may also be adopted for the more general direction of the gas and to assist in dispersing it but it may not be possible to discern the extent to which such measures are effective. Persons engaged in these or any operations should remain outside the danger zone, if necessary after lashing branches in position.

The most likely occasion for any person to enter the area of a gas cloud would be in consequence of a decision to attempt to stop the leak. Such a decision would normally be taken by the management or senior staff of the premises in consultation with fire brigade officers. In these circumstances, the operatives making the attempt should be given the utmost possible protection, including heavy coverage by water spray or fog from the moment of their entering the area. If necessary, the branchmen providing this coverage should, in turn, be similarly protected. In the rare circumstances when personnel may have occasion to work within the area of a gas cloud, breathing apparatus should invariably be worn.

Once the leak has been arrested, an interval of time must be allowed to elapse for the ultimate dispersal of the gas before normal movement within the area can be resumed. Tests for gas should be made in any basements, pits or low lying parts where additional ventilation may be necessary.

(h) Pre-planning and training

Preceding paragraphs on accidents involving bulk quantities of L.P.G. deal with the main contingencies in broad terms only. The success of operations at the time of an accident may largely depend upon the adequacy of fire-planning and training directly related to each of the premises concerned. This necessitates much more than good layout and a sufficiency of equipment for fire-fighting and cooling operations, important though these are. Both the staff of the premises and members of the fire brigade should have a clear understanding of the potential hazards and, in any given circumstances, know what action must be taken at the onset and thereafter; the arrival of the fire brigade should be merely one phase in the carrying out of a carefully prepared and well rehearsed combined operation, leading to the eventual restoration of safe conditions with the least practicable damage. The closest possible co-operation between managements and fire authorities is essential in pre-planning of this order.

It is, of course, necessary to balance the quantity and type of fire protection equipment against the facilities for its being brought into use at each stage. For example, if the least number of staff who may be available at the time of an emergency is small, automatic installations and good communications will be the more important. Again, if the estimated peak rate of water usage is not immediately available, the supplies which are available should be reasonably sufficient to contain the situation during the mobilising of further supplies from more distant sources. Considerations of this kind are familiar to fire officers and it is not the purpose in the *Manual* to suggest fire protection standards; however, it may be remarked that, for fixed water spray installations on containers, a discharge rate of 900 millilitres per minute per 0·09 square metres of surface area is widely accepted. Therefore, when assessing the total need of water, an allowance of this order may be made for each protected container which could be directly exposed to heat from a fire, the requirements for the application of water by other means being additional.

Once the details of a satisfactory emergency plan have been agreed, all persons liable to be concerned should be instructed and trained in its application. Fire brigade officers and men must be made familiar with the premises at risk and the facilities available, and should have a general understanding of the normal working of the installation or plant and any special measures of control which may be possible. Staff of the premises should have regular training to enable them to play the roles required of them with assurance and efficiency. Both fire brigade and installation personnel should be brought together in combined training and exercises so that their operations may be closely integrated and mutual confidence be established.

(j) Occurrences involving L.P.G. tankers

Spillage of L.P.G.—as of other flammable liquids—is always a possibility during loading and unloading operations. When spillage occurs at a loading point, filling operations should be stopped immediately and the same basic techniques be applied as for other L.P.G. occurrences—cooling in the event of ignition or assisting safe dispersal of the gas if ignition has not occurred. Where there is no fire, the engines of road tankers or other vehicles should not be started in order to remove them from the scene.

In the event of a tanker being involved in an accident or overturned upon the road—especially in a built-up area—the situation is liable to be complicated by damage to the tanker and by the exposure of nearby property and the public to risk. One consequence of a tanker being overturned may be to bring pressure relief valves, which are normally in the gas phase of the tank, below the liquid level so that liquid will be discharged through these valves if the tank becomes heated. In all circumstances, where fire occurs, cooling should be undertaken to prevent a dangerous build up of pressure.

If the burning at a point of leakage should be accidentally extinguished, it should be instantly relit by means of a remotely held flame. Provision for this purpose should be made before the necessity arises.

Where L.P.G. escapes from a road tanker without fire, vehicular traffic should at once be halted and all persons be excluded from the vicinity. Urgent measures need to be taken to eliminate sources of ignition in buildings also, special attention being given to basements. The occupants of all buildings within the danger area should, if possible, be evacuated and others, less immediately threatened, should be warned.

Pipe-lines

Pipe-lines have long been used for transporting crude oil to refineries and to shipping terminals, and petroleum and its derivatives are still the main fluids moved by pipe-line. With the growth in the consumption of goods and the development of technology, however, pipe-lines are being increasingly used for transporting various industrial materials and products, in particular those of the chemical industry. Progress has also been made with the transportation of solid materials as slurries in water or other fluids. The advantages of pipe-line transport include the reduction of costs, the easing of congestion on roads, the maintenance of delivery 'round the clock' irrespective of season or weather conditions, the elimination of contamination during transport, the removal of dangerous cargoes from the roads and the reduction of stocks being held in crowded areas.

Pipe-lines are classified as 'cross-country' or 'local', and in law a cross-country pipe-line is one whose length exceeds 16 km.

1. SAFETY CONSIDERATIONS

The imposition of safety requirements relating to the construction and operation of pipe-lines is a responsibility of the Minister of Power, who has discretion to impose different requirements for different sections of a line. He can, for example, require an additional valve to be provided in a high hazard area so that a suitably limited section of the pipe-line could be isolated in the event of a leakage or other emergency. Before exercising these powers the Minister of Power will take into consideration any recommendations which may have been made to him by the fire or other authorities concerned along the pipe-line route.

2. MARKING OF PIPE-LINES

It is generally recognised that the greatest danger to any pipe-line is the liability of damage being caused by excavation work carried out by other authorities, and clear and adequate marking can do much to minimise this danger. The hazard is most likely to arise where a pipe-line crosses a street, railway, canal or river, as other underground services often follow the line of these features. It is, therefore, at such crossings that the Minister of Power normally stipulates the provision of approved markers. These are also required at valve locations, where the marker will include the telephone number of a responsible representative of the pipe-line company.

A standardised form of pipe-line marker plate has been adopted by the Ministry of Power after consultation with the Home Office. The information to be given on markers at the street, rail, canal and river crossings referred to above, is as follows:

First line	Initials or short word indicating the operator of the pipe-line.	
Second line	Nature of the fluid	
1st symbol	Liquid	L
	Gas	G
	Liquefied gas, (i.e., a liquid when in the pipe-line but a gas at normal temperature and pressure)	V
2nd symbol	Degree of flammability:	
	Liquid with F.P. 73°F (22·8°C) or over	+
	Liquid with F.P. below 73°F (22·8°C)	—

Note: If the liquid is non-flammable the space is left blank.

3rd symbol	Other hazards:	
	Toxic or corrosive	T
	Asphyxiant	A
	(If neither, the space is left blank)	
Third line		
1st symbol	Diameter of the pipe-line (in inches)	
2nd symbol	Direction of normal flow	
	Away from reader	↑
	Towards reader	↓
	To right	→
	To left	←
Fourth line	Any system of numbering which will identify a particular marker, *e.g.*, miles and tenths of a mile from the origin, serial number from the origin, etc.	

The following are three examples of standard marker plates, which should be not less than 150 millimetres square and not more than 250 millimetres square.

SRCL L — 12 ↑ 51-7	SMBP L — 6 ↑ 15-7	ICI G — A 8 ↓ 75-9

3. PIPE-LINE ACCIDENTS

The *Pipe-lines Act* 1962 requires the owner of every pipe-line to ensure that efficient arrangements exist to give immediate notice of the accidental escape or ignition of anything in the line to every fire authority who will have to provide an attendance. The owners must also provide, on request, maps and other such information as the fire authority may reasonably require.

4. FIRE-FIGHTING

(a) When dealing with fires involving pipe-lines, it should be remembered that the premature closure of a valve may give rise to certain difficulties, such as preventing the drainage of the affected section into tankage at either end of the pipe-line; where there is a fire adjacent to a pipe-line it may be desirable to maintain flow at the highest possible rate to effect cooling. In certain pipe-lines a change in flow conditions without prior warning may even give rise to hazard at installations served by the pipe-line.

Emergency procedures agreed between the operating company and the fire brigade normally provide for communication with a responsible representative of the company before any valve may be operated.

(b) If there is no fire and volatile products are involved, the area should be cordoned off to reduce the risk of ignition. Vehicles should not be allowed to approach within 15 metres of any dangerous area and 'no smoking' should be enforced.

(c) If there is a fire, it should be controlled, but should not be extinguished until an overall course of action has been agreed between the fire brigade and the pipe-line company. The premature extinction of the fire may increase the danger by allowing a hazardous gas or liquid to escape. Branches should be laid out to protect any adjacent hazard and to prevent the fire from spreading.

(d) Action should be taken to stop the flow of a leaking pipe-line and in other pipe-lines adjacent to it, if appropriate. The closure of isolating valves may not immediately stop the draining of the isolated section via the leak in the pipe-line, and the route profile should be studied.

(e) Where a major leak occurs in a pipe-line carrying liquid, it will often be necessary to form a pit or bonded area to collect escaping liquid. The collecting area should be so sited as to minimise interference with other services.

(f) It may be necessary to search the surrounding area and to install baffles or dams in any ditches or streams immediately threatened with contamination by escaping fluid. Steps should be taken to prevent liquid entering drains, etc. Should any streams, rivers or drains be contaminated, a message should be sent to the mobilising control requesting that the appropriate authorities be notified.

Part 6C, Chapter 45, Section 6

Fires in grain, hops and their derivatives

THE incidence of fire in grain and hops and the industrial processes and installations associated with them is not high. Nor are the dangers to firemen from such fires very severe or often serious in their effects. Successful fire-fighting in them is, however, an important technique, and without some prior knowledge of the materials, processes and buildings in the industry the fireman is severely handicapped. Quick and effective fire suppression is particularly important in this field, moreover, since grains, flours, meals and hops, in common with most foodstuffs, are very liable to contamination by smoke and water, and a very serious waste may occur from a relatively small fire if the correct action is not taken at once.

Nature and properties of grain and hops

1. GRAIN

Grain is the term used for any fruit or seed of a cereal. It is frequently used as a substitute for the word 'corn', but it strictly includes rice. Grains grown in the United Kingdom are wheat, oats, barley and rye; wheat is the staple grain used for flour. Oats are ground coarsely into oatmeal, and are used whole for both human and animal food. Barley is little ground or milled, being used partly for food but more widely for producing malt liquors (beer, stout) by fermentation and spirits by distillation. Rye is little grown or used in this country, but is ground into flour in other countries. Grains also include maize, a staple human and animal food in the U.S.A. and Continental countries, and rice. The latter is used in many foodstuffs and, in common with corn, produces starch.

All grains are flammable, though not very readily so, and their burning characteristics are not remarkable. When heaped, most grains burn sluggishly due to the lack of oxygen in the interior of the heap, and they tend to give off plentiful and thick smoke: if the husks are still on the grain these will burn off very quickly on the outside of the heap, with the issue of whitish smoke. Burning grain is extinguishable without difficulty by the use of water, though a heap of grain will smoulder internally and presents difficulty in getting the water to the seat of burning.

The other particular properties of grain which may effect the fireman are as follows:

(a) Spontaneous heating

This is likely to occur in both damp and dry grain, though it is rare for heating to reach a temperature at which ignition can take place. The possibility of spontaneous ignition is greater when the grain has been ground or milled into flour or meal; as with other materials, the moisture content and the temperature are important factors in this risk. Generally speaking it can be said that high moisture content is a determining factor in spontaneous heating, while a low moisture content lessens the risk. Grain which has been allowed to become wet in storage, through leaking roofs or condensation from stone or cement floors, is susceptible to spontaneous heating. Other factors which may increase the risk are the presence of foreign matter such as weeds, clover, poppyheads, etc., and the presence of insect activity.

(b) Water absorption

All grains have to a greater or lesser extent the property of absorbing water. This absorption is accompanied by the action of swelling on the part of the grain, again varying in extent with different grains. (*see* Fig. 23).

The effect in either case is to present a hazard to the building if large quantities of water are discharged, or allowed to percolate. into bagged or heaped grain or grain stored in silos (*see* below). If it is a characteristic of the grain to swell considerably (as is the case with rice, for instance), the mass will swell and tend to push the walls outwards and endanger the building. Even when the swelling is not important, however, water allowed to sink into grain storage will stay there, steadily being absorbed by the grain and increasing its weight. The added load on the floors may again therefore endanger the building.

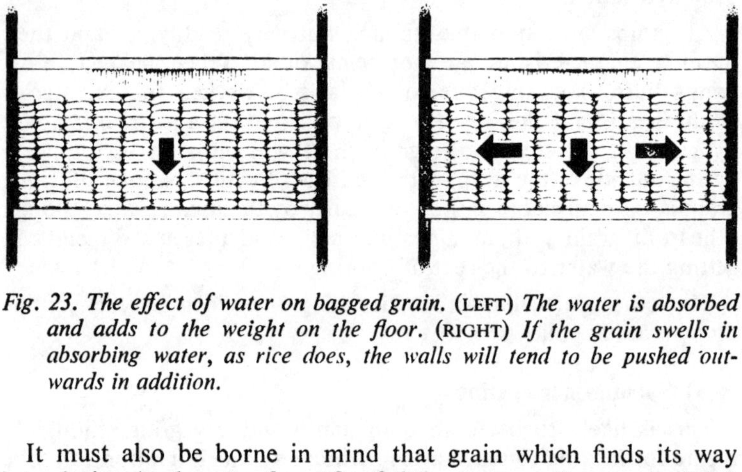

Fig. 23. The effect of water on bagged grain. (LEFT) *The water is absorbed and adds to the weight on the floor.* (RIGHT) *If the grain swells in absorbing water, as rice does, the walls will tend to be pushed outwards in addition.*

It must also be borne in mind that grain which finds its way into drains (and water from fire-fighting may well carry it there) is virtually certain to swell sufficiently to block them completely. This will always render salvage operations very difficult, and may sometimes increase the risk of water damage.

(c) Dust explosions

These have been dealt with at length in Section 1 of this Chapter. Two forms of dust explosion hazard are associated with grain and its derivatives: firstly, all flours and meals produced by the milling or grinding of all grains are susceptible to explosion under suitable conditions, and the usual dangers in the course of fire-fighting are therefore produced by these processes ar , by accumulated deposits of flour and meal. Starch dust is known to produce explosions of particular violence.

Secondly, even when the grain is being handled or stored whole, there is still a danger of explosion of the dust given off by the grain as it is handled; this dust is partly the product of pulverisation of the husk, etc., and partly of abrasion of the grain itself.

(d) Fumes from burning grain and starch

The quantities of thick smoke produced by burning grain will usually demand the use of breathing apparatus. In addition, particularly acrid fumes are produced by burning starch.

2. HOPS

These are the fruit ('cones') of a climbing plant, used for giving flavour to malt liquors, which include beer and stout. They are dried before use and, when dry, are flammable. They are easily extinguished by the use of water. In addition, they have the following properties:

(a) If bagged up in a damp condition, they may heat spontaneously. They are, however, unlikely to ignite from this cause.

(b) Hops absorb water in great quantity when dry, and will swell in doing so. They thus present the same dangers to the building as grains, and this is dealt with in par. 1 **(b)** above.

(c) Under certain conditions they may give rise to a dust explosion hazard. Hops carry a good deal of loosely adhering combustible matter, which, when dry, can form an explosive suspension in air.

Industrial processes and risks—grain

1. GRAIN SILOS

Large quantities of grain required for milling purposes are usually stored in silos. These have replaced, except for small stocks, the older granaries where grain was stored either in sacks or in bulk on the floor in partitioned compartments. (The word 'silo' in this connection should not be confused with the silo used on farms for storing green crops and fodder).

(a) Construction

As a general rule, a silo (Fig. 24) is of brick or concrete construction the roof being either slated or concrete, or occasionally of corrugated asbestos. Heights usually range from 30 metres to 45 metres. Generally speaking, the greater part of the wall area is devoid of windows, since practically the whole of the interior of the building consists of deep storage bins. There may, however, be a range of windows at the top and bottom of the silo where machinery is located. Common among older buildings are woodlined lantern light roofs, the window apertures being fitted with wooden frames and non-fire-resisting glass. Breakage of such windows results in direct ventilation being given to the floor below.

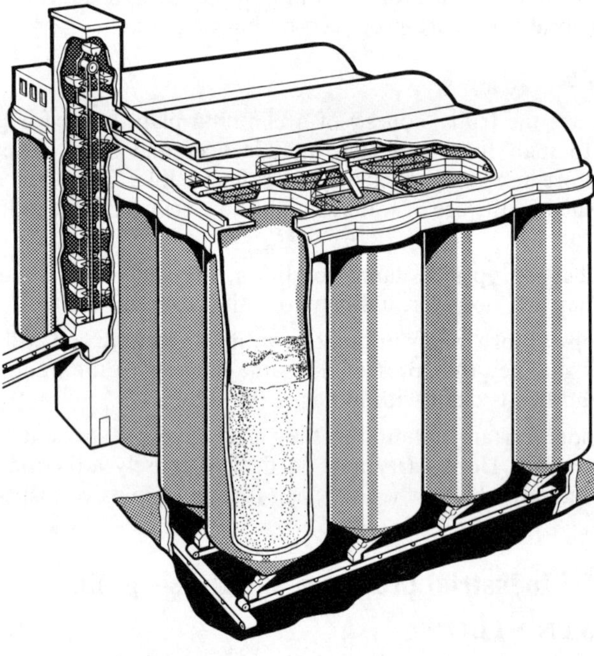

Fig. 24. Diagram showing the construction of a modern multiple grain silo, with an elevator at the side and conveyor serving all storage bins.

One small section of the building (usually in one corner) is left to provide room for a lift and staircase (though there may also be an external staircase) and for the elevator. The storage bins, which are 25 to 30 metres deep and 2.5 to 3 metres square, normally have concrete tops with metal traps to the inlets, and hopper-shaped bases of steel or concrete. Wooden bins are not uncommon in older silos, and are very strongly constructed to withstand the great pressure in the stored grain. The timbers used may be 150×75 millmetres thick at the bottom, where the pressure is greatest, reducing in size towards the top. The timber forming the bins is interlaced, and the planks are notched and spiked together. The tops of these bins are of wood and the bottoms usually of steel. Bins constructed entirely of steel may be found in some silos.

At the top of the building above the bins is a large room, extending as a rule over the whole area of the silo. The conveyor band is here. It is usually made of fabric or leather; it runs on rollers and is fed by the elevator, from which the material may be spouted to any desired bin. Trap doors in the floor give access to each bin for inspection purposes. When moving about above the bins it should not be assumed that the trap-doors are shut. At ground level below the bins is another large room which contains the hopper openings of the bins, conveyor bands for moving the grain, and grain mixers which let through a specific quantity from each bin. Cast iron, steel, brick or reinforced concrete columns may be found here supporting the bases of the bins.

In some cases two silos are built side by side. In this event it is possible that only one building has a staircase, access to the other being by means of a gangway at the level of the top floor.

In recent years construction has been simplified by making the building a single, circular bin of reinforced concrete, brick or steel, with the arrangements for loading and unloading grain already described. This circular bin silo may be used alongside a wharf for the temporary storage of grain taken from ships, the grain being later carried by a conveyor to larger silos similar to those already described.

(b) Fire-fighting

The principal cause of fire in silos is friction in the elevator or conveyor bearings causing overheating. Spontaneous heating is also possible if the grain is stored wet. The amount of flammable dust present depends upon the 'good housekeeping' of the silo concerned, but it is probable that considerable quantities of grain dust will coat many of the surfaces in the rooms both above and beneath the bins.

If the grain itself is on fire in the bins as a result of spontaneous heating, it should be worked out through the hoppers at the bottom, so bringing the fire with it. For this purpose it may be necessary to remove part of the lower opening through which the grain is fed to the bottom conveyor band; an official of the firm concerned should be consulted about this. When working out hoppers from the bottom, it must be borne in mind that there is a possibility of a dust explosion occurring within the bins. This may occur through the pressure of flammable gas from the slow combustion in the grain, or through the sudden collapse of the grain while being worked out.

Normally, it is possible to introduce water into a bin only through the opening at the top. Any considerable quantity of water put on to the grain will cause the latter to heat and swell and the swelling of the grain may eventually burst the walls of the bins or the silo itself. Where wooden bins are burning, grain is liable to escape into the small spaces between the bins and the walls, and swelling of this grain may similarly bring down the silo walls.

Fire in the older types of silo presents an unusually difficult problem, as the shape of the building offers a natural vent. A turntable ladder will probably be necessary to reach a fire in the room above the bins; but care should be taken to limit the amount of water used; in some cases the tops of the bins are raised and drains are provided to carry water away. After a silo fire at which water has entered a bin, the bin should be completely emptied and the grain dried as soon as possible.

In general, it is necessary to use breathing apparatus when fighting a grain fire in a confined space.

(c) Flour mills

(i) *Layout.* In the larger type of flour mill there are six departments. In some cases these departments may be housed under the same roof, in which case they will be separated by fire resisting doors; in other cases one or more of them may be in separate buildings.

The receiving house. This is usually situated near a railway, canal or river siding. Grain not taken in bulk direct to the silo is received in this department, weighed, and given a preliminary cleaning. A great deal of dust is usually present.

The silo. This has been dealt with under 1(a) above.

The screen house. Here the grain is thoroughly cleaned by means of shaking machines and other methods, including, for example, the use of magnetic separators.

The mill. Corn is converted into flour by rolling in the mill and may afterwards be bleached with chlorine dioxide (Dyox process) or some other gas. When nitrogen peroxide is used, it is generated on the premises; other gases are brought in steel cylinders.

The provender mill. Here offals and various cereals such as peas, beans and maize are prepared for provender. Much the same features will be found as in a flour mill.

The warehouses, where manufactured materials are packed,

stored and despatched. Here grain is stored in sacks, and also in some cases blending of one flour with another is undertaken.

(ii) *Special features and risks.* In old mills, where construction usually comprises stone walls, timber floors supported on unprotected cast-iron columns, and boarded and slated roofs, a hazard to the fire-fighter exists simply by reason of the age and inherent features of the premises and the large number of wall and floor openings which it possesses. These may cause rapid fire spread and quick collapse of the building.

Machinery on the top floor of old mills is very often tied to the ridge beam and/or other roof members for support. This adds to the risk of quick collapse of the roof and the danger of injury to firemen from falling machinery. The ridge of an old mill building is often noticeably bent, revealing the strain of this weight over a long period of years.

In a modern mill it is possible to see little evidence of dust except in the receiving and screen houses and the silos.

The principal points to note are as follows:

Height. Mills are usually of four or five storeys, shed buildings being used only for workshops.

Floor openings. A large number of these are commonly found in mills, screen houses, and warehouses. Openings for chutes, traps, elevators and stairs exist in all such buildings. Often a lift in the form of a continuous belt is fitted to raise and lower men from floor to floor and its unprotected openings are particularly dangerous. The chutes are likely to be lined with wood and constitute a high fire risk.

Defective construction. Sometimes there are no separating walls, or, if there are, they are not carried through the roof. Part of the roof of a mill may be raised to contain the heads of the elevators, the raised part being roofed with timber and glass.

Dust. Dust can be concealed in openings in spouts, elevators and floors. In some of the smaller and older types of mill an accumulation of dust is found on beams, crevices and other places.

Overcrowding. At some mills, though by no means in all, the floors are so crowded with machines, shafting, belting, elevators and spouts, that proper supervision and efficient lubrication of the machinery is very difficult, In such a case dust tends to accumulate and a fire may result from overheated bearings for example. If a fire does occur, the overcrowding may delay its detection and may subsequently cause difficulty in locating the seat of the fire.

Internal woodwork. In mills and screen houses there is a large amount of dry woodwork in the form of wooden floors, staircases, wood enclosed machines and spouts, and trunks of all kinds. Wood is used for these purposes to avoid condensation which would affect the flour. The trunks which draw dust from the screening machines are lined with dust and are particularly likely to spread fire rapidly, while if the dust is disturbed an explosion may occur.

Rope race. In many mills the machinery is driven by a steam engine and rope race of the type found in cotton mills (q.v.).

(iii) *Fixed fire protection.* Most modern flour mills are sprinklered. The installation conforms to the special rules for this class of risk, which specify a distance of 2·45 metres instead of 3 metres between heads, a floor coverage of 5·95 m^2 against 9·29 m^2 and the placing of heads not more than 1·2 m from walls. A sprinkler head must also be placed in the box head of every dust trunk or spout, inside each trunk or spout, and connected to every exhaust fan (with a few exceptions) on the delivery side. The heads should be examined weekly and dough removed.

(iv) *Fire-fighting.* The risk of a dust explosion is always present and it is, therefore, good practice to use the spray from a diffuser branch or fog nozzle to wet down any grain in hoppers or bins before emptying the latter. Piles of dust may safely be wetted down with a fine spray without danger of stirring up a cloud of material.

Ventilation should be shut off immediately and efforts should be made to prevent the fire travelling through ducts and chutes. Normally the extraction system blows the dust at high speed to canvas collecting sleeves. It should be remembered that when the system stops the dust will line the sides of the ducts. It will, therefore, be wise to use an inspection hatch to make a cut-off in case the fire should spread through the duct, and also to make sure that the fire is not travelling by this route to other floors.

It is usually wise to stop all machinery in the department affected.

3. BAKERIES

Premises coming under this classification vary from the small bakery in the basement of a shop to the huge mechanical plant of the large catering or biscuit manufacturing concerns.

(a) Risks

In small bakeries a coke-fired oven is the most usual, and fires sometimes occur as a result of the flue becoming foul. In large concerns the ovens may be heated by gas or electricity, oil fuel or creosote pitch. Fire may result from a gas explosion, an electrical fault, or a flashback from the oil burners when they are being lit.

Overheating of the ovens may lead to a fire started by conduction or radiation, especially in small premises adapted for use as bake-houses, where much dry wood is often included in the construction.

In the manufacture of biscuits and similar products, risks associated with sugar and with chocolate will be found. The principal ingredients used are flour, sugar, oleo oil, butter, cacao butter, cocoa beans, ammonia powder, soda, and tartaric acid. (The noticeable smell of ammonia in biscuit manufacture is due to small quantities of ammonia powder included in the dough, and not to refrigeration plant).

Large quantities of paper, cardboard wrappings, cartons and wooden boxes are used in the packing and despatching departments.

(b) Fixed fire protection

Large bakeries are often sprinklered, though in biscuit and similar manufacture some firms take the view that water damage may outweigh the protection given. If the oven furnaces are oil fired, the fronts are often protected by water spray, which may be mechanically or hand operated. In some cases a dual system is fitted, one mechanical and one hand operated.

There is generally no fixed fire protection in small bakeries.

(c) Fire-fighting

Fires in flues should be dealt with as described in Book II of the *Manual*, Chapter 5 for chimney and restaurant flue fires. Roof fires or fires involving flooring may occur as a result of an overheated flue combined with defective building construction.

When stocks of flour, sugar or cocoa are involved in a fire, a dust explosion may occur if sacks or containers fall in such a way that their contents are scattered and ignited, or if a carelessly used jet disturbs dust without wetting it. Raw materials used in bakeries can be greatly damaged by water and as these materials are often stored in the basement, salvage work should be undertaken with the least possible delay. Smoke, also, may readily contaminate foodstuffs.

In large mechanised bakeries, where there is a large number of travelling belts and conveyors, it will usually be necessary to have the machinery stopped and the heat turned off from the ovens. The staff will advise upon this.

4. STARCH WORKS

Starch is present in nearly all plants. It is found in various forms in industry, as powder, paste, and dextrine for example. It is also used in laundries, textile manufacturing and finishing, adhesives, explosives, certain food preparations, etc. and in the manufacture of dextrose and maltose which are used in brewing.

Many types of factory will be found where starch is processed in one way or another, and often a starch department is incorporated in a bigger works, such as a soap works. The raw material is usually kept at the top of the building, working its way downwards by gravity from process to process.

(a) Processes

The rice (occasionally maize) is ground to remove fibrous matter and then soaked in a weak caustic solution in decanting tanks, whence the fibrous matter is drawn off and the glutinous water runs to a deposit tank. The starch thus deposited, after being ground between millstones passes through centrifugal driers to get rid of impurities. Caustic is again added and the starch is turned out in wet blocks, wrapped in paper to limit evaporation, and placed in wood-lined drying rooms at a temperature of about 120°F. (49°C.).

Borax and gypsum are added, and the starch leaves the works in sacks or paper bags.

The fibrous matter referred to above has sulphuric acid added to it, is put through deposit tanks, filter presses and driers, and, after being bagged, leaves the works as cattle cake.

(b) Fire-fighting

Rice in sacks tends to heat spontaneously if it becomes wet. After a fire, therefore, arrangements should be made to dry affected stocks.

In the event of fire in the store, the sacks will burn and the rice tends to run loose and clog the drains. Rice will smoulder: it is also capable of absorbing very large quantities of water (though this process is slow) and the possible effect of this on the construction of the building should not be overlooked.

Normally, starch dust is found in quantity only in the milling room. It can transmit fire with great speed. In its final form (when mixed with gypsum starch) it is less flammable.

Starch dust is liable to explode violently, and thereby start a fire. In addition, a number of small dust explosions may occur during fire-fighting operations. This is probably due to the dust being disturbed by falling debris or by sacks falling through from

Plate 23. The cutting room of a modern clothing factory. Note the large amount of stock in the background which is liable to lose much of its value if wetted. (Section 4).

PHOTO: *Montague Burton Ltd.*

Plate 24. A rope walk at a cordage manufacturers. The length of working space required is considerable. (Section 4).

PHOTO: *British Ropes Ltd.*

Plate 25. A road tanker carrying 5586 kg of liquid butane, which overturned on the road. The gas escaping from the fractured tank ignited immediately. (Section 5).

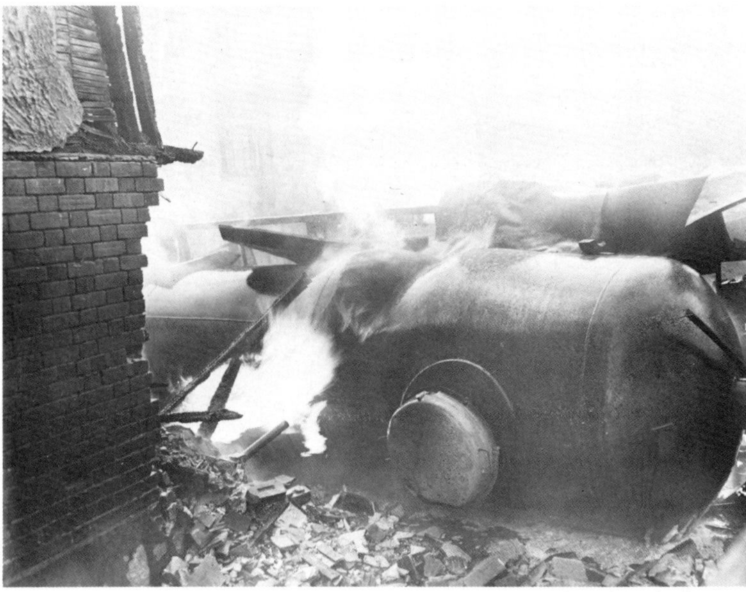

Plate 26. A close-up of the overturned tanker shown in Plate 25; the butane is burning at the points of fracture. The pressure bulges which developed in the tank can be clearly seen. (Section 5).

Plate 27. Fire resulting from an explosion at Feyzin refinery, near Lyon, France, where liquefied petroleum gas storage tanks were involved. 75 people were taken to hospital suffering from burns. (Section 5).

PHOTO: *Associated Press Ltd.*

Plate 28. An example of a well designed layout for L.P.G. bulk tank storage. A water spray installation is provided to protect the tanks and a raised sill to form a catchment area to contain any spillage of liquid (Section 5).

PHOTO: *Calor Gas Ltd.*

Plate 29. The results of an explosion caused by the overheating of a sodium nitrate salt bath, in which aluminium was being treated. The explosion was produced by the chemical reaction which followed the decomposition of the aluminium. (Section 7).

Plate 30. The interior of the factory shown in Plate 29; the many salt and plating baths, crowded together, can be seen. (Section 7).

one floor to another. The fireman should aim at laying starch dust rather than stirring it up, and it will often be found advantageous to use a water spray for this purpose before jets are brought into operation.

If the sulphuric acid used in cake manufacture comes into contact with starch in the presence of a piece of paper or similar material, a quick reaction will take place, and enough heat may be evolved to start a fire.

Starch in quantity, whether in bulk or in sacks, is not difficult to extinguish, though it burns readily. Acrid fumes are given off, but there is little smoke. The starch swells up in black globules as it burns, and continues to smoulder for a long time. The cake will also burn when sufficient heat is applied to it.

5. DISTILLERIES

Distilling includes the same processes as brewing (*see* below). The fermented wash, however, is converted into vapour by heat and the alcohol is separated by condensation. This alcohol is then diluted with water to reduce its strength, run to receiving houses, and stored in bond either in vats or casks.

The chief fire risks in distilleries arise in the grinding rooms and granaries, where the hazards are similar to those discussed under flour mills, maltings and breweries. In the still house there is also the danger of spirit vapour catching fire; this is especially so in the case of patent stills, where the concentrated spirit has a relatively low flash point. Spirit fumes in an empty still can also constitute a fire risk.

Fire-fighting

Generally speaking, stored spirit (see ethyl alcohol) is flammable; but its behaviour in a fire depends on its alcoholic strength. Some bonded stores possess a semi-basement, in which case spirit released from the vats is not likely to run out of the building. Once spirit has become ignited, the heat may cause spirit in other casks to vaporise and burst the casks. The resulting fire is likely to spread with great speed and fierceness.

One of the principal difficulties from the fireman's point of view lies in the construction of the buildings. Bonded stores have usually only one door, fastened with two padlocks, one of which is placed there by the Excise Department. Should the Excise lock be forced, the police must be notified and the premises guarded till their arrival. The fact that only one door exists, possibly to a range of interconnected buildings separated internally by fire-resisting doors, makes fire-fighting extremely hazardous. In some cases during the second World War a second emergency exit was provided. This door can only be opened from the inside, and is padlocked there. In case of fire, this padlock should be removed as soon as the building has been entered so that the door can be used in an emergency.

Bonded stores have heavily barred windows and ventilation is almost entirely absent. In case of fire, therefore, vapours are likely to increase in pressure, and the opening of a door may cause a back draught explosion. It may, therefore, sometimes be wise to ventilate the roof before entering.

Many bonded stores are of old, traditional construction. A risk of quick fire spread from one floor level to another exists, due to the slatted-type floors through which burning spirit can easily find its way from barrels or vats damaged by fire. Whole floor areas can be quickly involved in fire as a result, with the added danger of burning spirit flowing out to street level. Adequate steps should be taken to safeguard neighbouring basements, stationary vehicles, etc., and this point should always be borne in mind when siting fire appliances.

Generally speaking, once the seat of the fire is reached the best means of putting out the fire is to use diffuser branches and to dilute the spirit on fire until it is too watery to burn. Breathing apparatus will normally have to be used for fire-fighting in enclosed spaces, because the fumes from burning spirits have a rapidly intoxicating effect.

Industrial processes and risks—Hops

1. OAST HOUSES OR HOP KILNS

(a) Construction and processes

These buildings, in which the hops are dried by artificial heat, are usually circular and have a characteristic point cowl at the apex of the roof. They are to be found chiefly in the hop growing areas of Kent and Herefordshire, attached to farms or in separate groups. The usual type of construction provides a ground floor chamber containing a bricked-in coal stove. Above this is a wooden slatted floor upon which the hops are laid out to dry. In the wall of this chamber is fitted an exhauster fan, driven by an oil engine or electric motor placed outside the building. At the apex of the pointed roof is a revolving cowl which always has its back to the wind, thus preventing the wind blowing down into the kiln. Access to the hop drying floor is by internal or external staircase.

Sparks from the heating stoves are the principal cause of fire in kilns. The latest types of buildings, therefore, have the stove outside, the heat being carried within by pipes.

(b) Fire-fighting

Hops absorb water massively and swell when wetted, so that in fighting a fire there is always a risk of increasing the weight on the drying floor to danger point. Apart from the hops, which are themselves flammable, stocks of coal and of sulphur (which is burnt on the stove during the drying process) will be found at hop kilns. Sulphurous fumes must, therefore, be expected when entering the building. If the fans are electrically driven it is possible that the wiring may be affected by the fire and there may, therefore, be live cable ends.

After drying, hops are stored in bags known as pockets (Fig. 25). Some breweries compress the pockets of hops to about half their normal size, place ply-boards on their ends, and hold them in place with metal bands. This practice, which of course allows greater stocks to be carried in the same space, adds to the dangers resulting from water absorption should water reach the pockets. So far as possible, fires involving pockets should be dealt with by means of diffuser branches, while in the early stages of a fire, asbestos blankets or CO_2 are both of proved value. Pockets should be sheeted up at the earliest possible moment unless attempts are to be made to remove them. Hops are sometimes kept in cold storage for limited periods. From a fire-fighting point of view, this will add the hazards and difficulties of a cold store fire to the normal features of a fire in stored hops. Fires in refrigeration plants and cold stores are dealt with in Section 12.

2. MALTINGS

In these buildings barley is 'malted', i.e. germinated, as the first stage in the brewing process. Maltings may be found on the same premises as breweries, but they are normally in separate buildings in the barley growing areas, usually near a river, canal or railway and sometimes in groups of several buildings.

Most maltings are strongly built of brick or stone, the roof being tiled or slated with board beneath. The small wall openings which take the place of windows, are not generally glazed, and are fitted with wooden shutters. The building is often of long and narrow design, so that an even temperature can be maintained throughout with the least possible interference from walls and roofs.

At least 50 per cent. of the total area of the malt house is used

Fig. 25. Method of filling hop pockets. The dried hops are rammed down by the press until the pocket is tightly packed.

for the non-hazardous process of germinating or flooring, usually in the lower part of the building. Since large unobstructed areas are required there is an absence of internal separating walls. The working floors are usually covered with a non-combustible surface on open-joisted flooring, supported by beams or girders on naked cast iron columns. Unprotected floor openings exist, such as hoisting traps or stairs, while in modern labour-saving buildings floors are honey-combed with wood spouts, elevators, etc.

The top floor or floors, are often used for storing barley and malt, the barley in sacks or in bulk and the malt in bulk in bins. A large amount of timber is used in these stores for partitioning and for linings and ceilings (especially in attics). The attic within the slope of the roof, when used as a store, is generally lit by dormer windows.

In the bigger buildings, horizontal grain conveyors and vertical elevators and spouts will probably be found in the barley stores together with the barley cleaning plant, while a power-driven malt dressing plant will be situated in the malt store. Elevators and spouts are generally associated with malt and barley cleaning machines. In addition there will probably be one or two timber-built dust rooms whence dust is led in wooden trunks from the machines.

Adjoining the malt house, often with a common wall and roof, is the kiln. The latter is similar in construction to a hop kiln, the drying chamber being above the heating chamber. In some kilns there are two drying floors, one above the other, the green malt first being placed upon the top floor and then being brought lower to complete the process.

The hazards lie in the great internal size of the buildings, their proximity to one another and to the malt house and kiln, the large amount of woodwork and floor openings, the flammability of the dried grain and the amount of dust, which may give rise to a dust explosion.

3. BREWERIES

Brewing is a process of malting, grinding, mashing and fermenting. Malting involves steeping in water, couching (when water is drained and heat is generated in the grain), flooring (when germination takes place) and kiln-burning. The malt is ground and elevated to the brewing room by means of conveyors. The brewing is done in pear-shaped copper kettles heated by steam, and in these the malt is boiled with water. This forms the wort, which is strained, clarified, fermented and finally bottled. During the flooring process at maltings, heat is evolved as the grain absorbs oxygen and gives off carbon dioxide. Good ventilation is essential; if this is lacking, overheating may occur.

The following points should be noted:

(a) Brewery buildings are usually high up to 30 metres and are often built with a central, open shaft, topped with glass skylights, in order that plenty of air can reach the yeast and the vapours given off by the fermenting liquor can be carried away.

(b) The water tanks, malt stores and grinding machines are usually placed at the top of the building, the various processes taking place successively on lower floors.

(c) The main fire and explosion risk occurs in the mill rooms where the malt is ground. During this process much malt dust is produced, and a spark resulting from the passage of a foreign body through the rollers may produce a dust explosion. This will normally dictate the use of water sprays. (*See* Section 1).

(d) One or more refrigeration plants will be found in the bottling department on the ground floor or in the basement (*See* Section 12).

(e) Breweries are not usually sprinklered.

(f) When vinegar is brewed, carboys of acetic acid may be found.

Part 6C, Chapter 45, Section 7

Metal fires

METALS are one of the fields in which recent industrial development has added to the risks of fire, by which is meant not only the risk of causing an outbreak of fire but the risk of accident or injury to firemen during fire-fighting operations. Many uncommon metallic elements have been brought into wider industrial use over the past 20 years in an effort to obtain better strength-to-weight ratios and higher working temperatures in various engineering fields, and new processes have been introduced with the same objects. Firemen are well aware that the commonly held idea of the non-combustibility of metals is a false one, and that the majority of metals, while proof against being ignited in normal constructional forms and at temperatures normally encountered in everyday life, are readily combustible when in finely-divided forms and when very high temperatures prevail. The respects in which metals should be studied by the fireman are, broadly speaking, threefold. In the first place, metals can be the cause of fire outbreaks in various ways: in the second, they can offer very serious and varied dangers to those engaged in fighting fires, and, in the third, they can prove exceedingly difficult to extinguish when once burning.

As a cause of fire, perhaps the commonest, and most obvious ignition agent among metals is the metallic spark. Frictional sparks from ferrous metals are a very common cause of fire starting in readily flammable materials, and account for a large number of fires every year in such processes as milling, grinding, separating, screening, etc.; hence the need for substituting, where possible, non-ferrous metals such as copper, brass, bronze, zinc and aluminium in cases where sparks might be produced by ferrous metals and give rise to fire. Less well-known but more insidious is the property which many metals have (especially those termed 'pyrophoric') of self-heating, whereby they may, in certain forms and in given temperature and humidity conditions, experience a rise in temperature through rapid oxidation, or combination with the oxygen in the air. (Book I of the *Manual*, Chapter 10). This process of oxidation is particularly likely to occur rapidly when metals are finely divided, as in powders and dust, and it frequently results in an outbreak of fire. Metals may also combine chemically with other substances, again particularly when the metals are in a finely divided form, to cause fire and sometimes explosion. A particularly severe form of chemical reaction is that whereby certain metals known as 'alkali metals', and including caesium, lithium, potassium, rubidium and sodium) come into contact with water. They react violently, liberating hydrogen in the process, and generating sufficient heat to ignite the hydrogen. So marked is this reaction that some of these metals are capable of igniting spontaneously in air.

The most serious hazards in fire-fighting among metals arise from these same properties. The power of extremely rapid oxidation makes most metals in powder form capable of ignition as a dust cloud; this may produce an explosion of great violence, which has been known to destroy whole buildings. (Fuller details of dust explosions and fires in dust are included in Section 1 of this chapter). The metal/water reactions already referred to above are also capable of producing violent explosions, as are equally the reactions between many metals and other forms of extinguishing agent. This means that in many cases the use of any of the conventional fire extinguishing methods, such as water, foam, vaporising liquid, etc., would be extremely dangerous. .It follows from this also that even if all due care is taken in fire-fighting to avoid the use of unsuitable and dangerous extinguishing media, there are still severe risks, in all fires involving metals, of unexpected chemical reactions which may give rise to sudden increases in the rate of burning and to explosions.

Another important group of hazards to the fireman arising in the course of fire-fighting among metals is the risk of toxic fumes. Many metals, such as cadmium, give off poisonous vapours when raised to the high temperatures which can easily occur in fire. The effects of some of these poisons are more severe than others, but the risks in general are such that whenever fires are fought which are known or thought to involve the metals concerned, breathing apparatus is indispensable. Another highly specialised form of fire-fighting hazard in the presence of some metals is the radiation hazard from radioactive isotopes and radiation sources generally: this is dealt with fully in Section 11 of this Chapter.

Most combustible metals reach very high temperatures indeed in the course of burning, and it is a historical fact that the familiar 1 kg magnesium incendiary bomb burns with a temperature of about 6,850°F. (3,787°C.). Some mixtures of metals also contain within themselves the oxygen required for combustion, and this presents obvious difficulties in extinguishment. When to these properties are added the many possible physical and chemical reactions which may be produced by the use of normal extinguishing methods, it will be seen that burning metals, perhaps above all other materials, present difficulties in extinguishment and call for the use of special techniques and agents.

Characteristics and hazards of metals

1. METALS IN GENERAL

The following paragraphs give details of important properties from a fire-fighting point of view of some commonly encountered metals. No mention is made in this list of a number of metallic elements, such as radium, uranium, plutonium, neptunium, strontium and thallium, which are either naturally radioactive or occur more commonly as the radioactive isotope than as the stable element. Since these metals are not necessarily found in the metal-working industries as such, but are used in many different fields for special purposes, they have been regarded as radioactive hazards rather than as metal fire hazards. (For the fire-fighting hazards of radioactive materials, *see* Section 11 of this Chapter).

Aluminium

A light, ductile metal of good electrical conductivity and good resistance to corrosion. It is used in a wide variety of industries including aircraft, motor and electrical engineering and utensil manufacture, and is often alloyed with copper, silicone, etc. Its melting point is 1,220°F. (660°C.). Heated aluminium gives off hydrogen when in contact with solutions of potassium hydroxide or sodium hydroxide, and reacts explosively with carbon tetrachloride and methyl bromide. It has been known to ignite spontaneously with sodium peroxide. It burns vigorously at 1,472°F. (800°C.). Aluminium dust is a severe explosion hazard, since it ignites in cloud form at 1,193°F. (645°C.) and also decomposes water, producing hydrogen. It is not liable to spontaneous heating and is not regarded as an industrial poison.

Beryllium

This is a light, steely metal used mainly in atomic work. Its melting point is 2,336°F. (1,280°C.). When in contact with water, beryllium gives off hydrogen, and it is a moderate fire hazard when exposed to flame. Beryllium dust offers a slight explosion hazard. The metal is highly toxic; the inhalation of dust, or vapours from combustion can cause death. The process of producing beryllium is shown diagrammatically in Fig. 26.

Brass

An alloy, pale gold to red in colour, mainly consisting of zinc and copper with some lead. The melting point is about 1,720°F. (940°C.). While the metal in normal form is not considered to have a fire hazard, brass powder has a dust explosion hazard. It reacts with oxidising agents, and offers a toxic hazard in contact with these, or with acids or acid fumes.

Cadmium

A soft white metal used for coating small steel articles against corrosion and a common constituent of fusible alloys, its melting point is 610°F. (321°C.). It reacts vigorously with oxidising agents,

---

Fig. 26 (OPPOSITE). *Beryllium ore mixed with sodium silicofluoride is crushed and pressed into briquettes. Kiln roasting produces beryllium fluoride which is crushed, leached with water and filtered. Treatment with caustic soda precipitates beryllium hydroxide which is filtered out. Carbon and tar are added and the mixture is extruded, baked to remove the tar binder and converted to beryllium chloride by chlorination. Electrolysis of the beryllium chloride to flake metal from which the salts are removed by leaching, filtering and drying, is followed by consolidation by hot pressing into ingots. The metal can then be forged and fabricated for atomic reactor use. Note: The dotted areas indicate hazardous processes for which respirators and protective clothing are required.*

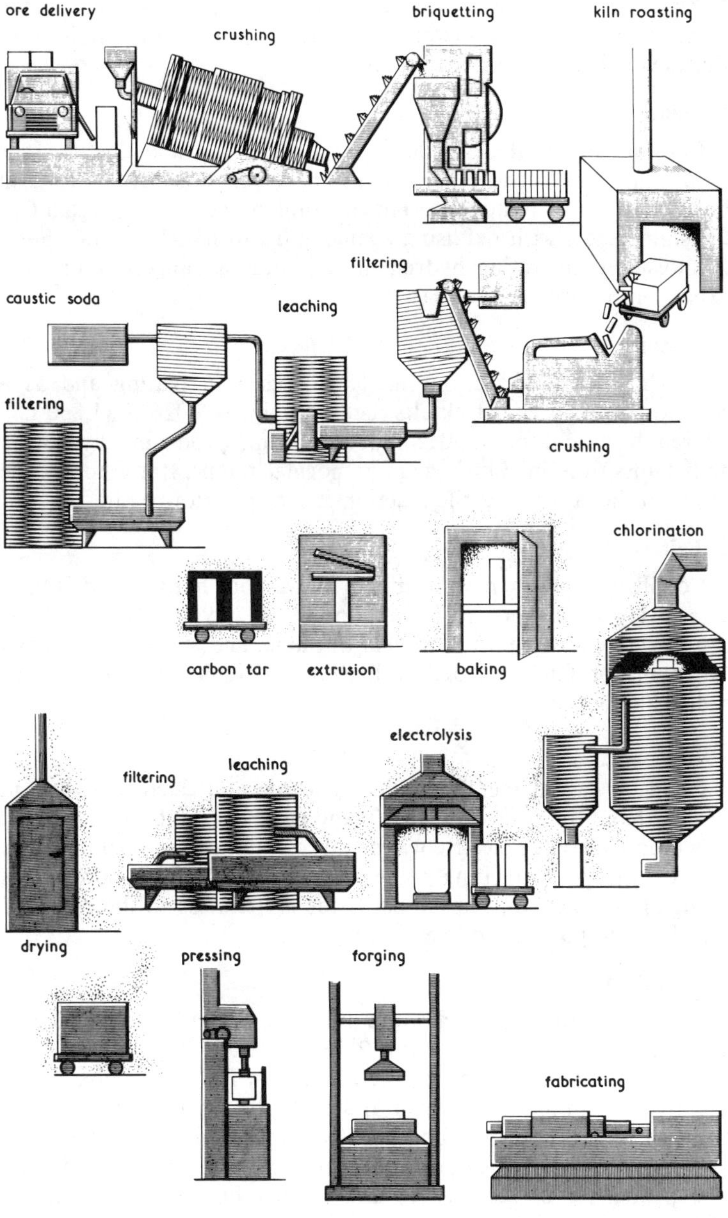
ore delivery
crushing
briquetting
kiln roasting
filtering
caustic soda
leaching
filtering
crushing
chlorination
carbon tar
extrusion
baking
electrolysis
filtering
leaching
drying
pressing
forging
fabricating

Fig. 26

releasing hydrogen, and will ignite when exposed to flame in the form of a dust. This metal, when heated, emits very poisonous fumes which can cause acute pain.

Caesium

Caesium is an alkali metal: it is silvery-white and ductile. It is used in photo-electric cells, and the radioactive isotope is used in hospital therapy treatment. The melting point is 84°F. (28.5°C.). Caesium reacts with oxidising agents and can ignite spontaneously in moist air, liberating hydrogen. It is thus a dangerous fire and explosion hazard.

Chromium

A white metal, hard but malleable, used in plating and as a constituent of special steels. Its melting point is 3,326°F. (1,830°C.). At red heat, chromium decomposes steam, producing hydrogen, and it dissolves in dilute acids at normal temperatures to do the same; in the latter case, the reaction accurs at a slower rate.

Cobalt

Cobalt is a silvery metal, similar to iron but harder, and is used extensively in alloys. Its melting point is 2,696°F. (1,480°C.). It can be ignited by exposure to heat or flame, and, especially in small fragments or finely divided, will heat and ignite by spontaneous chemical reaction.

Copper

Copper is a red metal, tough, malleable and ductile, with high electrical and thermal conductivity and good resistance to corrosion. It is used widely for electric wires and for piping; it is also the basis of many alloys. Its melting point is 1,985°F. (1,085°C.). Copper is a negligible fire hazard, but copper oxide is irritating to the eyes and is believed to give rise to 'metal fume fever'.

Iron and Steel

Iron in its pure state is a white metal, soft, malleable, tenacious and ductile. The melting point is 2,795°F. (1,535°C). Normally the ore is treated by roasting and then in a blast furnace to give pig or cast iron (combination of iron and carbon). Pig iron is very hard and brittle. Carbon may be almost entirely omitted from the pig iron to give wrought iron, or added in varying degrees to give steel. The properties of steel vary, and it may be treated in various additional ways to give combinations of strength and toughness: it melts at 1,515°C. Special steels are made by combining other metals (such as chromium, nickel, tungsten or manganese) with the iron and carbon. Contact between red-hot iron and steam or water liberates hydrogen, and the same effect is produced on contact between iron and strong acids, *e.g.*, hydrochloric or sulphuric. There is a fire and explosion hazard when iron dust is exposed to heat or flame. Iron pyrites has a spontaneous heating risk. There is no toxic hazard attached to iron itself but iron dust can injure the eyes and lungs.

Lead

This is a heavy, grey metal, soft and plastic. It is used for roofing, cable sheathing, lining chemical vessels, in storage batteries, metal bearings and for piping. The melting point is 621°F. (327°C.). Especially when in finely divided form, lead can react vigorously with oxidising agents, and it offers some fire and explosion hazard as a dust. When heated, lead gives off toxic fumes; inhalation of these or of lead dust is dangerous.

Lithium

A silvery, light, alkali metal which resembles sodium. It is used as an alloy in metal bearings and its melting point is 367°F. (186°C.). Lithium is a severe fire hazard when exposed to heat or flame and it reacts violently with water, acids and oxidising agents, producing hydrogen in reaction with water or steam. It offers a severe explosion hazard by chemical reaction. It is usually kept beneath the surface of paraffin or toluene, but can be stored with safety under inert gases such as helium or argon. (Nitrogen, recommended for some dangerous metals, is not suitable). Lithium offers a severe toxic hazard when burning.

Magnesium

Magnesium is a brilliant white metal, which is soft, ductile and malleable. It is used as a basis metal in light alloys for strength and toughness in the aircraft and motor engineering industries. Its melting point is 1,200°F. (649°C.). Dust or flakes of magnesium are very readily ignited, but in solid form it must be heated above its melting point before it will burn. In this state it burns tenaciously with a brilliant white light. When heated, also, it reacts violently with water and all moisture. In contact with water at normal temperature and pressure it reacts slowly and gives off hydrogen; the reaction is more rapid in contact with dilute acids. There is a severe risk of the ignition of magnesium in machine operation, while as a dust the explosion hazard is very great. When dispersed in air, magnesium ignites at 966°F. (519°C.).

Manganese

This is a hard, brittle metal, reddish-grey or silvery in appearance. It is mainly used in steel manufacture for purifying steel and as an alloy. Its melting point is 2,273°F. (1,245°C.). It decomposes water, producing hydrogen, and dissolves in dilute acids to do the same. It reacts with oxidising agents. It also offers a fire and explosion hazard when in dust or powder form. Dispersed in air the dust ignites at 842°F. (450°C.).

Nickel

Nickel is a silver-white metal, used in the pure form for electroplating, for coinage and for the linings of chemical vessels. It is used very widely as an alloy with steel and chromium, and with copper makes '*Monel metal*'. The alloys give high working temperatures. The melting point of nickel is 2,642°F. (1,450°C.). and in solid form it offers only slight fire hazards. It can be ignited, however, and offers an explosion hazard when exposed to flame in the form of dust.

Platinum

A rare, heavy, soft and ductile metal, immune to most chemical reagents and to oxidation up to high temperatures. Platinum wire is used in laboratory instruments but only the salts are in common industrial use. The melting point is 3,223°F. (1,773°C.). While normal fire and toxic hazards are negligible, platinum dust has been known to explode with a mixture of hydrogen and oxygen.

Potassium

This is an alkali metal, silvery in colour and highly reactive. As a metal, it has few applications, but the salts are used extensively. The melting point is 144°F. (62°C.). Potassium is a very dangerous fire hazard since it reacts violently with all moisture, producing much heat, and evolves hydrogen: there is thus a severe explosion hazard. In air, the reaction is slower, but it is violent with oxidising agents. Potassium may form an explosive compound in contact with carbon dioxide. The toxic hazard of potassium under reaction is very great. It can be safely stored under inert gases.

Sodium

This is a silver-white metal, light, soft, ductile and malleable. Sodium is an alkali metal: its melting point is 204°F. (95.5°C.). It is very reactive and easily ignited; heated, it will ignite spontaneously in air. By chemical reaction with water or any moisture, with air or with oxidising agents it will decompose violently, evolving great heat, and the hydrogen which is liberated will ignite. It also reacts violently with the halogens, with acids and with halogenated hydrocarbons. Hot sodium reacts violently with carbon-dioxide. It is not liable to spontaneous heating, but toxic hazards in all circumstances are severe. Sodium is usually stored below the surface of naphtha, paraffin or other liquids not containing oxygen, and may also be stored under inert gases. It may, however, be transported or kept for short periods in sealed airtight drums.

Tin

This is a white metal with a yellowish tinge, malleable but not very ductile. It is used as a foil in coating copper, lead, iron and steel and as an alloy with bronze, pewter, etc., its melting point is 449°F. (232°C.). Normally, it offers no fire hazard, but as a dust it can be ignited, and present an explosion hazard, when exposed to a flame or chemical reaction. Tin is not regarded as a toxic hazard.

Titanium

A strong white metal, lighter than steel, used in alloys with steel to give high working temperatures, mostly in the aircraft industry. Its melting point is 3,632°F. (2,000°C.). Titanium is easily ignited when in the form of thin chips, drillings or turnings, though larger pieces present only a slight fire hazard except in heat treatment baths. It offers a severe dust explosion hazard, however, being ignitable in cloud form at 896°F. (480°C.). It has also exploded with concentrated nitro-acids. Spontaneous ignition can occur when fine chips or dust are covered with oil, and spontaneous ignition has occurred in titanium dust when dispersed in air. There is no noteworthy toxic hazard.

Zinc

Zinc is a white metallic element with a bluish tint, and has very high resistance to atmospheric corrosion. It is used as a protective coating (*e.g.*, in galvanised iron) and is also used in sheets and as a constituent of alloys with aluminium, copper, magnesium, etc. The melting point is 786°F. (419°C.). The metal itself is not considered to have a fire risk, but it can be ignited. It also offers an explosion hazard when in the form of a dust. It is not liable to spontaneous heating, and toxicity hazards are low.

Zirconium

The melting point of this metal is 3,452°F. (1,900°C.). In most respects it may be regarded as similar to titanium, (q.v.) though the fire and explosion hazards are much more severe. It ignites as a dust cloud, for instance, at a temperature of 68°F. (20°C.).—The process of producing zirconium is shown in diagrammatic form in Plate 32.

2. METAL DUSTS

Metal powders have long been used as constituents of paints and coatings, and as explosive mixtures in fire-works, etc. Recent technical developments have led to their increasing use in the moulding and casting of small machine parts.

The dangers of fire and explosion in powdered metals (and other dusts) are dealt with in detail in Section 1 of this Chapter. The particulars of individual metals given in par. 1 above indicate those metals whose dusts are particularly hazardous, but it must also be remembered that exceptional hazards can arise when the dusts of different metals are mixed together. (*See* also Section 2 of this Chapter, 'Fires involving Explosives'.

3. LIQUID METALS

The only metal which is liquid at normal temperatures is mercury, and this is described in Section 16 of this Chapter, the list of dangerous chemicals. There is no hazard attached to it at ordinary temperatures, apart from its poisonous qualities when swallowed or in contact with the skin, but when heated it emits highly toxic fumes.

Other metals are heated to a molten state in foundries, engineering works and in any part of the metal-working industries where casting, moulding, the preparation of alloys or metal coating takes place. In a few highly specialised plants, moreover, liquid sodium potassium and sodium-lithium mixtures are used for cooling or heat exchange.

Water must at all costs be kept away from molten metals; even a small drop of any liquid may cause them to react violently and scatter. Escaping or scattered molten metal will set alight most flammable materials in its path.

Fighting metal fires

1. TECHNIQUE OF EXTINCTION

Most metal fires are difficult to extinguish, and they will normally respond only to the correct extinguishing medium in each case. For this reason it is a sound general rule that whenever a metal fire can be allowed to burn itself out with complete safety to lives and property and with no danger of increase in the extent of the fire, it should be allowed to do so. In some cases of metal fires it is necessary to extinguish the fire but this can be done at fairly long range, so that there is no risk to fire-fighting personnel even from an intensification of the fire: both these circumstances are more likely to occur out of doors than inside buildings, and the commonest example of these conditions is probably the magnesium fire which sometimes accompanies an aircraft crash (*see* Part 6B of the *Manual*, Chapter 43, 'Fires in Crashed Aircraft'). In these cases it is permissible to use water on the metal, in large quantities if possible; this will produce violent reactions and perhaps cause scattering and explosions, but the fire will either burn out the quicker or be extinguished by the strong cooling effect of the water. The risks of this proceeding should be borne in mind at all times, however.

This freedom to use water on a metal fire or to allow it to burn out will not often apply, and in the great majority of industrial metal fires it will be necessary to apply the correct technique to the burning metal. None of the extinguishing agents commonly used on other materials may be effective, however, and their use may be very dangerous; for example the use of water on aluminium or magnesium when in small fragments or powder form may be extremely dangerous. For this reason, the importance of previous liaison with factory managements on fire extinction methods cannot be overstressed. The fullest consultation is necessary, not only to ascertain what is the correct material to apply in the particular case but also to ensure that ample supplies of the extinguishing agent will be available on the firm's premises.

Among the substances suitable for use on the majority of metals are powdered graphite, powdered talc, soda ash, limestone and dry sand. (Dry powders based on bicarbonate of soda, such as are found in the commercial types of dry powder extinguisher, are not considered suitable for some metals, of which titanium is one, and carbon dioxide is of only limited suitability). The function of these powdered materials is to 'smother' the fire by the exclusion of air and to 'seal' it against causing further fire spread; the latter is done by the action of cutting off the radiated heat from the burning metal. In the case of powdered metals, for which they are also suitable, it would be unwise to apply them under pressure from an extinguisher: for these, the extinguishing agent should be carefully spread on to the fire, starting from outside the burning area and working towards the centre (*see* Fig. 27). The utmost care is necessary in avoiding any disturbance of the burning powder.

Fig. 27. Method of applying dry sand on to a metal fire.

The U.K.A.E.A. have developed two powders for use on fires involving radioactive metals, e.g., uranium, plutonium, etc. These are ternary eutectic chloride (T.E.C.), which is a mixture of potassium, barium and sodium chloride in such proportions to give a common melting point, and ternary eutectic fluoride, which includes lithium fluoride instead of sodium chloride. The lithium salt acts faster than the sodium one, but is more expensive. When applied to a burning metal, they form a 'frit' or semi-molten skin over its surface, thus effectively sealing off the metal from the surrounding air and the latent heat absorbed by the frit also helps to cool the metal.

Another powder which has been developed by J.F.R.O. is boron trioxide with added magnesium stearate to improve its flow properties, and this has successfully extinguished magnesium, aluminium and sodium fires. Research has also been carried out on trimethoxyboroxine (T.M.B.) for magnesium aluminium alloy fires, but this is not yet generally available.

It will be seen from the foregoing that powdered extinguishing agents for metals and metal powders are only suitable for small fires, since the technique of applying them could not easily be adapted to large fires. Fortunately, large fires in metals are not common, since the high thermal conductivity of most metals prevents any one spot in a mass of metal from remaining at a sufficiently high temperature to burn. In association with heat treatment baths, however, fires in relatively large pieces of metal are known, and here there is little or nothing that can be done except to prevent the spread of fire beyond the piece involved. Similarly, the necessity for fighting large fires in powdered metals must be avoided, since it cannot be faced, by restrictions on quantity in the storage and handling of such powders.

2. PERSONAL HAZARDS

Apart from the risks of explosion which have been mentioned, the danger to fire-fighting personnel of receiving severe and painful burns from metal fires should be guarded against, and there are many toxic hazards from the products of combustion and other reactions. Cadmium and beryllium are particularly dangerous in this respect. Breathing apparatus should be worn in fighting fires known or thought to involve these and other metals.

The radioactive hazards from certain metallic elements have already been referred to in earlier paragraphs.

Industrial processes and risks

1. ENGINEERING WORKS

(a) Construction

Heavy engineering works include foundries, heat treatment shops, machine shops for the heavier work, rolling mills and mills producing structural steel. Also classed under this heading are factories and yards which make heavy vehicles such as aircraft, cars, locomotives (including boiler manufacture) and ships, together with those manufacturing heavy machinery and the heavier equipment for agriculture and the electrical industry.

Light engineering works include light forge and machine shops and plants for galvanising, metal plating and metal coating and finishing. Together with these are light machinery and wire manufacture and works producing apparatus of the lighter kinds, for instance radio and electronic, photographic, plumbing, scientific (including instruments) and textile. Factories for lighter metal articles such as car accessories, household goods and appliances, jewellery, precision tools, metal toys, typewriters, cash registers, etc., are also included.

A large engineering undertaking may in some cases have a roof area of a million square feet or more, and the area occupied by the factory itself may be more than 40.47 hectares.

The works usually comprise a number of shops containing various engineering plants such as a foundry, with steel, iron and non-ferrous sections; a stamp shop or forge; a heat treatment shop; pickling plant; de-greasing plant; machine shops; press shops; assembly bays; welding plant; engine testing beds; saw mills; paint, or 'dope' shops; and electro-plating plant. There are also large quantities of stores such as steel, sheet metal, paint, trimming and timber, and consumable items such as stocks of oil and petrol, either grouped in one building or distributed in various places throughout the factory. In addition, there will be a general office and a drawing office, which in modern factories is usually a detached, brick-built building of normal construction; in older factories the offices may occupy separate compartments in one or more of the shops.

Owing to the wide range of industries concerned, it is difficult to generalise on the construction of engineering shops. Large engineering shops usually consist of long, rectangular-shaped buildings. These buildings may be steel-framed with brick panel filling, or may have solid brick walls. In the latter case, the walls may be load-bearing. There may be a considerable number of unprotected steel stanchions and trusses. The roof is ordinarily of large span, and may be of the Belfast (wooden) or steel truss type, or of the saw-tooth north light construction (*see* Book 8 of the *Manual*, Chapter 5 'Other elements of structure').

There may, especially in the newer industries, be very large and lofty buildings containing no dividing walls, so that fire, if it can take hold, is likely to involve the entire building: a particular danger arises when no adequate fire separation is provided between high-risk areas, such as wood-working shops and packing departments, and the remainder of the production space. The difficulties of providing fire separation are added to by the continuous flow production line methods, which are general in the industry. These factors have on occasions led to rapid fire spread combined with severe smoke-logging and great heat throughout the building.

In new buildings, the tendency towards unrestricted manufacturing spaces is even more pronounced, though there is less likelihood of flammable building materials being used: reinforced concrete structures are common and single-storey construction is becoming very general. Great use is still being made in engineering buildings, both new and when re-roofing older structures, of bitumen-protected metal sheeting. This material has excellent properties of resistance to corrosion, but it varies in its properties as regards surface spread of flame according to its constitution: not all varieties of sheeting are satisfactory.

(b) Processes and risks

In casting and moulding, molten metal is poured into moulds of sand (sand casting), thin shells of sand and resin (shell moulding), or, under pressure, into metal moulds (die casting). Metals commonly used in the first two processes are steel, iron, brass, copper and aluminium or magnesium alloys. In die casting, the molten metals are usually alloys containing tin, zinc, antimony, lead, aluminium or magnesium.

Fire may start through the ignition of wooden patterns or benches, or oils used in the processes, by split molten metal or by their coming into contact with furnaces or other sources of heat. The possibility of spread is increased where evaporated oil has settled on the roof and machinery. In die casting, fire may occur through the escape of flammable hydraulic fluids, and in shell moulding a dust explosion risk attaches to the resinous powder used in making the moulds. Where magnesium is the metal being cast, it presents fire risks due to its own properties. In some precision casting processes, flammable liquids are used as constituents of slurries, etc.

Other methods of forming metals into particular shapes are press forging, drop forging, machine forging and extrusion. In the first three of these processes, pressure is applied by hammers or presses, etc., to either hot or cold metal, while extrusion involves forcing the metal through a die by means of a ram. Either heated or cold metal may be used in extrusion processes. Forging and presswork are usually applied to steels, brass, copper, aluminium sheet, etc., and extrusion is used for copper, lead, aluminium, magnesium, zinc and alloys of these. The main fire risk arises when flammable hydraulic fluid escapes, though risks are also presented by thermal compression of lubricating oils and in static electricity.

Machining, which involves cutting metals in various ways, includes the operations of boring, broaching, drilling, gear cutting, grinding, milling, planing, shaping and turning. Risks from frictional heat with blunt tools are severe and a risk common to all these processes lies in the use of cutting oils, some of which may be highly flammable. Other risks vary with the different processes and with the metals being used. For example, there is a risk of dust explosion in grinding, and there is a considerable increase in all risks in any process using magnesium or titanium. In some works CO_2 extinguishing systems have been installed, with the object of securing immediate extrusion on the machine before any build-up of heat has had time to take place.

Metals are subjected to heat treatment in various ways. Furnaces are used for heating many different metals, and some of them, such as magnesium alloys, can cause explosions in furnaces if metal dust has gathered (*see* paragraph (e) 'Special hazards'). The fuel used may also cause fire, through an escape of oil, for example, or through mechanical faults when gas is used. Flame hardening, applied to fine-grain steels, is another method of heat treatment; this is often followed by metal quenching, in which there is considerable fire risk from overheating where the quenching medium is oil. The main cause of fire in this process is failure to immerse the red hot metal quickly enough in the oil, (through some fault retarding the process of lowering the metal), and this is liable to result in the immediate ignition of the oil. Induction heating, by high frequency current which may be used for treating steel, brass, copper or aluminium, is not particularly likely to cause fire, though there are electrical hazards in the apparatus.

There is a high fire and explosion risk in the final heat treatment method, salt baths, and this is described under 'Special hazards' below. Of the three types of salt bath, the low-temperature baths (nitrate baths), used for aluminium alloys and certain steels, are the most dangerous. Medium-temperature salt baths are used for steels and also non-ferrous metals. High-temperature baths treat steel alloys and copper.

Joining and assembling involve such processes as riveting, welding, soldering and brazing. Of these, welding is the most likely to lead to fires by igniting any oils, wood or other flammable substance within range of the operation. The special risks in oxy-acetylene cutting and welding are described among 'Special hazards' below; other types of welding also have certain dangers, including risks connected with electrical apparatus in resistance welding, and the great heat of the material in thermite welding. The metal being welded, aluminium, for example, may cause danger unless the work is done under an inert gas.

The fire risks in the various finishing processes vary greatly. The most usual methods of metal cleaning are de-greasing and de-scaling. In vapour de-greasing, trichlorethylene is generally used. It is not normally flammable (*see* Section 16), the only danger lying in the extremely poisonous nature of the vapours. In solvent de-greasing there is a fire risk where low-flash point solvents are used. De-scaling, on the other hand, is a type of salt bath process involving the hazards associated with salt baths and described below. It is essential that no water should get near the de-scaling apparatus or the stocks of sodium, which is one of the chemicals used in this type of salt bath. Fire may be started in this process by hydrogen explosion.

Electroplating is described in par. 3 below under 'Plating works.' The metals used range from silver and copper to zinc, chromium, lead, nickel and cadmium.

Hot dipping and coating, (in which tin, zinc and aluminium are among the covering metals,) involve the usual molten metal risks. In the case of coating, the methods vary greatly and may result in additional dangers in some cases: in hot spray coating, for instance, acetylene or hydrogen may be in use and the metal may be fed into the machine in powder form. In vacuum metallising, dip tanks and ovens are employed and the relative fire and explosion risks arise.

Paints and varnishes may be used for coating metals, and the fire risks connected with these and with paint spraying are covered in Section 9 'Paint and Varnish Fires.' In addition, fires are likely when the paint-coated metal is being baked and dried in ovens. The use of celluloid, also, is associated with the manufacture of cutlery (*see* Section 10).

Apart from the hazards arising from metal-working processes, which have been referred to above, there are many areas in most engineering works in which more normal causes of fire are common. Flammable materials such as wood, paper, textiles, fibres and rubber are used in most engineering factories including the manufacture of motor cars, tyres, aircraft, boats, tools, etc. In the parts of engineering works in which these flammable materials are used or stored, fires arise from careless handling of heating and lighting appliances, tools, electrical gear and smoking materials, from failure to maintain apparatus and from many other causes. In these areas, as in most industrial plants, the state of cleanliness and tidiness or otherwise and the extent of over-crowding of plant and workpeople govern to a great extent the likelihood of fire outbreak and fire spread.

(c) Fixed fire protection

Because of the wide range of processes carried on at engineering works, extensive fixed fire-protection systems are not general, though becoming more common. Reliance is usually placed upon an efficient works brigade; special risks, however, are occasionally protected by individual installations. Electrical generating or sub-station equipment is often protected by CO_2 installations, and particularly vulnerable sections such as paint stores and dip tanks may be similarly protected. In some factories, for example, the paint-drying ovens and the furnaces used for japanning are protected by manually or automatically operated fixed CO_2 systems, and the principle of applying fixed CO_2 protection to individual high-hazard machines and areas is being increasingly adopted. Although audible warning systems are installed in more and more engineering works, it is not common for these to be automatically operated by heat- or smoke-detectors and they are usually manually actuated. Hose reels are commonly found except where metals only are being protected.

The most usual form of protection is hand fire extinguishers and supplies of extinguishing agents (such as special powders for use where magnesium alloys are being machined).

(d) Fire-fighting

Because many large engineering undertakings maintain efficient private brigades, and because, owing to the comparative frequency of minor outbreaks of fire, the works personnel are usually familiar with the action to be taken in an emergency, the fire brigade is seldom called to a fire in time to deal with the immediate outbreak. There is, therefore, little point in describing here the methods of extinguishing very small fires such as those which occur in the individual machines.

If the private brigade has been unable to deal with an outbreak, their speed in subsequently asking for fire brigade help is of paramount importance, and may make all the difference to the conditions prevailing at the time of the brigade's arrival. Experience has shown that in many types of engineering works there can be a quick spread of smoke and a steep temperature rise throughout the building within a short time of the discovery of an outbreak. Very rapid reconnaissance (or enquiry made from fire-fighting personnel on arrival) will be very important, and early attention to ventilation will be necessary; this applies particularly in large undivided buildings of the single-storey 'shed' type.

The roof fire is one of the types of fire most likely to be encountered at engineering works, as this is the portion of the building which is most likely to burn. Such fires should be tackled with the greatest possible speed. Factory roofs present a serious problem to the fireman as, owing to their large span, they are seldom of sufficient strength to allow men to climb all over them. Fire-fighting operations will depend largely on the type of roof involved, but the best method is often to work from the valleys between the different roof spans, as these will usually provide a stable footing. The presence of cat walks indicates that the main roof area is almost certainly incapable of carrying the weight of a man, but the catwalks can, of course, themselves be used in fire-fighting. If a roof fire is not caught in its early stages, it may easily, owing to the extreme danger of using large quantities of water, involve the whole roof area of the shop. It is essential, moreover, that the fire should be extinguished as rapidly as possible, since the light steelwork of its construction is not usually protected and early collapse is always a risk.

In aircraft manufacture, particularly, the buildings are almost always large and tall. Many shops are 92 metres square and 15 – 20 metres high. A fire on the floor of the shop does not necessarily spread to the structure of the building at once, but is likely to spread rapidly from one aircraft to another, because of the presence of flammable materials and the absence of separating walls.

If aircraft in a hangar are endangered by fire, they should be pushed or pulled out of the hangar. The handbrake in the cockpit must be released before this can be done. Large aircraft can be towed out of the hangar by means of a tractor or a towing vehicle.

In general, one of the most important tasks in fire-fighting in engineering undertakings is that of preserving essential machinery intact as far as possible; every effort should be made to avoid damaging the machines by water. This applies particularly to the modern factory employing automation, which involves electric and electronic control gear of the utmost delicacy and value. The 'nerve centre' of such a system requires a high priority of attention. It is usually advisable in the early stages of the fire also to concentrate on the office block in order to prevent the records, on which the existence of the firm may well depend, from being affected.

The works manager should be contacted at the earliest possible moment. He will be able to advise on the hazards to be expected. Due weight should be given to suggestions he may make, particularly with regard to the danger of using water in foundries, die casting shops, heat treatment shops, and the like. The manager will also be able to advise on the priority to be given to the protection of the various shops from the point of view of production.

(e) Special hazards

The fireman's first impression on entering an engineering works is generally one of great complexity and, because of the presence of so much machinery and other plant, of extreme hazard. If the premises themselves are of reasonably sound construction, however, and good housekeeping is practised, the risk is in fact not unduly high, principally because a fire in a machine can often be isolated immediately and because there is little in the construction of the building itself to add to the intensity of the fire. Nevertheless, although the danger of an outbreak of fire can be kept within reasonable limits, the dangers to the fireman once a fire has occurred in these premises are considerable, particularly when space is congested and access avenues are narrow. Some of the principal hazards are briefly described below, but it must be remembered that industrial processes are changing continually, that the lay-out of a works is altered frequently and that it is extremely difficult for a person unfamiliar with the premises to know without expert advice, exactly what process is being performed by any particular machine, what dangers it presents and the best methods of fire extinction to employ.

(i) *Furnaces*. One of the more serious dangers to the fireman at a fire in an engineering works is the presence in certain parts of the works of numerous furnaces, ovens and cupolas for melting metal. These make it extremely hazardous to use water in any form, even on apparently cold objects. The fireman should take extreme care in all parts of the building where furnaces are situated. Water should on no account be used anywhere in the vicinity.

In some cases glass roofs have been boarded over. The boards may take fire from pieces of molten metal ejected by die casting machines, sparks from welding plants, etc. Despite the hazards of working on the roofs of engineering shops, which are often of wide span, it is the usual practice with works brigades for the men to get on top of the roof and cut roof fires out, or extinguish them as best they may with very small quantities of water from hand pumps, rather than take the risk of water from jets falling on the furnaces below.

The furnaces may be heated by gas (producer gas or town gas), liquid (fuel oil or creosote pitch) or solid fuel (coke or coal), and their existence is usually revealed by the presence of flues and chimneys. Electrically heated furnaces are now common, more-over, and may not be provided with flues where certain processes are concerned. The method of cutting off the supply of fuel or electricity varies in each case, the the advice of the workshop staff should therefore always be sought before interfering in any way with the furnaces. It should be remembered that a furnace may take many hours—or even days—to cool down after it has been extinguished.

In general, every precaution should be taken when working near furnaces. In particular, care should be taken not to pick up, or even to touch, metal parts which may be seen lying about, since these may be sufficiently hot to cause serious burns. Attached to the casting shop, for example, will frequently be found a series of ovens which are employed in casting the raw metal into ingots for use by the individual machines. When these ingots are withdrawn from the furnaces, they are laid out in rows on the floor to cool. They frequently cover the greater part of the floor space, and make it dangerous to move about unless great caution is exercised. Some furnaces, moreover, have oil baths attached in which small engineering products or metal strip may be quenched as they leave the furnace.

(ii) *Salt baths.* A method of heat treating metals, particularly aluminium alloys, by immersion in a bath containing a molten mixture of potassium and sodium nitrates and a small proportion of nitrites, is widely used. Various types of bath, employing different chemicals and different temperature ranges are used for varying purposes. Accidents have occurred at these baths from three distinct causes:

A. Explosion resulting from the reaction of hot nitrates with the metal of the bath or with the metal articles under treatment (*see* Plates 29 and 30).

B. Explosion resulting from the dipping of wet or greasy articles in the molten nitrates.

C. Gassing by nitrous fumes given off when the hot nitrates come into contact with oxidisable substances, e.g. oil, tar or soot, in the furnace, if the bath leaks or breaks down under the action of the nitrates.

In the low-temperature (300°—1,000°F.) (149°—538°C.) range, which embraces the common 'nitrate' bath, the mixture of molten nitrates is stable up to 900°F. (482°C.) the temperature at which it is generally used. If the temperature reaches 1,100°F. (593°C.) the nitrates begin to decompose and to react with the metal of the bath or with the articles under treatment. Above 1,100°F. this reaction becomes very violent, and cannot be controlled. Finally, the reaction attains explosive violence. Fatal injuries have in the past been caused by explosions occurring in this way.

While the contents of these baths are not in themselves flammable, they are strong oxidising agents and their temperature is sufficiently high to ignite any combustible substance with which they come in contact. The most important danger from the baths to the fireman, however, is the fact that if the water is allowed to come into contact with the molten salts, the instantaneous generation of steam will result in an explosion.

A fire in the vicinity of the baths, therefore, presents considerable risks. Every precaution should be taken to prevent water from entering the bath and, if the bath itself should become damaged, to prevent the contents of the bath flowing towards combustible articles. Because of this danger, some undertakings prohibit the installation of any type of fire extinguisher in the vicinity of the baths.

In modern heat treatment shops, medium-temperature (1,000°—1,750°F.) (538°—954°C.) cyanide hardening baths may be found. These are much smaller than nitrate baths but from a fire-fighting point of view present similar risks, and the use of water in their vicinity is exceptionally dangerous. This applies also to high-temperature baths.

(iii) *Pickling plants.* Pickling plants usually comprise twin vats, one vat containing steam-heated dilute sulphuric acid and the other containing clear running water. The size of these vats, some of which are very large, depends on the process being carried out.

The walls of the vats are usually constructed of wood, lined internally with sheet metal. Sometimes, however, the vats are constructed of other materials such as glazed brick, concrete, etc. The vats are usually partly sunk in the ground, with walls extending approximately 900 mm above the floor.

Great care should be exercised by firemen working in the vicinity of pickling plants, due to the danger of severe burns likely to be inflicted by contact with sulphuric acid.

Should a copious quantity of water be directed into the acid vat, the latter may overflow and, despite its dilution, inflict severe burns on the fireman's feet or legs, or any other part of his body with which it comes into contact.

(iv) *De-greasing plants.* In many works, particularly those such as motor or aircraft engine factories or repair shops, one or more de-greasing plants will be found (*see* Fig. 28). These are used to clean metal work prior to painting or coating. The plant generally consists of a deep metal tank, at the bottom of which is trichlorethylene. The latter is vaporised by heat (usually by gas, steam or electricity) and then caused to condense by means of water-cooled coils situated at the top of the bath. The articles to be cleaned are lowered into the bath on hangers or in metal baskets, and the trichlorethylene vapour condensing on them frees them from grease and oil. The vapour, although not normally flammable, is toxic, and will produce phosgene on being heated even by contact with a lighted cigarette, for example. The vapour is retained within the bath, however, and is not likely to be very dangerous. If any of the gas should escape it will remain at a low level. Anyone affected by it should be removed shoulder high. If a fire occurs in the vicinity of one of these baths, it will usually be advisable to ask the shop foreman to draw off the trichlorethylene.

(v) *Cutting and welding.* Oxy-acetylene cutting and welding are widely used in many branches of the engineering industry. The gas is dissolved under pressure in portable cylinders and these present a considerable fire and explosion hazard.

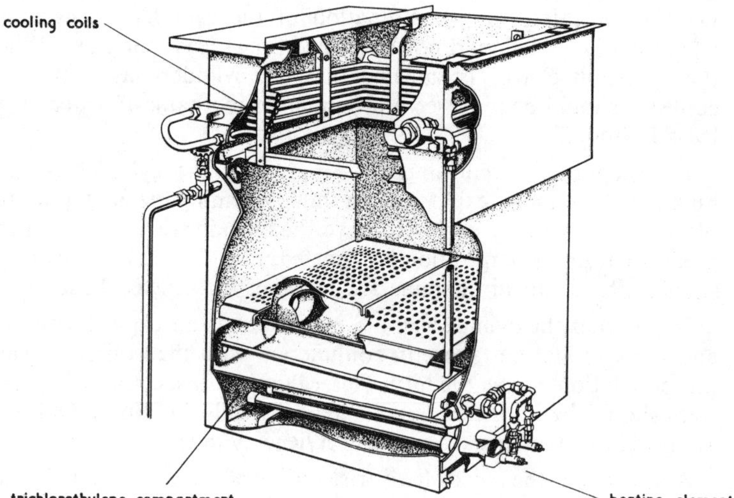

Fig. 28. Diagram showing the construction of a vapour de-greasing plant.

A dissolved acetylene cylinder which has been subjected to heating through being involved in a fire, or through a back-fire, or through careless handling when in use, is extremely dangerous; progressive decomposition of the acetylene and solvent contained in the cylinder takes place on heating and is liable to continue even after the cylinder has been cooled. Cylinders in this state should be treated as liable to burst or explode at any moment.

Before heat has reached the cylinders, but only before, it may be possible to remove them to a safe place, where they should be laid down so that they do not touch each other. Care should be taken during this operation not to subject the cylinders to shock or damage.

Once decomposition has begun, however, the cylinder should be left in position (unless circumstances make this impossible, see below) and cooled with a copious supply of water in the form of a spray. Where the main valve of the cylinder is found closed it should be kept closed; opening it will increase the danger. If the main valve is found open it should (if practicable) be closed, unless the pressure relief disc or plug has operated. In the latter case, no attempt need be made to close the main valve. The pressure relief disc is in the body of the main valve and there may be another at the base of the cylinder. A plug would be situated in the shoulder of the cylinder.

Cooling of the cylinder should be undertaken from behind suitable cover, the cooling branches being lashed and firemen withdrawn to a position of safety, and examination of the cylinder, to establish whether it is cool should not be started for at least one hour. When the officer in charge is satisfied that the cylinder has completely cooled it should be removed and immersed in a tank of water for at least 12 hours.

In exceptional circumstances, *e.g.*, where the cylinder is in a building containing explosives or highly flammable material, the officer in charge may decide to have the cylinder removed from the premises after decomposition has started. In such a case the cylinder should after removal be immersed in water for at least 12 hours.

In all cases the cylinder should be treated as an explosive missile and steps should be taken, in conjunction with the police, to warn the population in the vicinity. Her Majesty's Inspectors of Explosives should be informed immediately through the Home Office of all incidents involving dissolved acetylene cylinders.

Apart from the specialised risk of explosion associated with dissolved acetylene cylinders, the gas itself and the oxygen may cause or contribute to an outbreak of fire through leaks or backfires. The flame from oxy-acetylene cutting plant has a temperature of over 5,400°F. (2,982°C.), and conducted heat from cutting or welding operations can set fire to combustible material in the vicinity: sparks from blowpipes may be thrown great distances and can start fires if surroundings are not kept scrupulously free from material which they can ignite.

(vi) *Engine test beds.* In a factory producing internal-combustion engines, there will usually be a testing plant for measuring the power output of the finished products. These test beds present the usual risks associated with the internal-combustion engine, with the additional hazard that large quantities of petrol will be flowing into the engine from an external source, usually by means of a large diameter, gravity-fed, flexible piping, and any fracture of this may cause a considerable quantity of petrol to escape before the supply can be cut off. This applies particularly where aircraft engines are tested. CO_2 extinguishers are generally installed and will be used for attacking the fire in its early stages. When the outbreak has assumed larger proportions, however, the use of foam will almost certainly be necessary. It will be advisable to have all other engines shut off immediately and to cut off the supply of petrol to each.

2. FOUNDRIES

(a) Processes and risks

Metal castings are produced in a foundry. The principal sections of a foundry are the pattern shop, the foundry proper and the fettling shop.

(i) *Pattern shop.* In this shop are made the wooden patterns which are used to shape the sand moulds into which the metal is poured. Varnishes and methylated spirits may also be found here. These patterns are made to exact dimensions and often are of great value. It is, therefore, desirable not to use water for fire-fighting if it can be avoided as water may cause the pattern to swell and become distorted, thus making it useless. An extinguishing gas should be used, if available, and water should be used only as a last resort. In a fire, the first steps taken should be to protect the pattern shop.

(ii) *Foundry.* This may be a sand or a die foundry. In the sand foundry, molten steel, iron, brass, copper and aluminium or magnesium alloy as the case may be, is poured into a mould of sand which has already been shaped by a wooden pattern. In the die foundry, a molten alloy is injected, usually under pressure, into a metal mould. These alloys generally contain tin, zinc, antimony, lead, or other metals which have a low melting point. Sand casting is usually done on the floor of the shop. Sand castings occasionally weigh many tons; die castings are usually smaller, however, and this process is used principally for mass production work.

(iii) *Fettling shop.* Here the unwanted plugs of metal (called runners and risers) formed in the pouring process are trimmed off the castings by means of a bandsaw, a grinding machine or a hand file. Fine machining is done in the machine shop. The risks in these shops vary with the metal being used.

(b) Fire-fighting

It is most important to avoid using water anywhere in the vicinity of the furnaces or casting machines, or in places where it could run into the foundry and reach castings on the floor. If CO_2 extinguishers are available they may be used, as also can *dry* materials such as sand or special powders. (*See* 'Fighting metal fires', page 161). There is a danger that if water reaches white-hot metal in the moulds, a violent explosion may be caused and molten metal may be scattered in a dangerous manner. When fire-fighting is carried out near a foundry, men should be detailed to erect retaining walls or barriers to prevent water reaching the shop. Apart from the metal in moulds, the fire-boxes of the furnaces are often set below floor level where water may do much damage. Fire-fighting should otherwise be confined to cutting away and the use, if necessary, of small quantities of water, applied either by a hose reel or hand pumps. In view of the construction of foundries, the only part of the building likely to burn is the roof, and it will usually be possible, if the fire is caught at an early stage, to extinguish this without using a jet. Owing to the great heat always present in a foundry, all woodwork is likely to be extremely dry and therefore highly flammable.

3. ELECTRO-PLATING WORKS

(a) Construction

The smaller firms carrying on electro-plating of metals will usually be found in nondescript, converted premises which may be anything from an old shop front with tin sheds at the rear to an adapted office or other building. There is, however, a growing tendency for industries which depend on these processes to carry out the work themselves rather than to place it in the hands of smaller firms. In such cases the plating shop is usually a detached, single-storey building of brick construction; it may equally be found in a separate compartment within the main building, sometimes in a basement or sub-basement. Among the many industries in which plating shops are likely to be found are cutlers, die casters, fancy goods manufacturers and motor accessory manufacturers.

The interior of the plating shop is usually a single rectangular compartment with a series of cleaning and plating baths along each wall. In the middle is a long bench, or a series of separate work tables, for examining and checking the articles before and after plating. The floor is generally of stone or concrete with wooden treads or duck-boards running beside the tanks. If the shop is in a detached building, the roof is often wholly or partially glazed.

(b) Processes and risks

The article to be plated is first 'pickled' in an acid solution to remove impurities, and may also go through other cleaning processes such as de-greasing. It is then hung on one of a series of hooks in the plating bath (*see* Fig. 29), and connected by a short length of insulated wire and a clip to the negative terminal (cathode) of an electric circuit, A bar of silver or other metal, also immersed in the bath, is connected to the positive terminal (anode). A current is passed through the circuit, with the result that metal is removed from the bar at the positive terminal and deposited upon the article at the negative terminal.

The chemical contents of the bath vary according to the particular processes being carried out: copper sulphate, nickel chromate and silver sulphate are often used. Beside each plating tank is generally

FIRE CLASS SYMBOLS

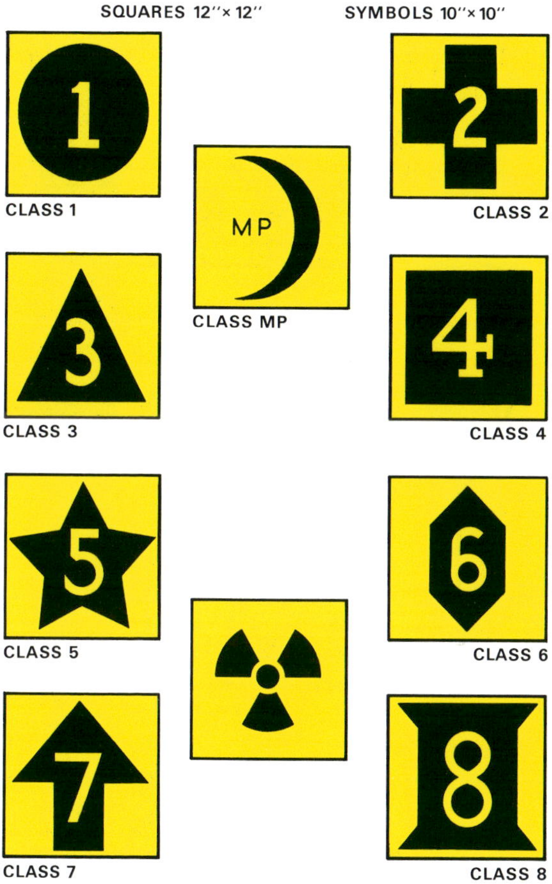

Plate 31. Fire classification symbols for explosives. For class 8 explosives, the trefoil symbol is also displayed to denote a radioactivity hazard. (For details see Section 2, pages 20 and 21.)

PRODUCTION OF ZIRCONIUM

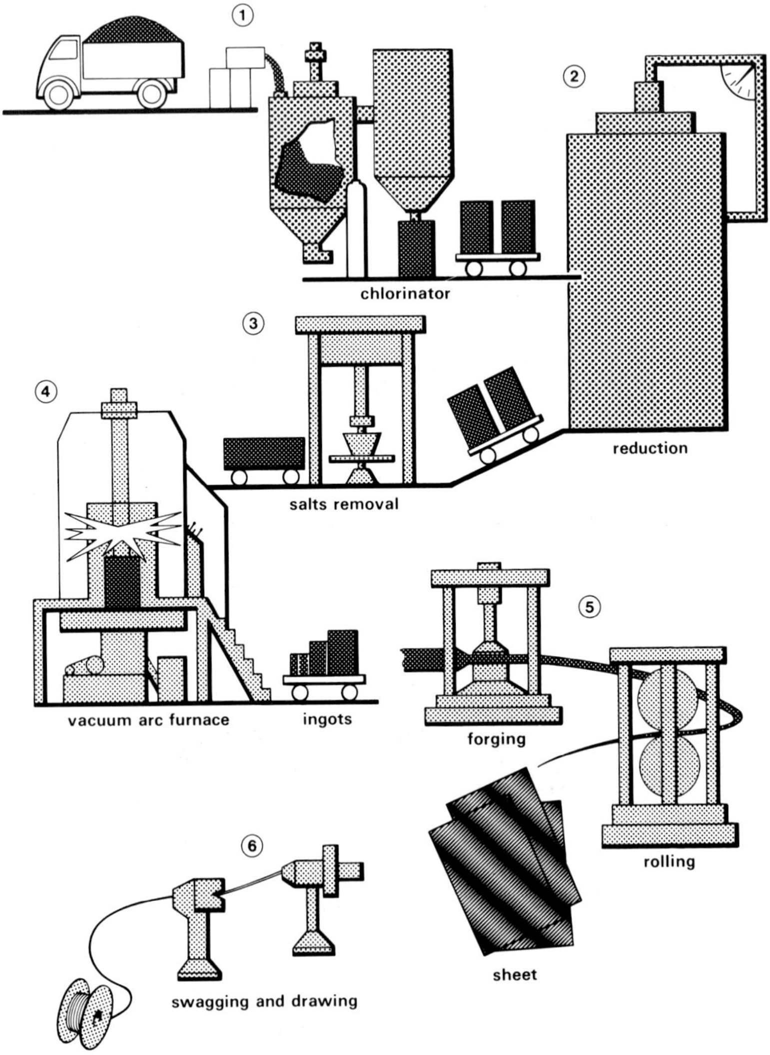

Plate 32. Zirconium silicate (from Australia) is mixed with carbon and heated with chlorine (1) forming zirconium chloride. Distillation over molten magnesium (2) produces a sponge-like material from which the magnesium chloride (3) is removed in a vacuum. A vacuum arc furnace (4) then produces zirconium ingots which are forged and rolled (5) into sheets or drawn into wire (6). (Section 7)

PRINCIPLES OF REFRIGERATION

DIRECT EXPANSION METHOD

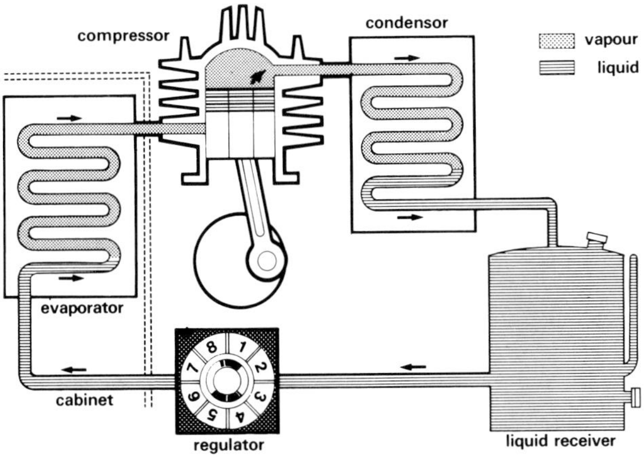

Plate 33. The liquid refrigerant passes from the receiver tank into the evaporator coils where it takes heat from the chamber to be cooled and consequently vaporises. The vapour returns via a compressor through a cooling condenser, where it is liquefied, back to the receiver tank. (Section 12)

ABSORPTION METHOD

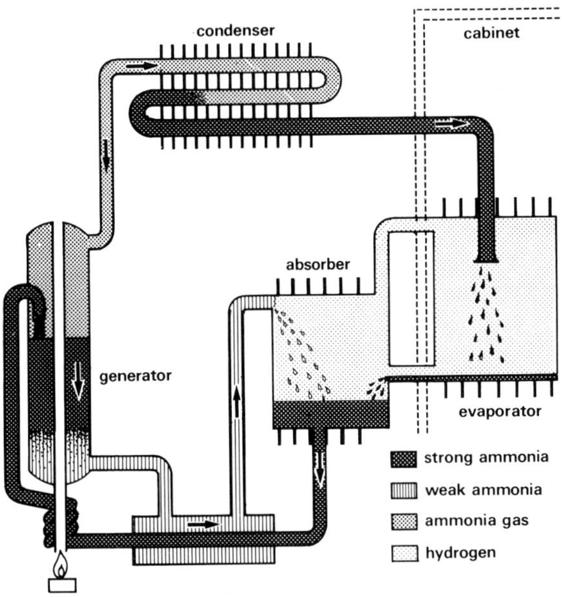

Plate 34. Diagram showing the principles of the absorption system of refrigeration, which are explained in detail in section 12 page 232.

THE MANUFACTURE OF SYNTHETIC RUBBER

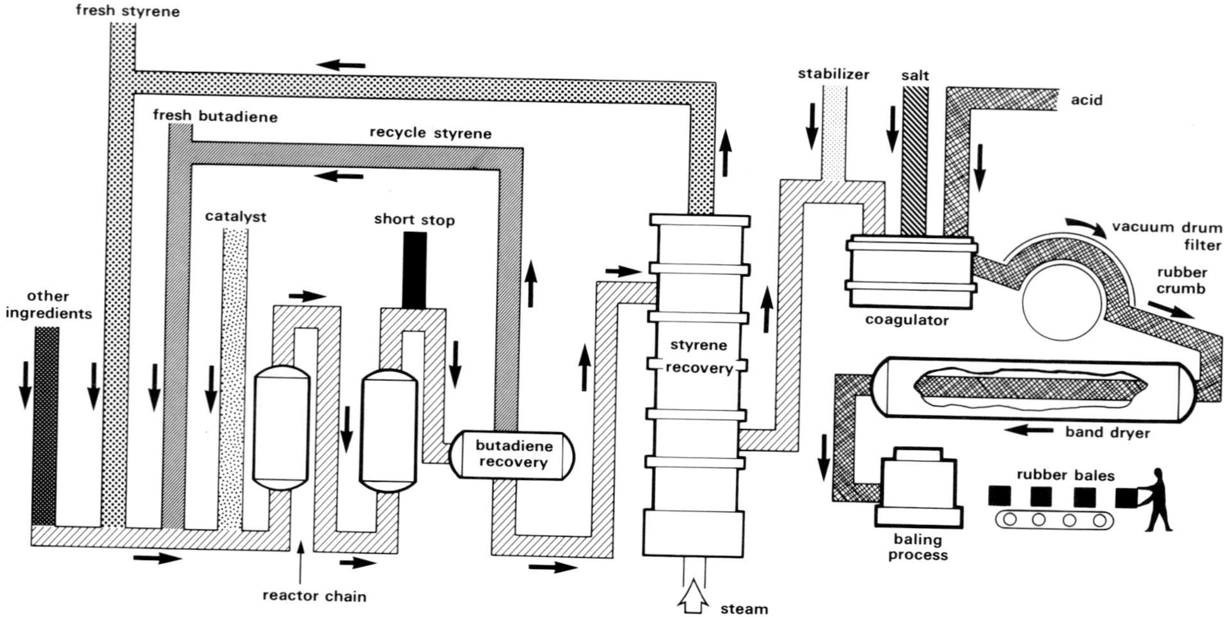

Plate 35. *Butadiene and styrene, in a 3 : 1 ratio, are mixed with small amounts of other ingredients. A catalyst is added to speed up the polymerisation process in a reactor. A reagent stops the process from continuing too far and some of the butadiene and styrene are recovered and used again. A stabiliser is added to the latex to prevent oxidation, and acids and salts coagulate the latex to a brown rubber-like mass. After washing it passes through a vacuum-drum filter and band drier ready for baling.* (*Section 14*)

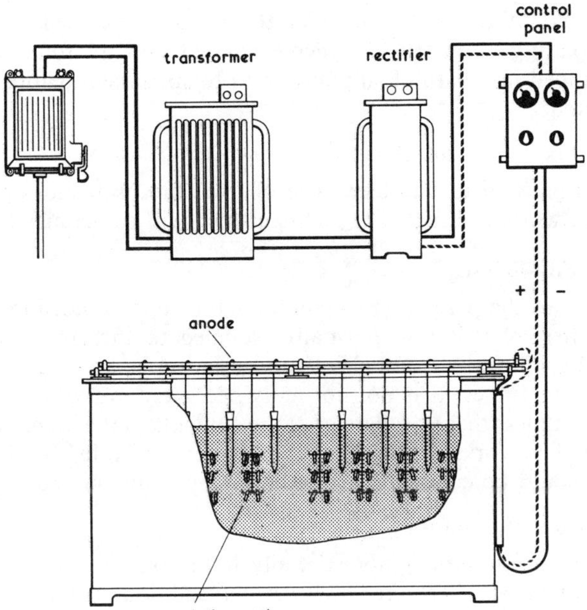

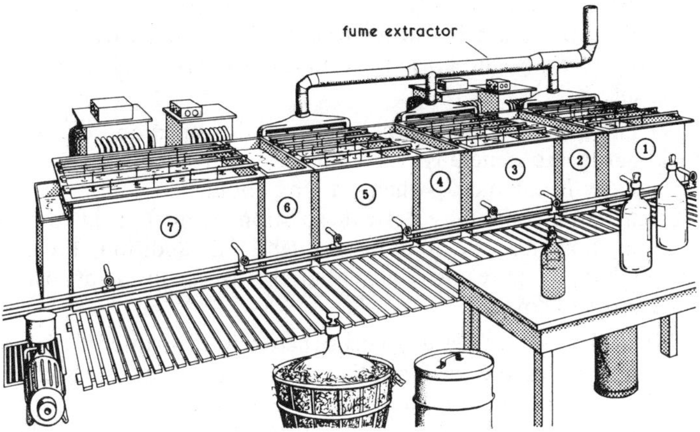

Fig. 29. Sketch showing the layout of a plating shop. Typical contents of tanks are: 1. Caustic potash; 2. water wash; 3. caustic soda; 4. water wash; 5. sulphuric acid; 6. water wash; 7. nickel sulphate.
ABOVE: *Diagram of electrical arrangement.*

a tank containing water, in which the articles are cooled and washed after plating. Each article is then examined, and any found defective are dipped in a bath of sulphuric acid to strip them before they are re-plated:

(c) Fixed fire protection

It is unusual to find any form of fixed fire protection in electroplating works; though CO_2 extinguishers will sometimes be found.

(d) Fire-fighting

None of the plating baths contain flammable liquid. Fire-fighting measures are therefore normally confined to the structure of the building. Immediate and effective ventilation is, however, necessary to remove fumes. It is not normally necessary to use breathing apparatus unless there is a danger of the contents of the baths becoming mixed. The works manager, or if he is not available, the shop foreman should be contacted at the earliest possible moment.

(e) Special hazards

The process itself is not specially hazardous from the fireman's point of view. There are, however, in addition to the usual risks associated with the chemicals employed, certain points to be borne in mind.

(i) Care should be taken not to damage or overturn any of the open baths and thus allow the contents to come in contact with each other. If cyanide is allowed to come into contact with sulphuric acid, for example, the poisonous fumes of hydrocyanic acid will be given off.

(ii) The sulphuric acid baths are invariably open and the wooden duckboards are generally very slippery. Care should be taken—particularly in the dark—when walking about.

Open baths containing a weak solution of nitric acid, used for cleaning, may also be found. Frequently in addition, there is a smaller bath of pure nitric acid, but this is usually kept in a safe place and well covered.

(iii) In the smaller plating works there will probably be a control panel on one of the walls of the shop, controlling the supply of electricity to the plating baths. Since low voltage direct current is employed, both transformers and rectifiers will be found. Though the electrical hazard is not great, the usual precautions should be taken; and it is advisable to open the main switch as soon as possible in case current enters the premises at medium or high voltages.

(iv) In general, plating shops attached to the larger and more responsible firms present few risks to the fireman if normal precautions are taken. Every possible precaution should, however, be taken against injury to firemen from acid burns, especially in the case of smaller firms. Many of them carry on their business in cramped conditions with little regard for safety.

Part 6c, Chapter 45, Section 8

Fires in animal and vegetable oils

VARIETIES OF ANIMAL AND VEGETABLE OILS

NATURE, PROPERTIES AND RISKS OF OILS

INDUSTRIAL PROCESSES IN SEED-CRUSHING MILLS

For purposes of this Section, a somewhat arbitrary distinction has been made between oils (other than mineral oils, which are dealt with in Part 6B of the *Manual*, Chapter 44 'Fires in Oil Installations') and fats; reference to Section 3 on 'Fat and Wax Fires' will confirm this. Most animal and vegetable oils have properties very similar in character to those of melted fats, and their industrial uses are also mainly similar. Much, therefore, of what is said in Section 3 on fats has some application to oils, and *vice versa;* since, however, fire-fighting technique may be said to be somewhat affected by whether the object is a solid or a liquid, the subjects of fats and oils have been treated separately for the convenience of the reader.

The flammability of animal and vegetable oils is well-known and they are universally accepted as a relatively high fire risk; they are, however, found less widely in industrial and domestic premises than they once were and their production, although a very old-established industry, has been contracting during this century. Industrial development has been such that in the nineteenth century both production and use of these liquids was at its highest: labour conditions in tropical countries allowed production of the raw materials to be done cheaply and abundantly, and the use of oils for illumination, foodstuffs and paint manufacture, for example, was widespread. These conditions have changed, other methods and materials having tended to replace those previously in use: oil is now seldom employed for lighting, and synthetic chemical products are displacing them in paint manufacture and in other industries. However, they continue to play a large part in outbreaks of fire, particularly in the old-established industrial centres and in ships, etc.; as far as manufacturing processes are concerned, the risks offered may be said to be highest in the processing stages, above average in the storage of raw materials, and average in the storage of finished products. The situation facing the fireman is liable, also, to vary according to the season: when, during the summer, demands for cattle cake are low, these products are liable to accumulate in storage spaces and even in the mills themselves. The fire risks to which these variations may give rise are such that a knowledge of the properties of the oils is important to the fireman.

Varieties of animal and vegetable oils

The oils under review may be divided into fatty oils and essential oils. The fatty oils, which may be of animal or vegetable origin, are glycerides and esters of fatty acids; essential oils, which are found in trace quantities in a very wide variety of plants, are volatile constituents with characteristic smells. A distinction can also be made between 'drying oils' and others; this will be dealt with more fully under 'Nature and properties of oils' par. 1(a) on page 183. The following are the properties of some of the principal members of each of these groups, but no attempt has been made to catalogue them all.

1. FATTY OILS, ANIMAL

Cod liver oil is a drying oil, very rich in vitamins, obtained from the fresh livers of cod fish.

Lard oil. This is produced by the rendering of lard, which is the fat of the pig. It has a flash point when pure of 464°F. (240°C.) which may be 100°F. lower in commercial grades. In common with all similar oils, lard oil is a carbohydrate consisting of glycerine and fatty acids chemically combined.

Menhaden oil. This is prepared from the menhaden fish (*clupeidae*). It is a drying oil and the flash points of this and whale oils (see below) are approximately the same as that of lard oil.

Whale or train oil. This is derived from rendering the blubber of the whale.

2. FATTY OILS, VEGETABLE

(a) Non-drying

Castor oil. Obtained from the seeds of the castor oil plant, (*ricinus communis*), originating in tropical Africa; a yellow or brown, syrupy, non-drying oil. It is slightly lighter than water. Used widely for medicinal purposes, and in India for illumination.

Groundnut or arachis oil. The fruit of the plant *aracis hypogaea*, originally from South America and now grown in most tropical countries. A non-drying oil, used in the preparation of foodstuffs.

Olive oil. A pale yellow or greenish oil obtained from the olive, the fruit of *olea europea.* Its flash point is approximately 425°F—500°F. (218°C—260°C.); it is widely used in the preparation of foodstuffs, especially in continental countries, and in cosmetics and pharmacy.

Palm oil. A reddish-yellow fatty non-drying oil from the fruit of *elaeis guineenis*, a variety of palm tree. Its flash point is approximately 421°F. (216°C.). Also known as *palm butter.*

Palm kernel oil. A yellowish oil from the nuts of *elaeis guineensis* sometimes referred to as *palm nut oil.* Its flash point is 398°F. (203°C.).

Rapeseed (or rape or colza) oil. This originates from the brassica plant known as *brassica campestris.* It is a non-drying oil, once used for lubricating and lighting, the residue being used for cattle food. The flash point varies from 325°F. to nearly 600°F. (163°C.—315°C.) according to constitution. This oil is refined by the use of sulphuric acid.

(b) Semi-drying

Cottonseed oil. Originally a by-product in the process of producing cotton; approximately 450 kilograms of cottonseed oil are produced to every 225 kilograms of cotton harvested. The flash point is approximately 486°F. (252°C.) and it is used for making refined oil for human consumption in foodstuffs; the poorer grades are used in soap, candles, etc. The residue (cottonseed cake) is an animal food-stuff also used for fertilisers, etc.

Maize oil (or corn oil). A pale yellow oil obtained from Indian corn and used largely for cattle feeding; it has only intermediate drying qualities. The flash point is 490°F. (254°C.).

(c) Drying

Linseed oil. This is a most valuable drying oil, produced from the seed of a flax plant *linum usitatissimum* and having a flash point, when raw, of about 432°F.—450°F. (222°C.—232°C.). Some of the oil is used in foodstuffs, but more in paints and varnishes, printing works and linoleum. Sulphuric acid is used in the refining of linseed oil.

Soya bean oil. This is derived from the bean of the *glycine soja*, originating in the East Indies and cultivated in China and Japan. The flash point is 540°F. (282°C.). It is used for foodstuffs and cattle feeding.

Tung oil (China wood oil). A yellow drying oil obtained from the seeds of *aleurites cordata.* It has a flash point of 552°F. (288°C.). It is used in paints and varnishes, especially enamels.

3. ESSENTIAL OILS (VEGETABLE ONLY)

The majority of plants yield, in addition to the non-volatile or 'fixed' oils already mentioned, which are found mainly in the seeds or fruit of the plant, some 'essential' oils in the leaves, stems or flowers. These latter are obtained by distillation with steam; they are used mainly in perfumery, being all characteristically odorous, but the most important is *oil of turpentine.* There are many variations of this, but all are lighter than water, and the flash point may be as low as 95°F. (35°C.). Turpentine is not miscible with water.

Nature, properties and risks of oils

1. NATURE AND PROPERTIES

(a) Fatty oils

Whether vegetable or animal in origin, it is characteristic of fatty oils that they are viscous liquids of high calorific value; they are flammable, but have relatively high flash points. It is also characteristic of all of them that they very readily combine with oxygen: this characteristic plays a part in the distinction which is made between 'drying oils' and 'non-drying' oils. The former have a particularly strong tendency to congeal into a resinous solid when exposed to the air, making them very suitable for use in paints. Tung oil, for example, will dry out to a solid in about 24 hours; linseed oil in 3—4 days, and others in longer periods. This characteristic of drying oils means, in practice, that these have an exceptionally strong tendency to absorb oxygen, and therefore a strong tendency towards self-heating. This tendency can be reduced, and sometimes is, by the admixture of mineral oil. However, in many of the industrial processes in which fatty oils are employed they are mixed with solvents which, unlike the addition of mineral oils, affects a sharp reduction in the flash point of the mixture and thus an increase in its fire hazard. Since fatty oils are liable to form the bulk of most of the products (such as paint, margarine, soap etc.) for which they are used, they are likely to be found in relatively large quantities in the places in which such things are manufactured.

As with mineral oils, oils of animals and vegetable origin are viscous at normal temperatures and become more fluid with a rise in temperature; they do not, however, become as freely fluid as paraffin, for instance, before their ignition temperature is reached. They are all slightly lighter than and immiscible with, water.

(b) Essential oils

As regards flammability, essential oils have lower flash points and are therefore of relatively higher hazard, as compared with fatty oils; they are sufficiently volatile in most cases to have the nature of a spirit. As opposed to these dangers, however, is the fact that, with the notable exception of oil of turpentine, they are normally extracted and handled in relatively small quantities, being concentrates of which only a very small quantity is required in manufacturing processes. Except therefore, in a few isolated instances, essential oils are unlikely to be found in bulk.

2. RISKS

(a) Burning characteristics

All fatty oils, while not readily ignited, burn fiercely with yellow flames accompanied by the emission of heavy smoke and particles of carbon. They are characteristically noisy burners, and their violence in burning increases strongly with the general temperature surrounding the fire. Since the oils are of high calorific value, great heat is evolved in the course of combustion. In common with fats, the flammable vapours produced can sometimes fail to be ignited at the moment of distillation and can collect in pockets, sometimes below the surface of the liquid, and result in explosions of great violence.

(b) Spontaneous heating

There is a severe spontaneous ignition hazard attached to fatty oils, particularly those referred to as drying oils. The tendency is at its strongest when cellulosic materials, particularly those in fibrous form, are impregnated with the oil; in the case of many of the vegetable oils, the residue, after the oil has been squeezed out or 'expressed' is still impregnated with a certain quantity of oil and is itself of cellulosic origin. This means that there is a particularly severe tendency to self-heating in the residues of cottonseed oil, linseed oil, etc. This tendency to self-heating is affected by the degree of moisture present; in the presence of moisture, fats are slowly decomposed, due to bacterial action. This accelerates the tendency to self-heating, and it should therefore be remembered that spontaneous ignition is more likely in these materials when a relatively high moisture content is present.

(c) Impregnation of woodwork

A further characteristic of all oils, increasing the hazard of their use in buildings, especially buildings of considerable age, is that they are absorbed very readily by wood and similar materials, and tend to soak and spread into every wooden part of a room or building which is attainable. In premises therefore in which these oils are used it is virtually impossible to prevent a condition in which all woodwork, partitioning, panelling, etc., is steeped in oil and therefore much more highly flammable than the original wood. This obviously adds greatly to the risk of fire spread in the building.

(d) Dust explosion hazard

There is a dust explosion hazard, in certain circumstances, attached to the residues of many of the processes associated with oil extraction. In many cases, after the oil has been extracted from the vegetable providing it, the latter is in a finely divided form, which, before drying, consists of a sticky mass. After drying, however, it will consist of a fine dust of basically cellulose material. This constitutes a severe fire hazard and at times a dust explosion hazard, depending on the conditions at which the dust is produced.

(e) Use of solvents

It should never be forgotten that the use and production of fatty oils in industrial processes are more or less inseparable from the use of solvents, and that the two forms of fire risk will nearly always be found together, the admixture of the solvents usually having a worsening effect upon the flammability of the mixture. For example, most of the fatty oil from vegetable sources is expressed by crushing in machines: in many processes, however, a further proportion of the oil is extracted by means of solvents, which are later separated from the oil which they have extracted, in a separate process. The fireman should therefore always expect to find solvents associated with animal and vegetable oils, and he should remember, moreover, that these solvents are of particularly low flash point, sometimes lower than that of petrol. There is a constant risk of the vapours coming within their explosive limits, and precautions should be taken accordingly.

(f) Sulphuric acid

A common method of refining linseed and rapeseed oils is by the use of sulphuric acid in strong concentrations, and this produces the hazards associated with this chemical. (*see* Section 16).

(g) Essential oils

Because of their relatively low flash point, these ignite more readily than the fatty oils, and burn with the characteristics of white spirit or benzene.

3. FIRE-FIGHTING

The proper fire-fighting medium for use on oils, in common with melted fats, is foam. The application of water to these fires is liable to produce reactions of explosive violence, the instant evaporation of the water into steam, with consequent sudden expansion being liable to throw quantities of burning oil in all directions. The remarks, however, which are made on the subject of fat fires in Section 3 are also applicable to animal and vegetable oil fires, *i.e.*, oil fires can be tackled in some circumstances with water sprays especially in the early stages of their development, and, if they are small enough, with carbon dioxide or dry powder.

Flowing fires in animal and vegetable oils should be dammed at the earliest possible opportunity: the application of foam may extinguish the fire, but it will not prevent the continued spread of the flowing oil. Water sprays may also have the effect of extinguishing the fire, but, unlike fat fires, oil fires will not normally respond to a cooling effect by solidifying. Steps must therefore be taken to check the flow of oil.

These oils are sometimes involved in fire when packed in tins, and these may be expected to burst after reaching temperatures of 400°F. (204°C.) and upwards. The tins should be cooled by water sprays before they have had time to burst.

Vegetable oils are frequently used for fish frying, and are therefore often found in fish frying ranges; the burning material in fires in fish frying ranges and in domestic cookers is, indeed, as likely to be cooking oil as cooking fat. For a description of the proper technique to be used on such fires, reference should be made to Section 3, 'Fat and wax fires'.

Due to the heavy concentration of smoke which is to be expected from this type of fire, breathing apparatus will nearly always be necessary.

Industrial processes in seed-crushing mills

The outline of processes used may be seen in the table on page 188 and is also shown in diagrammatic form in Plate 10.

1. BUILDINGS

The silo and expressing plant are generally housed in the same building, and the oil house and refinery in another. The solvent extraction plant is also separately situated.

The oil house and refinery is usually a lofty, one-storey building, probably with numerous stagings or part floors.

Seed silos resemble grain silos in construction. In smaller seed mills, however, the seed is often stored in bulk or in sacks on warehouse floors, in which case stocks are turned over periodically to avoid self-heating. Buildings used for storage are usually four to six storeys high, with substantial floors supported by iron columns. The floor may be divided into compartments by heavy wooden partitions. Milling machinery is not infrequently installed on the floors above those used for warehousing. The ground floors of seed silos are usually a network of chutes, conveyors and elevators. It is thus simple to transfer seeds from one bin to another in the event of suspected overheating: this tends to reduce the risks from spontaneous heating.

Hoist shafts, stairs, sack elevators, and belt and other openings are found in all storied buildings in the oil seed industry.

2. PROCESSES

After screening to remove refuse (and, in the case of cottonseed, the removal of the cotton linters (*see* Section 4), the seeds go to the mill for the oil to be driven out under pressure (called 'expressing'). In some cases the residue is treated in a solvent extraction plant (*see* Fig. 30), where further oil is obtained. The oils, after storage in tanks, may then be refined. The seed residue is used for cattle cake.

3. RISKS AND CAUSES OF FIRE

(i) *Dust*. Great quantities of dust are created in the receiving house, silo, seed warehouses, rolls house, grinding rooms and cake mills. In the mill, meal is likely to be scattered and to collect in corners.

(ii) *Spontaneous heating*. All seeds are to some extent liable to spontaneous heating, especially when damp. This applies particularly to linseed and cottonseed and also to palm kernels stored in jute sacks. If heating is suspected, the whole stack must be broken down and examined. If moreover, the oily jute sacks are used for some other commodity, the risk of spontaneous heating remains.

THE PROCESSES IN A SEED-CRUSHING MILL

Seed from silo to rollerhouse

Crushed seed from rollerhouse to extracting pots

Pots sealed and flushed with solvent from solvent storage tanks

- Solvent and oil pumped to evaporators where it is boiled
  - Solvent vapours to condensers. Condensers vented by trunking to solvent recovery plant
    - Solvent gravity fed or pumped from condensers to solvent recovery plant.
    - Solvent recovery plant to storage tanks ready for use again.
  - Oil pumped to refinery
    - From refinery to oil storage department
- Meal steamed and put into dryers Vapours through trunking to condensers
  - Dried meal from dryers to storage bins and/or into bagging department
  - Bags of meal to warehouse for storage

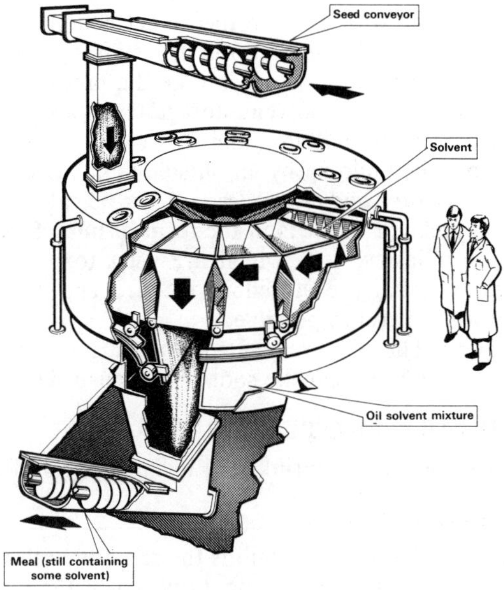

Fig. 30. Diagram of a rotary continuous solvent extraction plant. (See also Plate 10).

(iii) *Oil-soaked woodwork.* Throughout the buildings wood tends to become soaked in oil, thus assisting the spread of fire.

(iv) *De-linting rooms.* The machines used for de-linting (*see* Plate 37) resemble those in a cotton mill. It is possible for a flash to occur which will spread through the extraction ducts to the dust room. The de-linting rooms usually contain a considerable amount of cotton fluff.

(v) *Oil storage*, usually in large (50,000 kg approx.) open tanks. The building is often of brick and iron construction, with wooden roof louvres.

(vi) *Finished product storage—cattle cake, etc.* The chief risk here is that of an undetected excess of oil in the meal or cake, though the risk is less with meal that has been produced by the solvent process. The majority of firms now instal thermometers in the storage stacks, arranged to give readings in the foreman's office. This reduces the risk of undetected overheating.

(vii) *Solvents.* Most of the oil is extracted from the seeds by high pressure. The remainder is extracted by treating the crushed seeds with solvents. The principal solvents used are petrol, naphtha, carbon disulphide, carbon tetrachloride, and tri- or tetrachlorethylene. When dealing with a fire in the oil extraction plant, therefore, it is important to ascertain from the mill staff whether solvents are likely to be involved, and, if so, to take the precautions applicable to the particular solvent. Solvent storage tanks are usually situated outside the extraction building, and it is usually possible to cut off the tanks from the building by an outside tank valve.

(viii) *Cake storage and provender mill.* Cattle cake is stored either in blocks on racks or in sacks in the form of nuts. Stocks of wheat, oats, maize, molasses, etc., which are used to mix with cake to produce cattle meal, are stored in the provender mill, and dust explosions are sometimes possible.

(ix) *Refinery.* The oils are treated with various chemicals in the refinery. Hot water and caustic soda are extensively used.

4. FIXED FIRE PROTECTION

Seed mills are usually sprinklered in most parts of the building.

5. FIRE-FIGHTING

The chief fire risk in an oil mill is the rapidity with which the fire grows, and the difficulty in putting it out. Once alight, oil seeds and oil soaked materials (including, *e.g.*, the wooden structure of the mill) burn quickly. It should be remembered that cake or meal which is tainted with fire or smoke is unmarketable, as cattle refuse to touch it.

(**a**) Stocks of nuts or seeds which have been wetted should be ventilated as soon as possible to avoid self-heating. Where self-heating has occurred, it may be found at more than one place at the same time in a stock of bagged nuts. Heating will usually be localised, and if the fire has been detected quickly enough the heated bags can be lifted out.

If there is sufficient heat, the nuts will begin to give off vapours which may result in the development of a serious fire. Only the smallest possible quantity of water should be used, as not only may the water itself do damage, but it may also cause self-heating in the seeds.

(**b**) Burning seeds give off large quantities of pungent smoke. When much water is applied to them they swell, and may overload the floors or push out the walls.

(**c**) In case of fire in the de-linting machinery, the machine should be stopped, and after the fire has been extinguished a search for fire should be made in the ducts and dust rooms with which it communicates.

Part 6c, Chapter 45, Section 9

Paint and varnish fires

NATURE AND PROPERTIES OF PAINTS, VARNISHES, ETC.

1. **Paint**
2. **Varnish**
3. **Enamel**
4. **Lacquer**
5. **Cellulose lacquer and enamel**
6. **Metallic paints and enamels**
7. **Fire retardant paints**
8. **Bituminous paints**

FIRE AND OTHER HAZARDS

1. **Fire hazard of liquid paints, etc.**
2. **Flammability of dried paint coatings**
3. **Fighting paint and varnish fires**
4. **Personal hazards**
5. **Spontaneous ignition**

PAINT SPRAYING AND DIPPING

PROCESSES AND HAZARDS IN PAINT AND VARNISH MANUFACTURE

1. **Paint**
2. **Varnish**
3. **Enamel**
4. **Synthetic varnishes**
5. **Cellulose lacquers and enamels**
6. **Metallic enamels**
7. **Fixed fire protection**
8. **Fire-fighting**

CARBON BLACKS AND BONE BLACK

1. **Nature and properties**
2. **Fire-fighting**

PAINTS and varnishes have always been quite rightly regarded as substances of unusual fire hazard. Many flammable materials are used in their composition, with the result that the manufacturing processes present considerable risks. Similar risks, moreover, are duplicated to some extent in many thousands of other establishments, chiefly those engaged in wholesale and retail distribution, from the manufactured products themselves. A fireman, therefore, is liable to encounter paint or varnish fires in many places in his area and must be in a position to make a quick and effective attack upon them.

Special hazards associated with paint and varnish include cellulose nitrate solutions, flammable solvents, dust explosions and spontaneous ignition, all of which are dealt with in the following paragraphs. Reference should also be made to Section 8, 'Fires in animal and vegetable oils' and to Section 13, 'Fires in resins and gums'.

Nature and properties of paints, varnishes, etc.

1. **PAINT** is a general term for many types of decorative and protective coating. *Oil paints* are those in which the base material is an oil, with the addition of a solvent to give fluidity in application, a pigment to provide colouring and other additives for specific purposes. *Water paint* is a term sometimes used for *distemper*, which is a mixture of dry pigment with water, with occasionally a small amount of oil added. *Cellulose paint* is a loose term used for *cellulose lacquer* or *cellulose enamel* (see below). *Emulsion paints* are water-based synthetic plastic emulsions, commonly of polyvinyl acetate or polystyrene with a plasticiser, such as dibutyl phthalate, and a pigment.

2. **VARNISH** is a solution of gum or resin in a spirit, (*spirit varnish*) or an oil, (*oil varnish*), with the addition of smaller quantities of other materials for specific purposes.. *Spirit varnishes* are often referred to as *lacquers* (see below). Varnish is clear and contains no pigment. It may be of natural or synthetic manufacture, in that in the latter case it contains synthetic chemical resins, while in the former, natural resins are used. (*see* Section 13, 'Fires in resins and gums'). Varnish is applied for imparting a high gloss to paintwork, woodwork, etc. and for the protection of the surface.

3. **ENAMEL** is a type of high gloss oil paint which is essentially a combination of paint and varnish, since the ingredients include those of oil paint with the addition of a resin. Enamels are normally specially constituted to provide simple application with a minimum of mixing and a maximum of quick drying properties. (For cellulose enamel see below).

4. **LACQUER** is a term usually used to denote the same as *spirit varnish*. It consists essentially of a gum, such as shellac, dissolved in a spirit such as methylated spirit (denatured ethyl alcohol); this is, of course, in effect French polish. Like varnish, it is mainly used to provide a high gloss protective coating. Also like varnish, lacquers may be of natural or synthetic materials. (For cellulose lacquer see below).

5. CELLULOSE LACQUER AND ENAMEL are synthetic products prepared from cellulose nitrate (nitro-cellulose) or cellulose acetate with solvents, synthetic resins and plasticisers. When required clear, as a varnish, no pigment is added. Cellulose solutions, which include nitro-cellulose and acetyl-cellulose lacquers and enamels, are controlled by the *Cellulose Solutions Regulations*, 1934, under the *Factory Acts* or by the *Petroleum Mixtures Orders* under the *Petroleum (Consolidation) Act*, 1928.

6. METALLIC PAINTS AND ENAMELS are, in spite of the use of the word 'paints', largely enamel forms in which the pigment consists of a metallic dust, commonly aluminium, zinc or bronze powder. They have special properties in preventing corrosion in metals and resistance to heat up to about 350°F. (177°C.). They also present special hazards in manufacture.

7. FIRE RETARDANT PAINTS are not normal paints and are therefore often referred to as 'coatings'. A standard flame retardant coating quoted in British Standard 476 can be made from sodium silicate, kaolin and water, but proprietary coatings may be based on a phosphate resin or other chemicals.

8. BITUMINOUS PAINTS include straightforward prepared tar and pitch, as well as paint consisting of pigments combined with a bituminous medium. Some bituminous paints will withstand temperatures up to 600°F. (316°C.).

Fire and other hazards

1. FIRE HAZARDS OF LIQUID PAINTS, ETC.

With the exceptions of water paints (distemper) and the majority of plastic emulsion paints, all paints and varnishes must be considered high fire risks in liquid form before application. While the oils which form the base of oil paints are not themselves very readily flammable (linseed oil for example, has a flash point of over 400°F. (204°C.)), the solvents commonly employed are readily flammable and may have flash points as low as 90°F. (32·2°C.). The same considerations apply to enamels and oil varnishes, but spirit varnishes or lacquers are more flammable still: methylated spirits (denatured ethyl alcohol), for example, has a flash point of 65°F. (18·3°C.). The fire hazard of cellulose lacquers and enamels is also high, that of the cellulose nitrate ones being very high; the flash point of cellulose nitrate lacquers (40°F. or 4·4°C.) is considerably below room temperature and there is a severe explosion risk attached to them.

In addition, normally all the other constituents of most paints and varnishes are combustible. Some pigments are not, however, and when these are included in the paint its flammability is thereby reduced.

2. FLAMMABILITY OF DRIED PAINT COATINGS

As a coating after drying, most paints and varnishes (with the exceptions mentioned in par. 1 above) are still combustible, though once the solvent has disappeared the flammability is much reduced. In practice it depends upon the flammability of the oil or base used. Thus normal oil-based paints are not readily flammable, but the dangers attached to nitro-cellulose coatings are severe and a very rapid fire spread upon the surface is to be expected.

The ability of aluminium and other metallic enamels, when struck by a steel object, to produce sparks capable of igniting flammable vapours should be borne in mind. The hazard is especially severe with nitro-cellulose-based metallic enamels.

3. FIGHTING PAINT AND VARNISH FIRES

Once ignited, liquid paint burns fiercely and gives off a great deal of heavy black smoke. It also, obviously, flows and presents many of the features of an oil fire. Because of the presence in the paint of low flash point solvents, water is not a suitable extinguishing agent, and foam must be used if any substantial quantity of paint is involved. Cooling of surrounding materials by the use of water may also be required. On small quantities of paint or varnish it is possible to use a carbon dioxide or dry powder extinguisher as an alternative to a foam extinguisher. Water is the correct technique on dried paint coatings.

4. PERSONAL HAZARDS

Because of the dense smoke produced and also of the toxic fumes given off by some of the constituents of paint and varnish, breathing apparatus will always be necessary on a paint fire indoors. There is also a hazard to firemen from explosions: since paint in store is normally always in tins or drums (varying from 570 ml or less to 180 to 225 litres in capacity) which are tightly sealed, any fire in any kind of paint store is liable to heat up the drums and burst them. The contents are likely to ignite instantly on exposure to the air, and the effect will thus be explosive; the dangers, moreover, explosive; the dangers, moreover, from the burning liquid are obvious. There is also a subsidiary danger of explosions from trapped pockets of flammable vapour.

5. SPONTANEOUS IGNITION

Apart from the fires which can be produced by paint and varnish in association with an external means of ignition, there is a fire risk from spontaneous ignition. Due to the nature of the oils employed in paint manufacture, all paints and oil varnishes are capable of self-heating when they are closely in contact with cellulosic materials and particularly fibres, and sometimes even when they are not. Rags, therefore, and other materials which are impregnated with paint or oil varnish will self-heat if left in heaps or otherwise packed or heaped without provision for heat dispersal and this self-heating can easily develop to the point of self-ignition. Spontaneous ignition is also possible in cellulose lacquers and enamels which are sprayed alternately with oil paints, and when they are painted on to steam pipes.

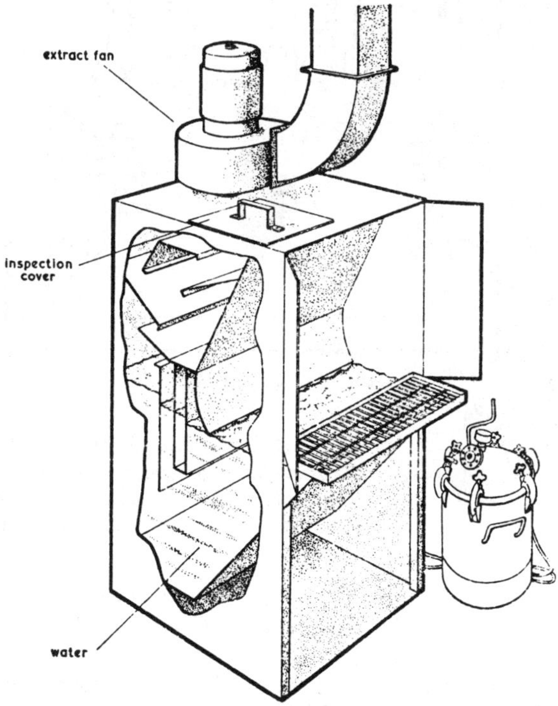

Fig. 31. A small spraying booth, with a water tank and fume ducting at the top.

6. PAINT SPRAYING AND DIPPING

Although these processes may be carried out in the application of any type of paint coating, they are used extensively to apply industrial lacquers and enamels designed for the purpose, and, of these, cellulose nitrate solutions are less common than they once were. In spraying (*see* Fig. 31), compressed air is used to project the paint from a pistol-like tool, while in dipping, which is normally used on small articles, it is common to use an overhead conveyor in conjunction with a dip tank and drip trays. The vapour-laden atmosphere surrounding the processes is removed by forced draught by extractor fans direct to the open air or through ducting (*see* Fig. 32). The hazards involved are very high, and the layout of the plants

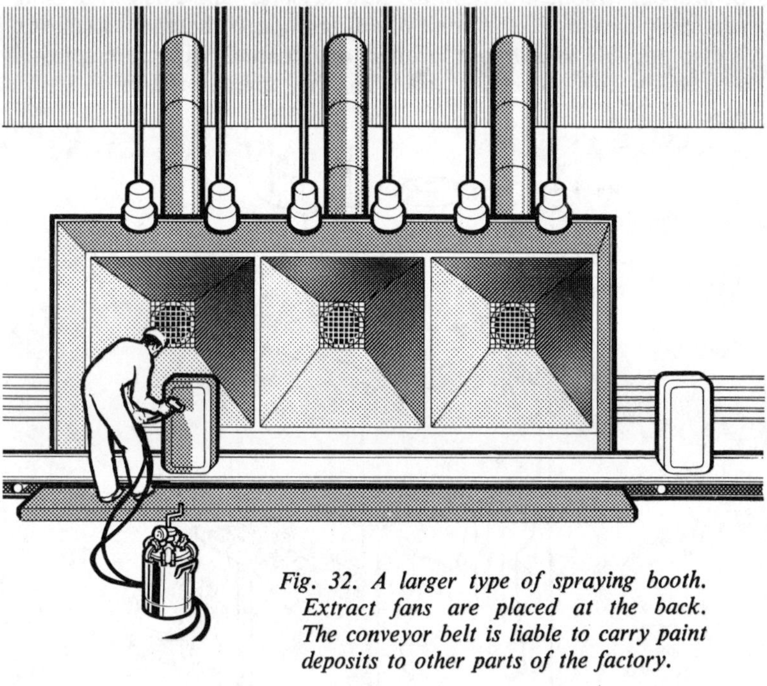

Fig. 32. A larger type of spraying booth. Extract fans are placed at the back. The conveyor belt is liable to carry paint deposits to other parts of the factory.

used, and the storage and handling of the liquids within the spraying booth, are controlled by Regulation, as mentioned in par. 5 on page 193.

In spite of these controls, fires are liable to occur, due to the accidental production of sparks from metallic objects, to electrical faults or to other breakdowns in safety procedure. In spraying processes, after the ignition or explosion of the vapour-air mixture, fires normally take the form of burning deposits of dried cellulose constituents in and around the spraying booth and ducting (*see* Fig. 33). These deposits are likely to be quite heavily caked and

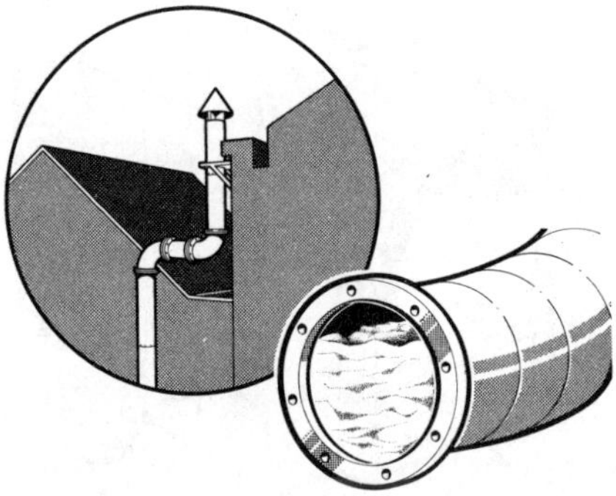

Fig. 33. (LEFT) *A hazard arising from bends in the extract duct from a cellulose spraying booth. Cellulose deposits block the duct at the bends.* (RIGHT) *Sectional view of the interior of the pipe, almost completely blocked with highly flammable residues.*

encrusted and they may contain air spaces due to the irregular pattern of their formation. Since, however, the volatile solvent constituents will have largely evaporated, the deposits do not burn as fiercely as the liquid nitrate cellulose and can usually be extinguished by means of water, those in the further reaches of the ducting giving most trouble. If forced draught ventilation is not at once cut off when a fire occurs, flames are likely to be drawn by the fan up the flue and cause the metal to become red hot, thereby encouraging the fire to spread. On dip tanks and drip trays, foam will be necessary, except where the outbreak is small enough to be dealt with by carbon dioxide, dry powder or vaporising liquid.

It is common in some industries for the spraying of even large objects, such as motor cars, to be done by fixed installations working on conveyors in sealed tunnels. In those cases it is usual to provide a total-flooding carbon dioxide installation.

Associated with many spraying and dipping processes is a drying oven (*see* Figs. 34 and 35) in which the fire hazards may also be severe. At almost any time in the cycle of operations in a solvent drying oven there may be a vapour/air mixture which is near its explosive limits and can reach its spontaneous ignition temperature if faults occur such as over-heating, electric sparks, or incorrect operation.

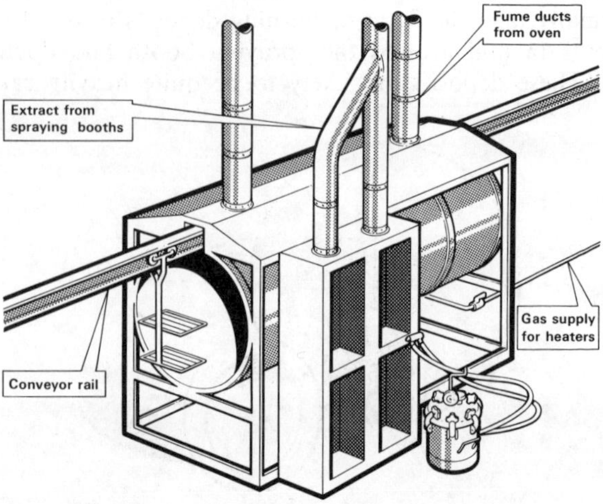

Fig. 34. Continuous-flow drying oven associated with spraying and dipping processes. Electricity or gas may be used. In this case a sample spraying booth is placed close to the oven.

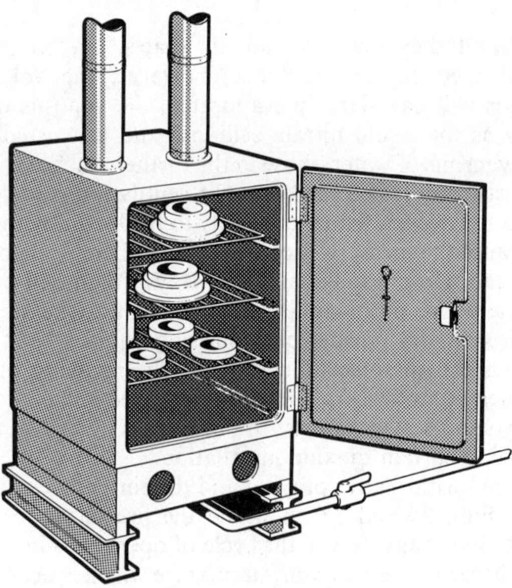

Fig. 35. A simple type of non-continuous drying oven.

Processes and hazards in paint and varnish manufacture

1. PAINT

Paint is a viscous liquid made up of a pigment or colouring matter, a medium or vehicle, a diluent or thinner, a drier, and sometimes an 'extender'. It is applied to surfaces as a protection and for decoration and dries by a complicated process of evaporation and chemical action.

(a) Pigments

Pigments are generally made at a colour works and are brought to the paint works in sacks or drums or in paper bags. They do not present a high fire risk except in their manufacture. Some, like Prussian blue, are flammable (Prussian blue if heated evolves cyanogen); others give off asphyxiants (Paris white evolves carbon dioxide); some, like carbon black, burn without smoke or flame (*see* page 205); whilst still others like chrome yellow are oxidising agents and in certain circumstances can present a risk of spontaneous combustion. This risk arises particularly in mixtures of blues and yellows ('chrome' green or 'Brunswick' green).

(b) The medium

Until recently the principal medium was boiled linseed oil, but now some kind of varnish or mixture of China wood oil, dehydrated castor oil or fish oils may be used. The purpose of the linseed oil is to form a hard surface by absorbing oxygen from the air. All oils used as media are drying oils (*see* Section 8). Of these oils, China wood oil (tung oil) oxidises most rapidly and is, therefore, the most dangerous from the point of view of spontaneous heating. Linseed oil and certain other oils used as media are subjected to a heating or cooking process. This treatment is carried out either in closed immovable pots (*see* Plate 47) from which fume extractors lead to condensing chambers on the roof, or in open pots (*see* Fig. 36) which are usually mounted on trolleys and pushed over a coke or gas fired furnace for the required length of time.

(c) Diluent or thinners

The most common diluents are turpentine, turpentine substitute (white spirit) and light grade coal tar naphtha. These are used to make the paint easier to apply.

(d) Driers

The driers are usually inorganic salts and fatty soaps of cobalt, lead, manganese, calcium and zinc. Among those employed are oxides of manganese, which are oxidising agents. The drier is usually added to the linseed oil or other medium in bulk, without greatly changing its fire risk; it acts as a catalyst and accelerates the drying and hardening process.

(e) Extenders

The function of an extender is to give weight. The most commonly used are white lead, zinc barytes, titanium oxide, calcium carbonate, iron oxides and lithopone (zinc sulphide with barium sulphate).

2. VARNISH

The preparation of a varnish involves bringing together a gum, (or a resin), a medium, a drier and a thinner.

(a) The gum (e.g., Congo copal) is subjected to a high temperature in a process known as gum running, *i.e.*, dry distillation. It is heated to 600°F. (316°C.), at which temperature volatile oils and spirits are given off, leaving behind a residue which is soluble in linseed oil. Gums and resins present a fire risk, as the vapours given off when they are heated are flammable. In addition, certain resins may produce a dust which can form an explosive mixture with air. (*See* Section 13).

(b) The medium usually consists of linseed oil, which is added to the gum when hot, the mixture then being cooked in pots with the addition of a drier. The pots, which normally contain about 140 litres, are wheeled to and from a coke fire pit on trolleys (*see* Fig. 36). For oil boiling, *set pots* are used, which are built into a brick furnace (*see* Plate 47). These contain anything up to two tonne of oil and/or varnish and are fired by coal, oil or gas. Ducts are installed to remove fumes. Linseed oil is capable of absorbing oxygen to 30 per cent of its own weight.

(c) The thinner is added, usually in the open, after which the varnish is pumped to a tank room for maturing. This room in modern factories will normally be of fire-resisting construction.

3. ENAMEL PAINTS

These are paints already mixed with varnish so that on drying they leave an enamelled glossy surface. They are prepared by a combination of processes described in pars. 1 and 2 above.

4. SYNTHETIC VARNISHES

Instead of natural gums or resins, resins made from chemical substances may be used to prepare synthetic varnishes. The two main groups are known as the phthalic alkyd resins and the phenol-

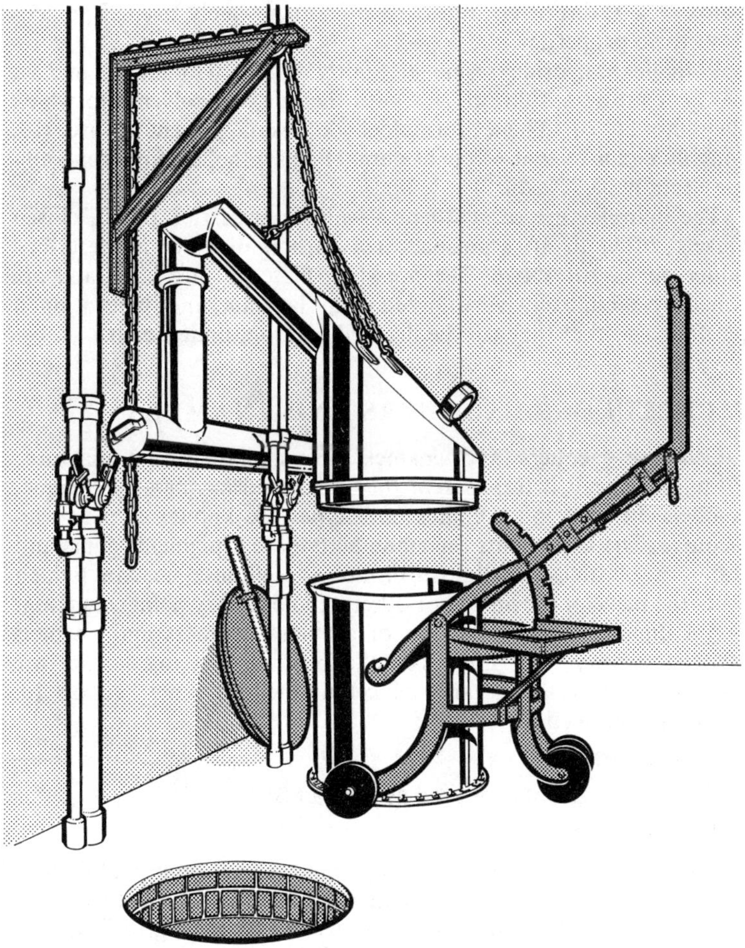

Fig. 36. The type of pot used in 'gum running,' ready for moving by means of the trolley. The pot is placed over the coke fire pit and the ducting is lowered to the top to carry off the volatile oils and spirits as they are evaporated.

formaldehyde resins, though a group of synthetic resins known as ethoxylene resins or epoxides are becoming widely used. Epoxides have toxic properties.

Phthalic anhydride resins are fairly highly flammable, and in fact, constitute a high fire risk with the oils and solvents to which they are added. Phenol-formaldehyde resins present a fire risk roughly equal to that of copal gum. (For synthetic resins generally *see* Section 10, 'Plastics').

The manufacture of synthetic varnishes is carried out in closed vessels, often in an atmosphere of carbon dioxide, as described under par. 2 (a) above. The pots (*see* Plate 46) are large and are generally completely surrounded with lagging, and placed on a platform, so that they can be drained off from underneath.

5. CELLULOSE LACQUERS AND ENAMELS

Cellulose lacquers and enamels are similar in manufacture: if the final result is transparent they are called lacquers; if opaque, enamels. In the United States both types are known as pyroxylin lacquers. The ingredients for manufacture are a pigment, gums and resins, a plasticiser, cellulose nitrate or cellulose acetate (which takes the place of the linseed or other oil in ordinary paint), a solvent and a diluent. The production of cellulose lacquers and enamels involves a very high fire risk as almost all the ingredients are flammable. (For the hazards of cellulose nitrate, *see* Section 10, 'Plastics' and Section 16).

(a) Pigment

The same pigments are used as for paints.

(b) Plasticiser

This is usually castor oil or an oily liquid such as dibutyl phthalate, which has a very high flash point, and therefore does not present a high fire risk. The function of the plasticiser is to give flexibility to the paint, thereby making it less brittle.

(c) Cellulose nitrate or cellulose acetate

Cellulose nitrate may be delivered to the works in heavy steel containers, the contents being damped with approximately 30 per cent denatured alcohol, such as methylated spirit or butyl alcohol, or occasionally with water. The damping reduces the risk of explosion, but the mixture remains flammable. In this state it resembles snow-white cotton-wool in appearance. The main store of cellulose nitrate will probably be in a separate building, but sufficient quantities are generally kept on the premises for jobs which are being done. Cellulose nitrate may also be obtained from scrap cinema film after it has been cleaned. Scrap film stored for this purpose presents a fire risk, and adequate precautions are unfortunately not always taken.

For the less hazardous cellulose acetate *see* Section 4 under 'Rayon', and for its manufacture *see* Section 10, 'Cellulose acetate manufacture'.

(d) Solvents

These are used to dissolve the cellulose. In larger works the solvents are fed from underground tanks to the mixers by pumps. Solvents in common use are acetone, amyl acetate, butyl acetate, ethyl acetate, butyl propionate, methyl ethyl ketone, and solvents of the alcohol group, the commonest probably being butyl acetate.

(e) Diluents

These are usually light coal tar derivatives such as benzole, toluol or xylol.

(f) Gums and resins

Small quantities of gums and resins are added to improve the lustre and tenacity, after which the finished lacquers or enamels are stored in drums, usually of 22·5 litres capacity.

6. METALLIC ENAMELS

These present severe hazards in manufacture. There are various ways of producing the metal powder, such as by grinding, by spraying and cooling the molten metal or by condensation of the metal vapour; each of these processes has characteristic hazards. They are more likely to be carried out in metal works, the powders being purchased in bulk by paint manufacturers. It is probably true to say, however, that aluminium powder is now mostly purchased as a powder in white spirit. The dangers attached to metal powders and dusts are serious, and are mentioned in Sections 1 'Dusts' and 7 'Metals'.

7. FIXED FIRE PROTECTION

Most paint and varnish works are usually well equipped with fixed installations such as sprinklers, water spray, and carbon dioxide systems. These installations should be used where possible, as they will be the most suitable for the process concerned. CO_2 should always be used when available, as it does less damage in this case than, for example, foam.

8. FIRE-FIGHTING

The chief fire risk in the production of paints and lacquers is associated with the heat treatment of the media, which will boil over if they are over-heated. When fixed pots for boiling oil are used, a flash from the pot may ignite the caked resin and oil and fires sometimes occur in this way in the ducts. It may be possible to shut off the duct, otherwise it may be necessary to open up the duct and use carbon dioxide or foam. A fire in the pot can usually be extinguished by closing the bottom end of the duct, thus excluding air. In modern works movable open pots are usually protected by carbon dioxide fixed installations; this gas does not affect the materials. If carbon dioxide is not available in this way, however, it may be possible to move the pot on its carriage outside the shop and there to extinguish the fire: CO_2 or vaporising liquids are more suitable than foam on fixed pots or ducting, and may make it possible to avoid spoiling the contents of the pot.

In general, it is almost always necessary to use foam if the fire involves the articles which are being processed. If the use of water on heated oils and spirits is unavoidable, the greatest care should be taken to avoid the latter being scattered or slopped over.

The use of breathing apparatus will almost always be necessary in a paint works fire, due to the acrid smoke given off: in the case of cellulose lacquers, moreover, the nitrous fumes are highly toxic, and for this breathing apparatus is essential.

Sawdust has been used with success by works personnel on small lacquer fires. Its efficacy is increased by pre-mixing with bicarbonate of soda in the proportion of 4·5 kilograms of soda to forty litres approx. of sawdust. Foam can also be used. Sand may be used if the lacquer is spread in a thin film on the floor or table but is otherwise not suitable for fighting this type of fire. Its weight causes it to sink to the bottom of the container; it does not have the blanketing effect of sawdust.

Many of the finely ground, dry pigments cannot easily be wetted. A jet will scatter them and may possibly create a risk of a dust explosion. It is, therefore, better to use a spray in such circumstances, although foam is often the best means to employ.

Carbon blacks and bone black

1. NATURE AND PROPERTIES

Carbon blacks consist of finely divided carbon produced as a result of the imperfect combustion of various solid, liquid and gaseous products and appear as fluffy, velvety black substances.

They are known by many different names which usually indicate the way the black was made, *e.g.*, acetylene black, gas black (obtained by cracking natural gas), lampblack (usually obtained from oil), etc. They are sometimes manufactured in factories engaged solely in their production; more often, however, they originate as a by-product in some other process.

Carbon blacks are used in the manufacture of rubber, printing inks, polishes and blacking, black paint, carbon paper, gramophone records, brushes for dynamos, carbons for arc lamps, electrodes, insulating materials, imitation leather and many other articles.

Bone black is obtained by greatly heating clean, dry bones in a retort out of contact with air, grinding the resulting product and treating it with acids to remove the calcium. It is used for sugar, glycerine and fat refining and for taking the colour out of various liquids.

2. FIRE-FIGHTING

Black does not ignite easily and burns very slowly without smoke or flame; and combustion is really difficult to detect, usually being recognised only by the diminution in the size of the heap. The rise in temperature is very slow, especially in a well ventilated store. Since carbon blacks are thought to present a slight risk of dust explosion, fires should be attacked with a fine spray or be flooded through open-ended hose, care being taken to turn over the heap thoroughly.

Part 6c, Chapter 45, Section 10

Fires involving plastics, including celluloid

NATURE, PROPERTIES AND RISKS OF PLASTICS GENERALLY

1. **Characteristics of plastics**
2. **Classification of plastics**
3. **Raw materials and composition**
4. **Burning characteristics**
5. **Personal hazards**

MANUFACTURING PROCESSES AND RISKS

1. **Manufacture of plastic materials generally**
   (a) buildings
   (b) processes
   (c) risks
   (d) fire-fighting
2. **Fabrication of plastic products generally**
   (a) buildings
   (b) processes
   (c) risks
   (d) fire-fighting

CELLULOSE NITRATE AND CELLULOSE ACETATE PLASTICS

1. **Cellulose nitrate plastic**
   (a) manufacture
   (b) fire-fighting
2. **Cellulose acetate plastic**
   (a) manufacture
   (b) fire-fighting

THE growth in the use of plastics which has taken place in every walk of life over the past twenty years has been phenomenal and promises to continue so. Perhaps because of this very rapid development, there is still much confusion in the public mind as to the nature and composition of the various types of plastics, and their burning characteristics and manufacturing hazards are not yet familiar in detail to the average fireman. This is hardly surprising, since new formulae and developments occur almost daily in this rapidly expanding industry.

In the early days of plastics the best known name was that of celluloid, and this material has always been regarded by all authorities as one of the most hazardous, both in manufacture and use, that exist, its basic ingredient being cellulose nitrate. Since that day, none of the more recently introduced plastics has possessed such dangerous qualities, and indeed many of them are either to all intents and purposes non-combustible or have slow burning rates. While these qualities are of the greatest possible service to the public in that they tend to reduce many domestic fire hazards, they do not assist the fireman greatly. This is because the materials used in the manufacture of plastics are in many cases of higher fire and explosion hazard than the end product, and also because of the toxic dangers presented: many plastic materials, while quite non-combustible or very slow burning, give off extremely dangerous products of decomposition when involved in fire.

In any survey of the fire hazards of the plastics industry, it is extremely difficult to pinpoint where the various dangers are likely to be found. The older types of plastics, chiefly the semi-synthetics such as cellulose nitrate (celluloid) and cellulose acetate, will usually be found manufactured in works of a somewhat different character from those in which the newer synthetic materials (of which polyethylene, polyvinyl chloride, the polyesters and the fluoro-carbons are typical examples) are handled. The development of the industry in the case of the latter types is progressing on different lines from those of the former: a small number of very large firms manufacture the constituents in large modern plants and sell them in bulk (in the form of moulding powders, for example) to the fabricators who produce the finished articles. These manufacturers of plastic articles are for the most part smaller firms, many of them housed in modern factories but some in a very small way of business indeed, whose processes are mostly mechanical as opposed to chemical and thus involve a very different type of risk. Because of these varying activities within the industry it is not always possible in the following paragraphs to generalise on the characteristics of plastics factories, and the fire brigade officer is advised to make prior contact with all types of firms to obtain the necessary information.

Many of the terms used in this new and specialised industry are highly individual and technical; they can only be defined in terms of advanced organic and inorganic chemistry. Any reader of the *Manual* who desires to pursue the chemical aspects of the subject in greater detail, is referred to the 'Glossary of terms used in the Plastics Industry', published by the British Standards Institution as B.S. 1755 and amendments.

Nature, properties and risks of plastics generally

1. CHARACTERISTICS OF PLASTICS

Plastics as a term may be said to cover a very wide range of materials which are plastic at some stage in their manufacture and can be shaped by application of heat or pressure or both, but which are stable in shape and form at normal temperatures. Certain inorganic materials might be said to conform to this description, such as glass, ceramics and portland cement, but these are excluded in the following paragraphs since they do not conform to the generally accepted meaning of the term "plastics"; there are other groups of plastic materials which are based upon such substances as bitumen and shellac, which are also excluded. Synthetic rubbers are dealt with in Section 14 of this Chapter.

Within the narrow limits laid down above, therefore, 'plastics' comprise organic materials chiefly composed of carbon in combination with other chemicals such as hydrogen, oxygen and nitrogen. They are semi-synthetic or synthetic in origin: the former are largely derived from cellulose, (the older forms, derived from casein, are no longer commonly found and are omitted from the present classification) and the better-known of the principal plastics are listed below, grouped into classes where appropriate:

Semi-synthetic

P CELLULOSE DERIVATIVES:
- Cellulose acetate
- Cellulose acetate-butyrate
- Cellulose butyrate
- Cellulose nitrate (celluloid)
- Benzyl cellulose
- Ethyl cellulose
- Methyl cellulose

Synthetic

S AMINOPLASTICS:
- (S or P) Aniline-formaldehyde
- (S or P) Melamine-formaldehyde
- Phenol-formaldehyde
- Urea-formaldehyde

P ACRYLIC ESTERS:
- (e.g. polymethyl methacrylate)

S FLUOROCARBONS:
- (e.g. polytetrafluoroethylene)

P POLYAMIDES:
- (e.g. nylon)

S & P POLYESTER RESINS:
- (e.g. terylene fibre)

P POLYSTYRENE

P POLYETHYLENE:
- (e.g. alkathene)

P POLYVINYL ACETALS:
- Polyvinyl acetal
- Polyvinyl butyral
- Polyvinyl formal

S SILICONE RESINS

P VINYL DERIVATIVES:
- Polyvinyl acetate
- Polyvinyl alcohol
- Polyvinyl chloride
- Polyvinyl chloride-acetate
- Polyvinylidine chloride

NOTE: S denotes thermo-setting
P denotes thermo-plastic
(*See* paragraph 2 below)

Plate 37. De-linting cottonseed in an oil and cake mill. The accumulation of cotton fly can be seen on some of the horizontal surfaces. (Section 8).
PHOTO: *Rose, Down and Thompson Ltd.*

Plate 36. A battery of high-pressure continuous oil expellers in a modern seed-crushing mill. (Section 8).
PHOTO: *Rose, Down and Thompson Ltd.*

Plate 38. An example of a modern, well-planned and efficiently maintained engineering workshop.
(Section 7).

PHOTO: *Fox Photos Ltd*

Plate 39. Men of the London Fire Brigade at work during the fire at the Union Cold Stores, Smithfield Market, on the 23rd January, 1958. (Section 12).

PHOTO: *Planet News*

Plate 40. Results of a fire in an old type of cold store. The space between the brickwork and the wood panelling was filled with granulated cork. The fire spread from the compartment below up the chase left for the pipework. (Section 12).

Plate 41. Machinery for filling sugar bags. (Section 15).
PHOTO: *British Sugar Corporation*

2. CLASSIFICATION OF PLASTICS

Plastic materials may be classified according to whether they are semi-synthetic or synthetic in origin, as was shown in the table above. A more common basis for classifying them is, however, the division into those which are termed '*thermo-setting*' and those which are '*thermo-plastic*'. Thermo-setting plastics are plastic during the course of their manufacture but are then heated, and become permanently set on reaching quite moderate temperatures. Any further application of heat does not have any softening effect, and they char at a temperature of about 650°F. (343°C.); being non-combustible, they do not ignite at any stage in the temperature rise, but decompose at something above this temperature.

Thermo-plastics, on the other hand, are of such a composition that no chemical change takes place on an increase in temperature, and they may be softened on heating and hardened on cooling as often as may be desired. They are, therefore, plastic during manufacture, stable in form at ordinary temperatures, and become plastic again when temperatures in most cases exceeding 250°F. (121°C.) are reached, or higher according to the type of plastic. The division into thermo-setting and thermo-plastics is indicated on the table above, and it will be seen that in a few cases plastic substances can be made so as to be thermo-setting or thermo-plastic. The plastics industry, however, is not solely concerned with the manufacture of finished articles made in plastic; many of the compounds listed above are, in fact, used in the form of synthetic resins. As such, they may form the basis of finished articles, or they may be combined with other materials such as wood to produce composite boards, or they may be used by other industries (paint and varnish, for example) as substitutes for natural resins. (*See* Section 9, 'Paint and varnish fires', and 13, 'Fires in resins and gums').

3. RAW MATERIALS AND COMPOSITION

The plastics industry relies for its raw materials on the chemical industry, and a wide variety of products is employed: it would be impracticable to give an exhaustive list, but this would include, for example, phenol, cresol, benzene, (which can be converted into phenol) and coke, which are obtained from coal by distillation. Coke, air and water supply the gases required for the syntheses of methyl-alcohol and ammonia. Formaldehyde, once made exclusively from wood alcohol, is now made from synthetic methyl-alcohol. Urea is made from carbon dioxide (derived from coal) and synthetic ammonia. Phenol (and cresol), urea and formaldehyde form the basic materials for the phenol-formaldehyde and urea-formaldehyde groups of plastics. Acetylene (the basis of the vinyl group) is made from calcium carbide, prepared by heating limestone and coke to a high temperature in an electric furnace. The cellulose-based plastics are largely composed of cotton linters while cellulosic materials, such as wood flour, wood pulp, paper and cloth play a large part in the manufacture of many types of plastics.

Generally speaking, a plastic is a combination of a number of ingredients, each of which has its own function and effect in the qualities of the finished product. These qualities can be varied at will by the manufacturer by varying the types and proportions of the ingredients: these latter may be classified as binders, fillers, plasticisers, solvents and pigments, though all of these are not necessarily required in the making of all plastics. The function of the binder may be loosely said to be that of cementing the constituents together, and it will normally consist of a synthetic or natural resin; the filler may be said to provide bulk and reduce shrinkage, and is commonly found from cellulosic materials such as wood flour; the plasticiser gives flexibility and makes moulding possible, and may be one of a variety of synthetic chemical compounds; and the function and nature of the solvents and pigments is self-explanatory. Many other substances in smaller quantities are also liable to be used for special purposes.

4. BURNING CHARACTERISTICS

The oldest well-known form of plastic is the most dangerous: *celluloid* is another name for cellulose nitrate plastic, which is certainly hazardous and is described in detail below. It decomposes with an exothermic reaction at temperatures from 250°F. (121°C.) and upwards, and the reaction progresses with great rapidity without the addition of oxygen from the air; flammable vapour is produced by the reaction, and, if this is allowed to accumulate before ignition, it can be violently explosive. In any event, cellulose nitrate will burn vigorously, with the emission of large quantities of toxic and flammable vapour, accompanied by thick yellowish smoke, and is difficult to extinguish.

No hard and fast division is possible between one plastic and another in describing burning characteristics; the manufacturers vary the specifications of their products in detail, and one manufacturer's sample of a material may not have the same characteristics as regards flammability as those of another. It is also very important to remember that, in common with other materials, the state of division of a plastic material is a vital factor in determining how it burns: a substance which is to all intents and purposes non-combustible when in a mass such as that of its normal manufactured form may burn quite readily when in thin sheet, and even more so when in powder form.

Within these limits, however, it is possible to suggest burning rates for plastics in three roughly-divided groups: in the first are those materials which will either not burn at all or which will cease to burn if the source of ignition is removed: in the second group are those which are combustible but have a relatively slow rate of burning, which may or may not cease on removal of the source of ignition: and in the third are those which burn away without difficulty and are capable of being more or less completely consumed after the removal of the source of ignition. (The hazardous materials referred to above in the nitro-cellulose group are excepted, being in a class of their own).

In the first group will be found those of the phenolic plastics which are asbestos-filled as opposed to wood filled; some of the polyvinyl chlorides; some of the vinyl derivatives; the silicone resins; polyamides, such as 'nylon'; polyester resins such as 'terylene'; and the fluorocarbons. In the second group will be found the wood-filled formaldehydes (urea or phenol); some vinyl derivatives; melamine-formaldehyde; and glycerol-phthalic anhydride. In the third group polystyrene; acrylic esters such as polymethyl-methacrylate; some types of cellulose acetate; cellulose acetate butyrate; polyvinyl alcohol; polyethylene; aniline-formaldehyde; and polyvinyl acetals.

The remainder are difficult to classify owing to the possible variation in their composition: it should be borne in mind that even specimens of a definite type of plastic are capable of being composed of varying materials, and this will vitally effect the burning rate. This is particularly notable in the case of those which sometimes include mineral materials in their composition as fillers, which would place them more or less certainly in the non-combustible category. It is important to remember that flammability is not the only hazard attached to plastic materials, many of which have severe toxic hazards associated with them; these are dealt with in the following paragraphs.

5. PERSONAL HAZARDS

Many types of plastic materials will emit considerable quantities of smoke when burning or when subjected to high temperatures;

the most notable of them are probably those based on cellulose nitrate, but most other forms of plastic emit smoke. In some cases, the only dangers attached to this are the normal dangers of carbon dioxide and monoxide; in many other cases, however, the products of decomposition are additionally poisonous, and very severe dangers indeed can arise from these. It is not a practical possibility to ascertain and learn the toxic characteristics of all plastic products, and in these circumstances it is as well to bear in mind that whenever plastic materials are burning or subjected to very high temperatures breathing apparatus should be worn for fire-fighting purposes.

Tests have shown that chlorinated plastics, such as polyvinyl chloride, are capable of evolving hydrochloric acid in addition to carbon dioxide and carbon monoxide; very small quantities of phosgene may also be formed. Polymethyl methacrylate plastics may liberate hydrocyanic gas when burning.

Among the newer types of plastics particular dangers are associated with the fluorocarbons, and in particular polytetrafluoroethylene or 'P.T.F.E.'. This may, at temperatures above 400°C. (752°F.), give off traces of vapour on the depolymerisation of the material into its monomer and other fluorine-containing compounds. The vapour can produce unpleasant effects, sometimes referred to as 'polymer fume fever' and produce the symptoms of a severe feverish chill. There is also evidence that the fine dust which may be given off during disintegration may give rise to similar symptoms if inhaled. It is particularly dangerous if particles or dust from P.T.F.E. find their way on to the burning tip of a cigarette and are inhaled by the smoker. Breathing apparatus is essential for personnel engaged in fighting a fire in which this material is involved.

It is necessary to emphasise that, as has been said earlier, it is of no use to attempt to distinguish during fire-fighting operations between those plastics whose products and decomposition are not poisonous and those which are; in practice it will seldom be possible to be sure that only one material is involved or exactly what the properties of that material are. The general rule should therefore be observed that breathing apparatus is always necessary in the presence of plastic materials which are burning or at high temperatures.

A study of later paragraphs in this section dealing with industrial processes in the plastics industry will show that many of the materials employed in the industry also offer toxic or other personal hazards to the fireman; as far as possible these will be noted in the text but the reader should consult all possible sources to ascertain the dangerous properties of any materials concerned.

Manufacturing processes and risks

1. MANUFACTURE OF PLASTIC MATERIALS GENERALLY

(a) Buildings

The materials side of the plastic industry is of comparatively recent growth, and tends to trade in large quantities; the buildings are therefore usually of relatively modern construction and often of considerable size. They are likely to be of steel-framed construction and the fire risk arising from the construction of the building is likely to be low.

(b) Processes

Again on the materials side, the majority of processes are chemical and the commonest process is that of polymerisation. This is a process producing chemical changes in large quantities of materials, usually accompanied by solidification, and is commonly done in autoclaves or polymerisation tanks. The essential factors in the processes are mixing, the addition of catalysts, and controls of temperature and pressure. Other chemical processes include distilling, blending, 'digesting' and sintering; some or all of them may involve the application of heat and/or pressure in varying degrees. Mechanical processes may include grinding, pulverising, milling, filtering, weighing, etc. The mixing processes may involve liquids, including solvents, or solids in powder form, and mixing rooms are usually at the tops of buildings; in the majority of plants many of the liquid constituents are handled in bulk by pipe lines. Further details of specific manufacturing processes are included below under 'Cellulose nitrate' and 'Cellulose acetate plastics'.

(c) Risks, including toxic hazards

In materials production, hazardous liquids and gases are handled, sometimes in large quantities, and present the usual fire and toxic risks if strict safety routines are not adhered to. Associated with some of the powders involved in plastics manufacture is a dust explosion risk, since many of the materials, while not hazardous in the normal solid form, may be so when finely divided. Dangers also arise in the storage and handling of hazardous chemicals, and in the processes in which those chemicals are involved; temperatures and pressures may be critical in some cases. Apart from these specialised risks, ordinary hazards arise from faulty process operation, faulty temperature controls and faulty maintenance of equipment;

electricity is extensively used for driving motors in process machinery for heating the moulding presses and for other purposes, and fires may result from incorrect use or maintenance of such appliances. Because of the dangerous or flammable nature of so many of the materials used, any unsuitability of the equipment (to withstand pressure or corrosion for example) or any failure to maintain good equipment maintenance and good housekeeping are liable to produce outbreaks of fire. Toxic hazards may be associated with any such outbreak.

Processes involving the irradiation of plastics constituents from radioactive isotopes exist, but they are new and comparatively rare; they are likely to be found in association with research laboratories more than in production plants, and only a very few manufacturers have these facilities at the present time. In such cases, prior contact with the factory concerned is all the more important, and guidance from works' chemists or physicists should be sought, both before the fire and during operations. (For the hazards to firemen associated with radioactive materials *see* Section 11).

(d) Fire-fighting

It is impossible to draw a hard-and-fast line between the character of fires in materials manufacturing plants and those in fabrication works, and the paragraph headed 'Fire-fighting' under 2 (**d**) below should also be studied in this connection.

Due to the nature of the processes, there is little likelihood of fixed extinguishing installations being found in plastics manufacture, other than in special high-hazard process machinery, though hose-reels or internal hydrants may sometimes be installed. Since the majority of materials plants are large and well-staffed, the probability is that the fire brigade will be faced with a well-developed fire on arrival, or at least a situation in which heavy smoke-logging has taken place. Breathing apparatus will almost certainly be necessary; toxic gases may well be present in addition to smoke, and narcotic fumes may be released such as formaldehyde and chloroform. The most generally useful fire-fighting medium will be water sprays; particular care should be taken to avoid the use of jets on heated pressure vessels, heated piping and stocks of powdered chemicals. Jets may, however, be the best attack on burning cellulose nitrate plastics if circumstances permit.

If substantial quantities of flammable liquids are involved (especially in tanks), such as aniline, styrene or ethylene, foam will be required, though small fires in these liquids, perhaps arising from leaks in piping, etc., will usually respond to the use of water sprays, carbon dioxide or dry powder if they are too small to justify getting a foam-making branch or a high expansion foam unit to work. It should be remembered that aniline, which has a specific gravity of 1·02, will rise to the top of a tank if water is introduced into the aniline when the latter is heated. Since these tanks may be placed at a high level, the dangers from an overspill are obvious. This applies generally to any liquid with a specific gravity of between 1 and 1·1. Water in the form of spray is the best medium for use on butadiene, and water sprays may also be useful in keeping down heavy vapours such as those from urea.

Moulding powders and granules are completely ruined by admixture with water, so that efforts must be made to avoid this and to salvage existing stocks as far as possible. The probability of water damage will be much reduced if early contact can be made with works chemists, who will know from past experience whether special fire-fighting techniques (such as smothering powder fires by shovelling on more of the same powder) can be adopted. It is important to know in any case what plastic materials are involved, if any, and a watch should be kept (and early enquiries should be made) for the possible presence of special dangers such as cellulose nitrate stocks, the corrosive phenol, the fluorocarbons or (very rarely) radioactive materials. The help of the works staff may also be very urgently needed to have piped liquids and electricity supplies cut off and to release rising pressure in sealed tanks, etc.

2. FABRICATION OF PLASTICS PRODUCTS GENERALLY

(a) Buildings

A very great number of firms have begun production in the plastics fabrication trade in the last twenty years. The larger articles, such as baths and radio cabinets, require the use of large moulds and quite heavy machinery, and this involves substantial capital investment; firms in this class of business usually have large modern buildings in which large-scale production is carried on and materials are stored and handled in bulk. Many thousands of small articles however, are produced in small factories of recent growth: the majority of these may be of modern construction, but many are converted from their original purpose and quite a large number are partly converted domestic premises. They undoubtedly include a number of buildings of unsuitable or high-hazard construction and many more which offer risks through over-crowding, improvisation, poor maintenance and indifferent housekeeping.

(b) Processes

These may be divided into moulding processes on the one hand and sheet production on the other, since most of the techniques other than those of sheet production are in effect moulding processes. Materials are received from the manufacturers in the form of powders or granules in most cases, though some firms purchase sheet or rods, etc., ready made; plasticisers, solvents and pigments will be added to the pure resins to make moulding materials in some factories. Moulding may be by compression, injection or extrusion: in compression moulding the powder is loaded into a heated steel mould and pressure is applied, while injection moulding (*see* Fig. 37) involves pressing the material in a liquid or

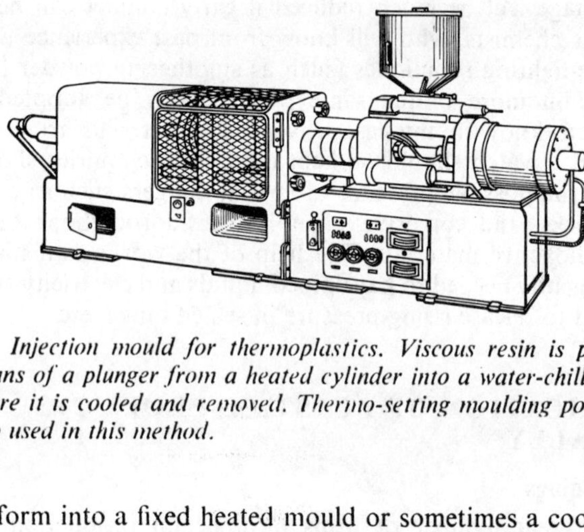

Fig. 37. Injection mould for thermoplastics. Viscous resin is pressed by means of a plunger from a heated cylinder into a water-chilled mould, where it is cooled and removed. Thermo-setting moulding powders are also used in this method.

ductile form into a fixed heated mould or sometimes a cool mould. Extrusion moulding (*see* Fig. 38) is done by passing the ductile plastic through a heated tube by means of a screw pump, the die which forms the extrusion being placed at the end of the tube. Some moulding processes are done, particularly with thermo-plastic types, by vacuum, whereby a sheet of heated plastic is caused to take the shape of an internal mould by the extraction of air from the mould, and a similar principle is employed in the 'rubber-bag' technique, in which pressure is applied over the surface of a mould by inflating a rubber bag against the mould within an autoclave. Other processes include drilling and machining, 'potting' (the insertion of small metal parts into a plastic mass), the cementing of parts together and various forms of coating.

Sheet plastics processes nearly always employ rolling machines, either heated or cold. They are used to produce laminated sheets, resin-bonded plywoods and various types of faced boards, in addition to the making of sheets and the coating of paper, card, metal and plastic sheet itself.

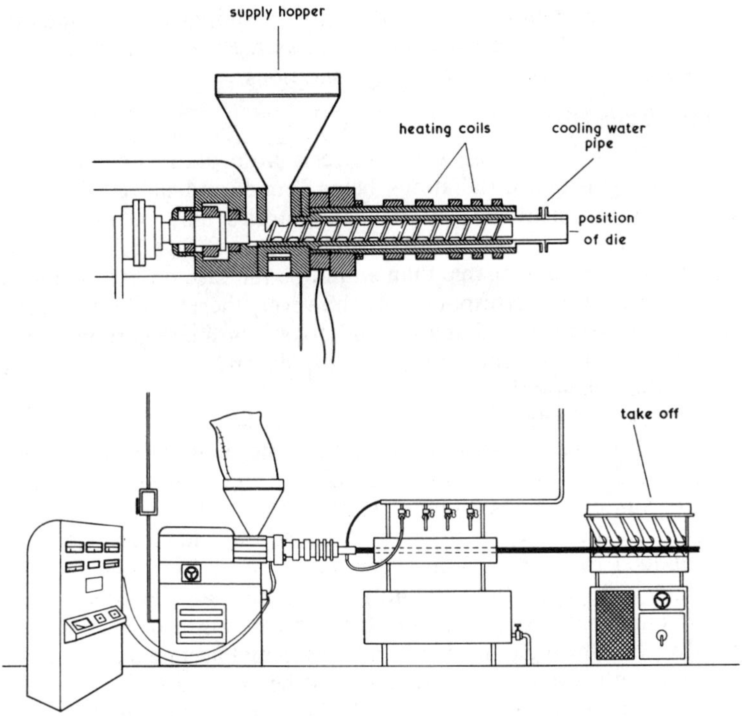

Fig. 38. Plastics extrusion. Tubes and rods are made by forcing the plastic through a suitable die by means of a screw. Controlled heating is applied at the extrusion nozzle followed by water cooling. ABOVE: *Sectional view of the screw.*

(c) Risks

In fabrication work, the risks are less highly specialised and chemical in nature; they have more the character of normal factory risks. Few of the plastic materials themselves are highly flammable (always excepting cellulose nitrate), but flammable liquids are widely used in the form of cements, solvents and plasticisers, and present the risks to be expected of them. Toxic hazards may arise from the moulding powders or from the finished products. The dust explosion hazard is more likely to be present in milling and grinding processes than in moulding, though waste powder is also sometimes present in quantity where moulding is done. Presses and extrusion machines are usually worked hydraulically, and therefore present dangers from hydraulic fluids escaping at high temperature and pressure. Static electricity in the presence of volatile flammable liquids is a serious hazard, and electrical equipment is abundant. Apart from these, the usual factory risks are present and may be aggravated by unsuitable working conditions.

(d) Fire-fighting

In plastics fabrication works there is little or no likelihood of fixed extinguishing installations being found, and indeed fire protective arrangements in the smaller factories may be only of a rudimentary kind. There is also much less chance of a works' chemist or physicist being available than would be the case on the materials side of the trade; correspondingly, however, there is a greater probability in these works that a fire will involve ordinary carbonaceous materials such as roof linings, partitions and packing materials rather than chemicals.

In most incidents water jets or sprays will be called for, therefore, though it should be borne in mind that one or more of a number of flammable solvents may be involved, and that this may call for foam unless the quantities are relatively small. Foam is the correct medium for benzene, naphtha and the ketones, and alcohols should be subdued by dilution with water unless alcohol-resistant foam is available; ethyl alcohol (ethanol) and methyl alcohol (methanol) are irritant and toxic, especially the latter, whose toxic effects can be absorbed through the skin and lungs. Amyl acetate also requires dilution with water, and acetone should be extinguished with water sprays: further details of these and other solvents will be found in Section 16, 'Dangerous Chemicals'. (Small fires in flammable liquids may, of course, be successfully attacked with carbon dioxide, dry powder, vaporising liquid, etc., in their early stages if these media are available.) Stocks of powdered or granulated chemicals should not be attacked with jets, which might tend to produce a dust cloud, and sprays are preferable; the use of water on moulding powders should in any case be avoided as far as possible, since they are spoiled by the admixture of water. Machinery should be protected against sudden localised cooling (which might be produced by the application of jets) and against the action of water generally; while this is a normal machinery point such as is encountered in any factory, the fireman should remember that moulding and extrusion machinery is heated, and in addition may contain flammable hydraulic fluids at high temperatures.

If plastics or plastic materials are burning, breathing apparatus will be needed against poisonous fumes and narcotics. The information given under fire-fighting in plastic materials manufacture (above), and that under cellulose nitrate and acetate plastics (below), should also be studied in this connection.

Manufacturing processes and fire-fighting hazards of cellulose nitrate and cellulose acetate plastics

1. CELLULOSE NITRATE PLASTIC

This, as *celluloid*, is the oldest of the plastics, and it is still in use; it is, however, not nearly so common as it once was. It is no longer used for cinematograph film manufacture, and film made from it has almost completely disappeared; it is worth noting that cellulose plastics of all kinds represented only 20 per cent of plastics production in the United States in 1953, and this proportion will be much lower now. However, a few household articles are still made from it, notably table knife handles, and it is essential for every fireman to be aware of its dangers.

(a) Manufacture

Cotton linters, which are nitrated with a mixture of nitric and sulphuric acids to make cellulose nitrate form the base. After purification, the cellulose nitrate passes to the mixing department, where it is compounded with plasticisers (camphor is chiefly used, a plant for manufacturing camphor from turpentine often being found in the same works), pigments or dyes, and fillers. Solvents are used to assist the plasticising action. As many as 140 different plasticisers and 240 solvents are used in the industry. Most of the plasticisers have high flash points with no special fire risks, but risks arise with low flash point solvents such as acetone and benzene. An outbreak of fire is often accompanied by small explosions of solvent vapours.

The surplus solvent is removed partly by rolling the plastic dough on steam-heated rollers, and partly by seasoning the sheets of celluloid in heated stoves (often in spaced, self-contained buildings at a distance from the main works). These sheets are made by pressing the rolled plastic dough into block form in steam or water-heated hydraulic presses, and then slicing from the blocks sheets of the desired thickness. During this process as much of the solvent as possible is recovered by means of a vapour recovery plant connected to the rolling machines and stoves.

The material is then ready for making up into articles.

(b) Fire-fighting

Celluloid is a highly flammable and easily ignited substance which burns fiercely. If heated to 212°F. (100°C.) it begins to break down, while above 248°F. (120°C.) sudden, and possibly explosive, decomposition takes place. Fumes are given off, and these can catch fire as a result if heat generated by the decomposition process. Fumes which have collected in a confined space may explode when air is admitted. The violence with which decomposition sometimes takes place may force gases a considerable distance and produce an explosive mixture some distance from the celluloid. The fumes are poisonous, flammable, heavier than air, and contain carbon monoxide, nitric oxide and hydrocyanic acid. It is almost impossible to extinguish a celluloid fire with water owing to the high oxygen content of celluloid, though a very hard-hitting jet may be successful. Large quantities of water should, however, be used for cooling, and preventing the spread of the fire. A small fire can be extinguished by the prompt application of CO_2. Considerable noise is made when celluloid burns, and flames are forced some way from the fire. Firemen should avoid inhaling the nitrous fumes, by using breathing apparatus, even in the open air. Fumes can be made less noxious and less liable to flame by the use of water spray.

Celluloid in sheet or other form which is anywhere near a fire should be cooled in order to prevent it decomposing. Where sheets are stacked in piles, they should be separated to assist the dissipation of heat.

Finely divided celluloid is liable to dust explosion. Cutting and scrap (used in the paint, oilcloth and other trades) are often stored in sacks which allow flakes to leak through the fibres. These flakes, which may be found spread over the floors and passages, facilitate the spread of fire.

A fire in the nitrating department is dangerous because of the presence of large quantities of acid. The latter is often propelled by compressed air from one part of the works to another by pipeline. If this pipe line is fractured, acid may be sprayed over a wide area. In the event of fire, therefore, it is advisable to see that the pumps are stopped.

Disastrous fires have sometimes occurred during the rolling and seasoning processes. The solvent vapour in the atmosphere presents a constant risk, and it is possible for fine particles of hardened nitrocellulose plastic, which tend to accumulate, to take fire as a result of becoming heated by friction in the machinery or in some other way.

2. CELLULOSE ACETATE

Cellulose acetate forms the basis of 'safety' (or 'non-flam') film and of many transparent wrapping materials, and cellulose acetate plastics are made to many varying specifications; they are widely used.

(a) Manufacture

Cellulose acetate is made by acetylating cotton linters with a mixture of acetic acid, acetic anhydride and a little sulphuric acid. The resulting product takes the form of a syrupy solution in acetic acid. This compound is ripened in a warm aqueous solution. It is then soluble in acetone and other solvents.

Cellulose acetate is made up into plastics by means of the celluloid technique. The solvents and plasticisers commonly employed are acetone, methyl-ethyl-ketone, alcohol, triphenyl phosphate, triacetin, diethyl phthalate, dimethyl phthalate and naphthalene products for cellulose acetate; and alcohol, benzene and various phthalates for ethyl cellulose.

Cellulose acetate plastics are spun in liquid form to make rayon (*see* Section 4, 'Fibrous materials, rayon'), made into film by special processes, or made into moulding powder. In the latter case the sheets of plastic are granulated in disintegrators, and there is a risk of a dust explosion.

(b) Fire-fighting

Although never highly flammable like the corresponding nitrate, cellulose acetate plastic is produced in different grades (varying according to the plasticisers used). The 'non-flammable' grades are difficult to ignite and will cease to burn if the source of ignition is removed, but the other grades, especially if in thin sheet or powder form, burn fairly readily and will continue to do so after removal of the source of ignition. All grades tend, when heated, to melt, form bubbles and char. The fumes given off have a very acrid smell of burnt cellulose and acetic acid.

Part 6c, Chapter 45, Section 11

Radioactive materials

THE rapid development of atomic energy as a science has taken most ordinary people somewhat by surprise. In spite of the fact that many of the fundamental discoveries about radioactivity were being made as long ago as the turn of the century, most people knew little of the subject until the first atomic bomb was dropped on Hiroshima in 1945, and nuclear weapons have tended to predominate in the public mind ever since. (Radium and X-rays are, of course, familiar manifestations of radioactivity, but the man in the street knows little of the working principles, and not much more of the dangers, even of these). Only eight years after Hiroshima the United Kingdom Atomic Energy Authority was established, and within six years of that event two atomic power stations were working and two more were substantially complete. This startling rate of development has outstripped the ordinary man's understanding of atomic energy technology, and has also very quickly produced a specialised and unusual form of fire-fighting hazard, with which firemen were not until recently familiar.

It is the purpose of this Section to explain this hazard. It should be made clear at the outset that it is not with fire dangers that the Section is concerned; there are no inherent fire hazards associated with radioactive materials in general. Some radioactive materials are themselves flammable or combustible in varying degrees and to this extent they have a fire hazard attached to them, but this hazard is no greater because of their radioactive properties; the effect of the latter, with which we are here concerned, is to produce a risk of personal injury to firemen (and perhaps to the public) during the course of fire-fighting operations. A study of the following pages will show, moreover, that the danger of personal injury is in general very slight: the uses to which radioactive sources are put, the strength of the sources used and the precautions which are taken are such that it is very unlikely that any fireman will be in serious danger from radiation in the ordinary course of his duties or be in a position to create serious danger to others by his actions. There are, however, a few commonsense precautions which everyone in the fire-fighting services in peace-time should understand and apply; it is only by understanding the nature and uses of radioactive materials that the logic of these precautions can be fully grasped.

The increasing use of radio-active materials has led to the introduction of the NAIR scheme (National arrangements for dealing with incidents involving radioactivity). Details of this scheme have been issued to all Brigades.

Properties and hazards of radioactive materials

1. PROPERTIES OF RADIOACTIVE MATERIALS

(a) Composition of matter

The reader will be familiar with the accepted theory that all matter consists either of one of about 100 chemical elements which are known to exist or of compounds of one or more of these elements, and that the atom is, or was once considered to be, the smallest particle of matter and indivisible. It is now accepted that the atom consists of smaller particles, the most important of which are the proton, the electron and the neutron: the proton is a relatively heavy particle which is positively charged, the neutron is a heavy particle also but without electrical charge, and the electron is a very light particle with a negative charge. The proton and the neutron are held closely together in a central mass known as the nucleus of the atom, while the electrons revolve round the nucleus in one or more orbits resembling the orbits of the planets round the sun. The number of protons in any atom is the same as the number of electrons (and the electric charges are thus balanced) and the difference between one element and another is accepted as being the difference in the number of protons (and, of course, of electrons, since this is the same) possessed by each. The 'atomic number' of an element is the number of protons that there are in the nucleus, and hydrogen is well known as having the atomic number 1; uranium has 92, and the other elements each have their own individual atomic numbers. The idea that electrons, the particles of negative charge, can be detached from their atoms and 'flow' in some materials from one atom to another and through the air, and thus form an electric current or discharge, is a familiar one; less so perhaps is the fact that ions, which are atoms or groups of atoms also with a characteristic electric charge, can exist and move about both within materials and outside them.

(b) Isotopes

Although an element is always one recognisable substance which is chemically the same throughout, because its atoms all have the same atomic number, its atoms may not, in fact, all have an identical nuclear structure, and indeed this is the case with many materials. This is due to the presence in the nucleus of the other type of particle in addition to the proton, i.e., the neutron. Although all atoms of the same substance contain the same number of protons and electrons, they do not all contain the same number of neutrons; different atoms of elements which are chemically the same are known as isotopes of that element, and vary with respect to the number of neutrons in the nucleus. Because, also, the varying number of neutrons affects the mass of the atom as a whole, (neutrons, like protons, being relatively heavy particles) the isotopes of an element each have a different mass number or atomic weight; this is shown, for instance, in the distinction between 'uranium–235' and 'uranium–238'. Since there are 92 protons in a uranium atom, it follows that there are 143 neutrons in an atom of uranium–235, and 146 in that of uranium–238.

(c) Stable and unstable isotopes

There can be other differences between isotopes. One of the most important from the present point of view is that some isotopes of some elements are 'stable' and some are not: those which are not are called 'unstable' or 'radioactive', and are referred to as radio-isotopes or radioactive sources. Stability in this connection denotes that the nuclear structure of the isotope is not changing of its own accord; radioactivity, on the other hand, means that the nuclear structure is undergoing change. The important feature of this change is the emission of either particles or rays, whereby the atom is steadily becoming transmuted, as it were: that is to say, it is steadily losing the characteristics of an unstable isotope and acquiring those of a stable one. The emission of particles or rays is termed 'radiation', and the isotope will continue to emit radiation until it reaches a stable condition and its transmutation is over. In the course of changing from instability to stability, an atom may pass through a number of intervening stages, these stages being represented by further isotopes in a series. Uranium–238, for example, becomes in course of time uranium–234, which becomes in turn thorium–230, which also in turn becomes radium–226. After passing through several other stages, all the isotopes in the uranium series become stable in the form of lead (Pb.–206). It may take a long time or a short time for stability to be reached: it is not realistic to state the full time taken for isotopes to cease giving off radiation, because in theory the radiation dwindles, but never actually ceases, for an infinite length of time. Therefore, the time taken for the radio-activity to fall to one-half is the figure quoted, and the proper term for this is the isotope's 'half-life'; the half-life, for example, of uranium–238 is 4,500 million years, while the half-life of bismuth–214 is only about 20 minutes.

Many radioactive materials have existed naturally since the beginning of time, some of them even being present in the human body in microscopic amounts. Most elements have both stable and radio-active isotopes, and the majority of natural elements as we know them consist of a mixture of the two, or perhaps more than two. For example, the atomic weight of chlorine is accepted as 35·457, the atomic number (the number of protons in the nucleus) being 17; this would lead one to deduce that there were 18·457 neutrons in the nucleus. This is manifestly absurd, since there cannot be a fraction of a neutron; the explanation lies in the fact that chlorine as normally found consists of a mixture of the isotopes chlorine–35 and chlorine–37, in proportions (roughly 25:75) such as to produce an average of 18·457 neutrons in each atomic nucleus. Mixtures of different isotopes can be separated in a variety of ways, so that substances can be produced which are pure forms of any of the isotopes. Moreover, some isotopes, such as strontium–90, do not occur in nature at all and can only be produced artificially.

(d) Effects of radio-isotopes

A reference has been made above to the characteristic of the atoms of radio-isotopes whereby they are emitting particles or rays.

The importance and value of these isotopes lies in the ability of their emissions to affect other materials; the effects are many and may produce physical, biological or chemical changes.

When, for example, material B is subjected to radiation from material A, material B may be caused to change from one isotope into another through the absorption of neutrons; alternatively, through bombardment by neutrons, the neutrons in material B may be caused to migrate from one atom to another. We are familiar with the fission process, whereby uranium atoms are split or divided into approximately equal parts by bombardment with neutrons, the process continuing spontaneously if desired until all the atoms are so divided. The possible variations in these effects is very great and their details need not concern us. It is sufficient to say, first, that obviously if human beings are subjected to radiation from a radioactive source the tissue of the human body will experience changes which may be harmful, and, second, that the possible industrial, medical, photographic and other techniques which radio-isotopes make possible are too numerous to mention here; they are referred to, however, in later paragraphs.

It should be noted that in no case is the radioactive isotope of a substance any more flammable—or any less—than the stable form, and that the radioactive properties of a material cannot be destroyed by fire or by any other known means.

(e) Types of radiation

There is a variety of types of radiation ('*alpha*' particles, '*beta*' particles, '*gamma*' rays, etc.), whose effects differ considerably, being dependent upon their penetrating power: *alpha* particles are very easily stopped and are only dangerous if emitted within the body (*e.g.*, by radioactive dust which has been inhaled), *beta* particles are of intermediate penetrating power and *gamma* rays are the most penetrative of all. This makes a great difference to the protective measures required, and is taken into account in the advice given in par. 2(c) on protection from radiation hazards. A table summarising the hazards of different types of radiation will be found in paragraph 2(b) below. There is little to be gained from a fire brigade point of view by a more detailed study of the differences in type of radiation, since in some cases specific instructions will be available from authorities on the spot as to the action to be taken, while in the others it may be impossible to know what type of radiation is involved. In the latter event, as a safeguard, the existence of the worst danger should be assumed and action taken accordingly.

(f) Source strength, dose rates and decay

Three factors are important in speaking of the radiation characteristics of a radioactive source: one is the rate at which radiation is being produced, or the strength of the source, the second is the dose a person would receive from that radiation at a given distance from the source, and the third is the rate at which the strength of the source dies away.

Quantities of most ordinary materials are determined by weight or size. With radioactive substances, however, the weights of the radionuclides involved are usually extremely small, and in practice radioactive materials are nearly always specified in relation to their 'activity' or source strength. The special unit of activity is the *curie*, which is defined as exactly $3{\cdot}7 \times 10^{10}$ nuclear transformations per second. Activity or source strength is therefore quoted in curies or sub-units (a millicurie=1,000th part of a curie; a microcurie= 1,000,000th part). Although in the past the majority of radioactive sources were measurable in millicuries (and were used in medical treatment), it is nowadays quite common for sources of several curies to be used, and there are a few sources of 1,000 curies or over. Even larger sources will be used in the future.

Three units of radiation dose are in common use and although in connection with most applications of radio-isotopes they are, in practice, numerically the same within a few per cent, it is useful to consider them in detail so that the fundamental differences between them can be appreciated.

(i) *Roentgen.* The unit for X- or *gamma* radiation. A measurement in roentgens (r). (exposure) expresses the capacity of a given beam of radiation to ionise air. It is not a direct measure of the effect of the radiation on substances other than air.

(ii) *The rad.* This unit (RAD=radiation absorbed dose) came into use because of a need to express the energy absorbed in a particular material from a radiation flux without reference to the type and energy of the radiation. The rad is strictly a measure of absorbed dose in a specified material and is equal to 100 ergs per gramme. However, a radiation exposure which would produce 1 roentgen in air will produce an absorbed dose of very nearly 1 rad in normal body tissue, so that for most practical purposes in relation to radiation protection, exposures in air expressed in roentgens can be taken as giving the absorbed dose in tissue in rads.

(iii) *The rem.* It has been found that the biological effects of equal absorbed doses, in rads, of different kinds of radiation are not the same. Heavy-charged particles, such as *alpha* particles, produce greater biological effects for a given energy absorption than do *gamma* rays or *beta* particles. Hence a unit of biologically effective dose is required. The *rem* (roentgen equivalent mass) is defined as the absorbed dose in rads multiplied by the quality factor (QF) of the particular type of radiation under consideration. However, for all X-rays, *gamma* rays and *beta* particles which are likely to be encountered, the QF is 1. For *alpha* particles, it can be taken as 10, so that conversion from rads to rems is quite simple.

So far as the fire brigade is concerned, the rad is considered to be the most suitable term to use to denote radiation dosages; 1,000th part of a rad is called a millirad and 1,000,000th part is a microrad. The term is usually used in conjunction with a unit of time to express the amount of radiation absorbed in a given period, such as 'five millirads per hour.'

The fact that a radioactive source loses its strength was referred to earlier. This decrease is known as 'decay', and is spoken of in terms of the isotope's 'half-life'. Knowledge of the half-lives of individual isotopes is not of practical value in fire-fighting, however, since few of the materials likely to be encountered decay significantly in less than two or three days.

2. PERSONAL HAZARDS, PROTECTION AND DECONTAMINATION

(a) Effects of radiation

The risks which may be encountered by firemen will vary considerably, and in the vast majority of cases where radioactive materials are involved in fire the sources concerned will be so small, and the dangers to be expected from them so slight, that no special precautions need to be taken.

Where precautions are necessary because the sources are substantial it may be assumed that the premises concerned will have staff available who are qualified to point out what the dangers are and what steps should be taken, and that the premises will be known by the fire brigade to contain a substantial source. In all cases, prior arrangements can be made as to the precautions necessary. The personal hazards and the precautions against them are explained in the following paragraphs, but it will be understood from what is said above that the dangers will only be significant in a limited number of places.

The damage to the human body is caused by the radiation. As has been mentioned in paragraph (e) above, the radiation may be of several different kinds, but the physical effects may be regarded as generally similar. It may cause delayed burns (which may be superficial or deep and are similar to normal burns except that they may well not become apparent for some days), and it may increase the probability of eventual serious illness.

The question arises when considering these hazards of how much radiation it is 'safe' for the human body to take, and thus what is a 'maximum permissible dose'. This is a reasonable basis on which to fix safety procedures for those working in close day-to-day contact with radioactive materials, since it is feasible for such people to be continuously 'monitored', i.e. kept under observation with instruments; reliable estimates have been made of the likely effects on the human body of different dose rates for varying lengths of time.

Firemen are regarded as belonging to the 'occupational exposure' group of workers and legislation has been laid down which limits such workers to a dose of five rads in any period of twelve months, or, in the case of firemen who may be regularly engaged in the handling of the radioactive sources supplied for training purposes or for testing radiation instruments (*see* the *Fire Service Drill Book*, Part III, 'Standard tests for fire service equipment'), to a dose of 0·1 rad (100 millirads) per week. Because these training radioactive sources are very weak, there is hardly any likelihood of any fireman receiving anything like 100 millirads a week when handling them.

The total dose of five rads in any period of twelve months applies to all firemen, but exceptionally a fireman attending a fire or other incident in the presence of radioactive substances may receive a dose of up to three rads at one incident, or in any period of thirteen weeks, provided that any dose received during the remainder of the twelve months does not exceed a further two rads. Any dose received during the testing of instruments or in training must be taken into account in calculating the further dose which can be received during the remainder of the twelve months period.

(b) External and internal radiation

An important feature of radiation effects on the human body is the difference between 'external' and 'internal' radiation. The former is radiation which is received by the body from some outside source. The latter is radiation from within, *i.e.*, emitted by radioactive material taken into and lodged in the body in the form of dust, by breathing ('inhalation'), eating or drinking ('ingestion'), or by penetration of the skin, for example through wounds. Generally speaking, internal and external hazards may be due to *beta* or *gamma* radiation, while internal hazards are also caused by *alpha* radiation. Table IV summarises in general terms the external and internal hazards from these different types of radiation:

TABLE IV
Summary of Hazards

Isotope emitting	*Hazard*	
	External	*Internal*
Alpha particles ..	None	Very serious
Beta particles ..	To skin and eyes only	Serious
Gamma rays ..	To whole body and internal organs	Not normally significant

Note: X-rays are identical with *gamma* rays but are generated by high voltage machines. Unlike *gamma* rays, they may be switched off when not required.

Some radioactive substances emit more than one kind of radiation, but with others only one kind of emission is significant; for example, the radiation from plutonium is almost entirely *alpha*, that from radioactive phosphorus is solely *beta*, whereas radioactive cobalt emits both *beta* and *gamma* radiation. The methods of protection required are not the same for external as for internal radiation: each hazard requires its own code of rules for personal safety, and these are outlined below.

(c) Personal protection

(i) *External radiation.* Protection from external radiation is to be found in a combination of three factors, viz.: time, distance and shielding. These can be more fully explained as follows:

A. The duration of exposure is of importance; the shorter the exposure, the smaller is the dose incurred. It is thus of the utmost importance that anyone who may be exposed to radiation should cease to be so exposed at the earliest possible moment.

B. The intensity is much greater at a short distance than at a long one. Protection from the emission of *beta* and *gamma* rays can therefore to some extent be obtained by keeping as far from the radioactive source as possible. For example, the dose rate from a *gamma* source is nearly 1,000 times as great at 300 millimetres distance as at 9 metres.

C. Protection can also be obtained by placing some form of shielding (*see* Fig.39) between the source and the person exposed. All radioactive sources capable of creating an external radiation hazard are normally shielded according to the strength of the source and the nature of the radiation, the object of the shielding being to absorb the emitted rays so that they cannot reach and harm people nearby. In emergency, radioactive sources may lose their shielding, and in such cases it is necessary to make use of any other shielding which may be available. The effectiveness of a shield depends upon its weight; for example the following materials in the thicknesses shown will reduce to approximately a quarter the *gamma* radiation from radioactive cobalt:

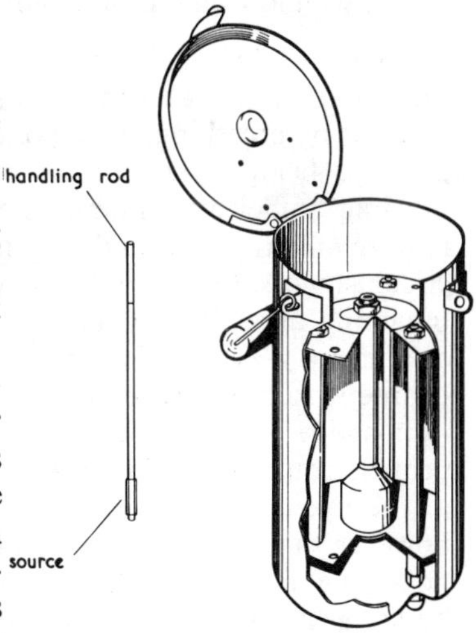

Fig. 39. A transport container for radioactive isotopes. The source, at the end of the handling rod, is stored in the centre of the container, which is generally lead-lined.

Lead	25 mm
Steel	38 mm
Concrete	150 mm
Earth or brick ..	190 mm
Water	330 mm
Wood	500 mm

It should be borne in mind also that it is more important

to protect the trunk and vital organs of the body than the arms and legs.

D. There is no form of clothing which will protect the wearer from an external *gamma* radiation hazard. Skin contamination hazard can occur from *alpha* or *beta* sources or from dust, smoke or water contaminated by them, and against these the ordinary fire-fighting clothing gives a large measure of protection. All exposed parts of the body should however be washed thoroughly after operations are over, particular attention being paid to the hair and nails and to any cuts or wounds, and a bath should be taken as soon as possible.

(ii) *Internal radiation.* Protection against internal radiation relies on avoiding the absorption into the body of radioactive dust, whether by breathing, eating or drinking or through wounds. When a fire is being fought in the open air, the possibility of firemen inhaling sufficient contamination to produce any harmful effect is so remote that it would only need to be considered in exceptional circumstances. In buildings believed to contain radioactive materials, breathing apparatus should be worn as a precautionary measure, even if not necessary for fire-fighting purposes, unless the officer in charge is told or has reason to believe that the hazard arising from radioactive contamination of smoke is negligible. Respirators provide equal protection against radioactive dust, and may be used in the open air where breathing apparatus is not available. Since however they do not provide protection against carbon monoxide poisoning or oxygen deficiency, they should be not used inside buildings if there is fire or smoke present.

Smoking, eating and drinking should be avoided in the suspected presence of contamination, and any men who suffer cuts or wounds should be withdrawn and instructed to wash the injury thoroughly.

(iii) *Radiation instruments.* All first-line fire brigade appliances which are likely to attend at premises where radioactive substances are used carry quartzfibre dosimeters and a survey meter. The quartz-fibre dosimeter will record a total dose of up to 5 rads, and will be worn by all fire brigade personnel who are required to enter a contamination control area (see par. (d) below). The 0–10 rads/hr. survey meter is used to monitor the area to verify whether there is any radiological hazard to firemen at the incident. As the scale is a logarithmic one, dose rates in millirads/hr. are easily discernible at the lower end of the scale.

The type of contamination meter recommended for use by fire brigades is one which will give a disintegration count rate of as low as 5 per second, so that a very small amount of radioactive contamination is detectable. A *beta* probe as well as one for *gamma* radiation is recommended, but it is not considered necessary to be able to detect contamination due to *alpha* particles. Contamination meters are normally carried on control units.

(d) Contamination control

It may be desirable at an incident involving large quantities of radioactive materials to fix the limits of a contamination control zone, so that, for example, men and equipment which may have become contaminated by radioactive dust, smoke or water can be segregated within the zone and further contamination be prevented. Steps should be taken to see that nothing—personnel, hose, equipment, clothing, food, etc.—passes out of the contaminated zone until it has been checked to verify that it is not contaminated with radioactive dust. Fire appliances should be kept outside the zone (and, if possible, upwind of the incident) as should also, of course, members of the public.

(e) Decontamination

After fire-fighting operations which may have involved radioactive materials, steps must be taken to ensure that no personnel, clothing, equipment or appliances remain contaminated by radioactive dust. Normally the operations will be conducted under the supervision of specialist personnel, who will give instructions as to a check and decontamination of the objects concerned. Removal of contamination from the person involves very thorough washing (see paragraph (c) (i) D above). As to appliances, since the object of decontamination is to remove radioactive dust, thorough hosing down should have the desired effect. Equipment should, as far as possible, be washed, and uniform should be carefully removed in a suitable place set aside for the purpose. All articles suspected of having been contaminated should be segregated pending a check in accordance with the paragraph below.

Any personnel and/or equipment which may have become contaminated should be examined before being returned to their station or resuming normal duty. Checking for contamination will normally be done at the time of the incident, and guidance will be given to the officer in charge before fire appliances leave.

3. FIRE-FIGHTING PROCEDURE

Any fire-fighting operations are by their nature an emergency, and if radioactive materials are involved the emergency may be more acute, particularly if rescues are to be carried out. It is accepted that in emergency some risks have to be taken and it may therefore be impracticable to prevent all fire-fighting personnel from being exposed to radiation. However, a study of the risks in different circumstances as detailed in later paragraphs combined with the knowledge he obtains from prior liaison with local officials will enable a fire brigade officer to have some appreciation of the risks involved in any operation, even when no officials are present who are qualified to give advice. Unless, therefore, the risks are known to be slight, he should act in the light of this knowledge and bear the following points in mind:

(a) Only the smallest essential number of men should be employed on any duty involving radiation exposure.

(b) Men so employed should be removed from the radiation area immediately it becomes possible to do so.

(c) Operations should be conducted as far as possible from behind any available cover which may provide shielding from radiation.

(d) The possible need for a contamination control zone should be borne in mind.

(e) Control should be exercised over any water used for fire-fighting (and which might therefore contain radioactive particles or dust) to direct it into the drains and thus protect members of the public from direct contamination. The Medical Officer of Health should be warned when this has been done.

(f) Scrupulous care should be taken after the operation is over to ensure that men, equipment and appliances have not been, or do not remain, contaminated.

(g) It is essential that burning materials, among which radioactive materials might be included, should be disturbed as little as possible in the course of fire-fighting. For this reason, it is always preferable to use spray branches as opposed to jets. 'Turning over' and 'cooling down' operations should similarly be conducted with care to avoid the undue disturbance of debris (which should not be touched with the bare hands) until the radioactive source, if one be present, has been identified. When this has been done it should be left alone. Salvage operations should not be undertaken in the vicinity of any radioactive material until guidance has been given by appropriate authorities that it is safe to do so.

(h) In the course of fire-fighting operations in these circumstances it should be remembered that the containers of radioactive sources are not themselves combustible but might be broken open or melted by the heat of the fire. The materials themselves (if they can be identified) should therefore be disturbed as little as possible by the action of spray branches or jets, which should be kept clear of the sources themselves and used only on surrounding combustible materials.

(j) It is important that an informative message be sent back at an early stage from any fire in which the presence of radiation is suspected. A form of wording which will cause as little public alarm as possible is preferable, and it is recommended that only the two words 'Radiation Suspected' be added to the text of the message.

Industrial, medical and other applications

1. ATOMIC ENERGY FACTORIES AND NUCLEAR RESEARCH ESTABLISHMENTS

(a) Radiation and fire hazards

Most fire hazards are the same as those found elsewhere in industry and in research establishments, such as from electrical apparatus and flammable solvents. It should be emphasised that only limited areas of establishments present radiation hazards; these are clearly marked and have been given special attention to reduce the risk of fire. Potential radiation risks associated with fire-fighting vary greatly from one establishment to another, and although the amount of radiation in certain areas may present a very severe hazard to the uninitiated in the absence of radiation safety staff, with the guidance of the latter fire-fighting can be undertaken without risk of harm to those involved. In addition to the general risks of fire-fighting these establishments may present some special hazards, on which some guidance is given in the following paragraphs.

A number of the metals handled are pyrophoric (*i.e.*, liable to spontaneous combustion), powdered uranium and thorium being the most common; fortunately these two materials are by nature only slightly radioactive and present little radiation hazard to those engaged in fire-fighting. (*See* Section 7: 'Fires involving Metals'). If, however, the metals have been irradiated in a reactor, the presence of fission products will make them extremely active and guidance from safety staff will be necessary during fire-fighting. Another pyrophoric material, plutonium, is handled in a number of establishments; the fire risk is normally minimal as the metal is always handled in an inert atmosphere, such as argon. However, should the material escape into the air and catch fire it is so extremely dangerous that fire-fighting operations would have to be undertaken under the most stringent protective control. Other more flammable substances, for example liquid sodium and sodium-potassium mixtures, are in use in some establishments but are at present only rarely associated with radioactivity. Flammable solvents are also employed but in only one or two establishments is a radiation hazard associated with their large scale use.

The release of energy in nuclear reactors is obviously a potential source of fire and great attention is paid in the design and operation of reactors to minimise the risk of an uncontrolled release of energy. Fire-fighting in the vicinity of a reactor must be undertaken only under the guidance of both reactor-operation and radiation-safety personnel.

High-voltage machinery for producing radiation does not in general present a major radiation hazard in the event of fire as the radiation would be switched off. A small amount of radioactivity may remain within the shielded parts of some machines but the radiation from this is unlikely to be more serious than that of sources commonly encountered in industry (*see* paragraph 3 below, 'Factories').

Some establishments contain large quantities of *beta-gamma* active substances, either in the form of fission products arising from the chemical treatment of spent fuel elements, or as materials intentionally produced by irradiation in nuclear reactors. These substances may be undergoing treatment to convert them to a form suitable for use—for despatch to hospitals or to industry, for example—or for storage or disposal as waste materials. The advice of radiation safety staff should always be obtained at the outset on the precautions needed for fire-fighting, as these will depend on the radioactivity of the material involved.

(b) Fire-fighting procedure

All establishments have their own fire services or fire-fighting teams, liaison between establishments and the fire brigade exists to ensure proper protection of fire-fighting personnel. Although an outbreak of fire may have no associated radiation hazard, close co-operation with establishment officials is essential during operations as radiation sources are not necessarily self-evident. In buildings which present a very high radiation risk it is vital to safety that the advice of radiation safety staff on fire-fighting should be obtained at the outset.

No general guidance on fire-fighting techniques is possible as these will depend on the particular materials and conditions encountered. It is, however, essential that strict operational discipline be maintained, and, in particular, that warning signs and notices be respected and the advice of radiation safety staff be heeded.

2. LABORATORIES

(a) Radiation hazards

Laboratories vary enormously in size from single rooms to large blocks, so that it is impossible to discuss them other than in very general terms. It may be stated, however, that apart from a few highly specialised laboratories coming within the control of statutory authorities, the quantities of radioactive materials handled at any one time are generally quite small and the consequent risks in fire-fighting similarly small. Any exceptions to this rule would be subject to the same detailed control as in the case of the specialised establishments referred to in (1) above.

Many of the laboratories in which radioactive materials are used are merely specialised chemical laboratories and the presence of small quantities of radioactive material, normally in solution, will add little to the hazards of fire-fighting. Some laboratories use radioactive materials in the solid form, but as these are frequently in hermetically sealed capsules (*see* Fig. 40), and are normally in some form of protected enclosure, the hazard will again be small.

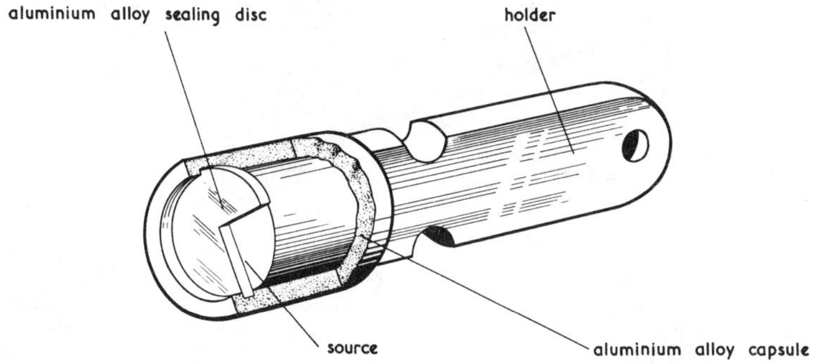

Fig. 40. One of the many methods of sealing a radioactive source into a capsule for industrial use. This particular example is about an inch long.

(b) Fire-fighting procedure

In view of the diversity of the types of the laboratories and of their work with radioactive materials, it is not possible to specify a detailed procedure. The importance of pre-fire liaison with the laboratory authorities and the setting up of a procedure to be followed in the event of a fire cannot be over-emphasised. This procedure should cover any special precautions which may need to be taken at the outset before the arrival of a member of the laboratory staff capable of giving expert advice.

3. FACTORIES

(a) Radiation hazards

The industrial use of radioactive materials is widespread and increasing, but in general the nature and quantity of the radioisotopes involved and the methods of their storage and use are not such as to present severe problems in fire-fighting. Common examples of the use of radioisotopes in factories are of curie amounts of cobalt, iridium, caesium and thulium in the radiography of castings and welds, of millicurie amounts of thallium and strontium, in instruments for thickness gauging of metals and plastics *see* Fig. 41 and 42), and in devices for static elimination in the textile

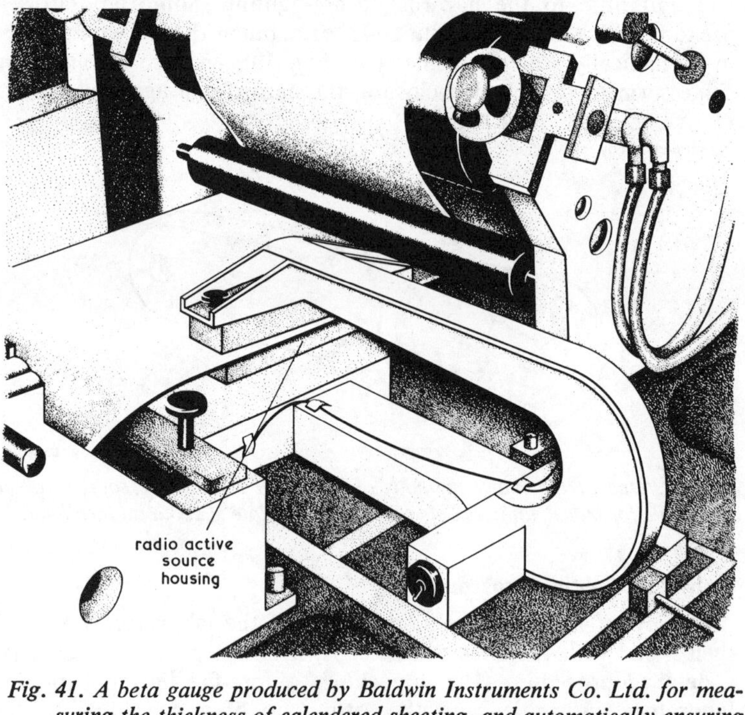

Fig. 41. A beta gauge produced by Baldwin Instruments Co. Ltd. for measuring the thickness of calendered sheeting, and automatically ensuring uniform thickness.

and printing industries, and of microcurie (1,000,000th curie) amounts of up to perhaps fifty different isotopes (mostly short-lived) for tracing the flow or transformations of material in works plant, or in experimental laboratory apparatus. Some naturally radioactive material also finds industrial uses, in particular the use of radium luminous compound for luminising watches, clocks and aircraft instrument dials and of thorium compounds in the gas mantle industry.

The *Factories (Luminising) Special Regulations* are in operation and further Regulations have been made under the *Factories Acts*

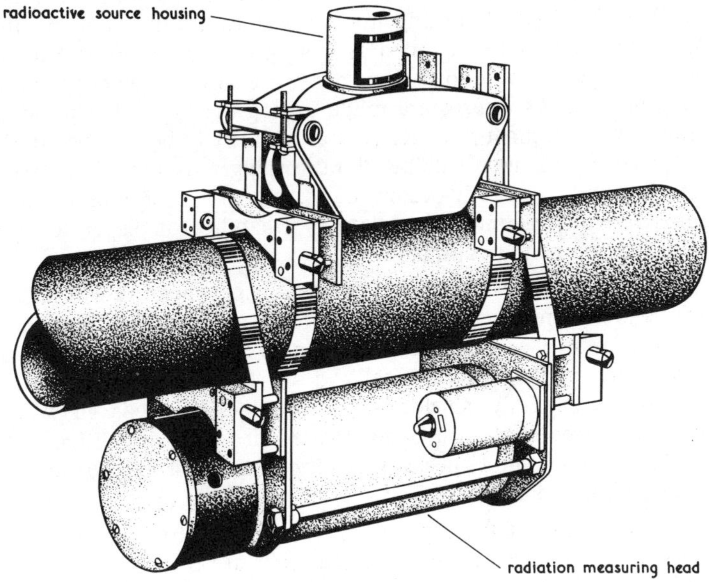

Fig. 42. A fluid density gauge, made by Ekco Electronics Ltd., using a radioactive source to measure the density of a liquid passing through a pipe.

which require that radioactive material is safely stored, transported and used within factory premises. The Factory Inspectorate will enforce observance of these regulations in the usual way. The radiation hazards in most circumstances will therefore be small. Against this, however, is the fact that radioactive materials are liable to be found in any building of the premises rather than confined to particular hazardous areas, though this does not normally apply to gamma-radiography or to luminising, which are usually carried out in special areas or enclosures marked by warning notices.

(b) Fire hazards

There are no fire risks as such from the presence of radioactive materials. The hazard is confined to the possible presence of radioactive sources in a fire of a nature normal to the surroundings. Most sources are enclosed in a metal capsule of foil which cannot easily be disrupted by fire, though in some cases the covering can be decomposed or melted. The capsules of those which are capable of creating a considerable external radiation hazard—the *gamma* radiography sources—are manipulated and transported in thick metal shielding pots. These protective containers bear orange-coloured markings and the British Standard trefoil warning notice (*see* Plate 31). After working hours they are usually kept in safes (which can usually be regarded as having some hours of fire resistance) though on rare occasions they are left in their operating positions while exposures are made during the night. It should be emphasised that most radioactive sources likely to be involved in the course of fire-fighting will be of little danger as regards external radiation, and could only become dangerous (from contamination by radioactive dust or internal radiation) if handled or dispersed after the disruption of the container.

A limited number of firms is at present experimenting with multi-curie sources (mostly of cobalt) for the irradiation of foodstuffs, chemical and pharmaceutical materials, etc. In all cases these sources are installed in specially designed and fire-resisting irradiation cells, fully shielded and interlocked. Cells in which combustible materials are being treated are normally fitted with sprinkler systems, and special precautions are taken to reduce fire risks in adjoining areas.

The only class of factory outside the Atomic Energy Establishments where unsealed radioactive materials are present of a nature and in amounts sufficient to constitute a possible hazard to firemen are the radium luminising factories. Though only millicurie amounts, at most, are in use during working hours as mixed or unmixed luminous compound or in the form of luminised dials being dried off, radium is a particularly dangerous radioactive element if it gets into the body. After working hours all stocks of luminous compound are normally stored in iron safes, which are frequently lead-lined.

(c) Fire-fighting procedure—before the call

Close liaison should be maintained with managements on such matters as the marking of rooms, storage, access, etc., and the making of predetermined plans for fire-fighting. This applies particularly

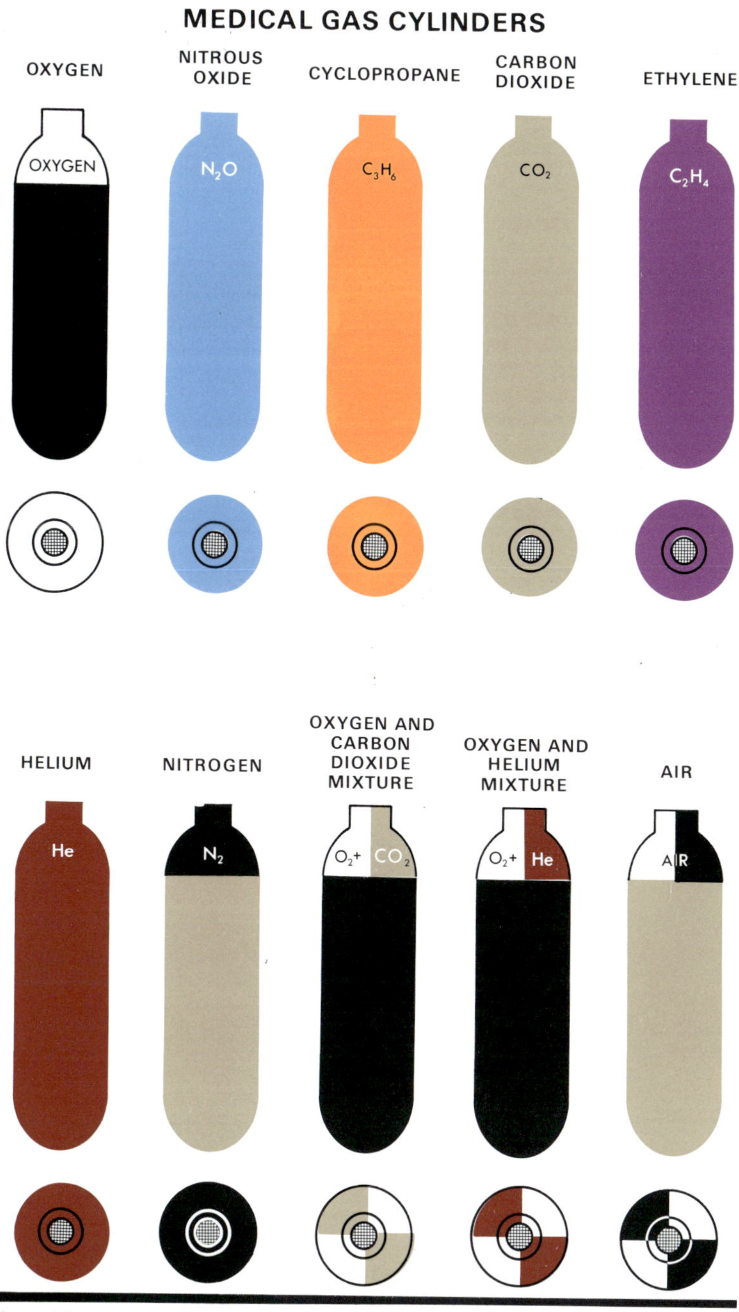

Plate 42. British Standard colours for medical gas cylinders. (In accordance with B.S. 1319: 1955.) (Section 16)

IDENTIFICATION COLOURS FOR GAS CYLINDERS

Plate 43. British Standard colours for commercial gas cylinders. (In accordance with B.S. 349: 1932.) (Section 16)

IDENTIFICATION COLOURS FOR GAS CYLINDERS

Plate 44. British Standard colours for commercial gas cylinders. (In accordance with B.S. 349: 1932.) (Section 16)

SPECIAL GASES OR MIXTURES

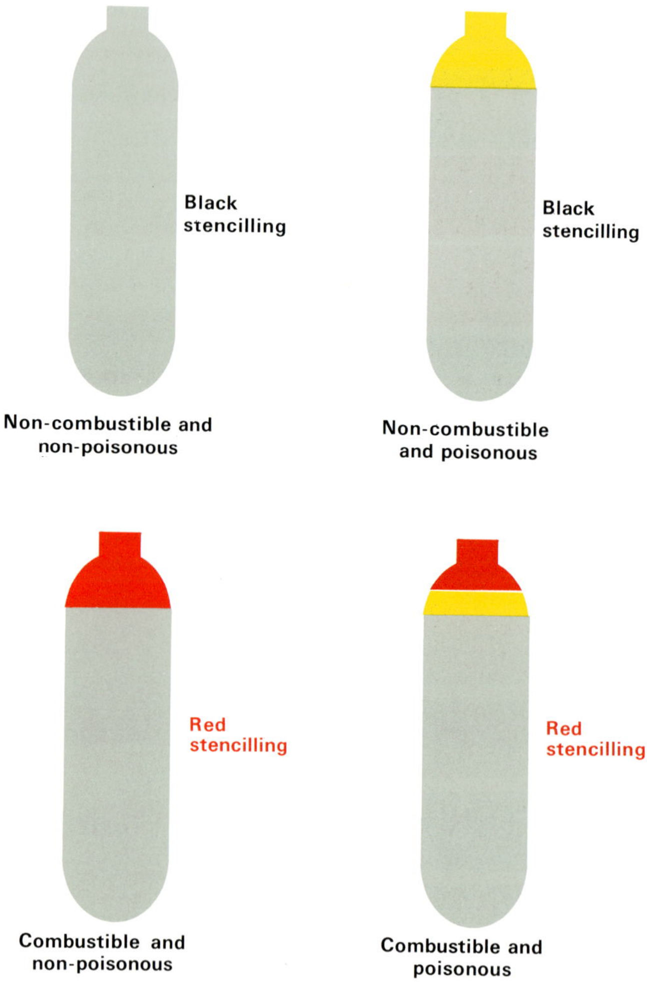

Areas of grey in illustration denotes silver paint on cylinders

Plate 45. For gases or mixtures of gases with no special colour allotted. The name of the gas or the words "Special Mixture" is stencilled on the body of the cylinder. (Section 16)

to firms using strong sources and to the radium luminising factories mentioned above.

(d) Fire-fighting procedure—at the time of the call

If predetermined plans have been made they should be carried out. If no plans have been made and the presence of radioactive isotopes is suspected, fire-fighting should be carried out with care to avoid undue disturbance of burning material, spray branches being used in preference to jets where possible. Unless the hazard is known to be slight, breathing apparatus should be worn at all radium luminising factories and at other factories if it is believed that an isotope capsule may have been broken open; and the appropriate protective measures detailed in paragraph 2(c) pages 230/3 should be applied. The debris should not be touched and contact should be made with factory officials as soon as possible to determine the location, and ultimately the disposal, of the radioactive sources.

4. HOSPITALS

(a) Radiation hazards

A radiation hazard additional to normal fire risks can arise in those hospitals where radiotherapy treatment is carried out and exceptionally in a very few others.

There are three types of treatment and of related appliances to be considered, namely (i) teletherapy units (using radioactive sources) (ii) radium containers and (iii) liquid radioisotopes.

(i) *Teletherapy units.* The units contain from ten to upwards of 1,000 curies of radioactive cobalt (or similar strengths of other isotopes, such as caesium). Most sources are permanently housed in a massive steel-encased lead treatment head fitted with various safeguards, though some of the weaker sources, when not in actual use, are transferred automatically to a thick-walled lead safe. Unusually high temperatures would be needed to disrupt or expose these sources. As the units are necessarily housed in rooms having unusually thick walls of concrete or brick, such structural shielding would form an effective heat shield against fires originating outside the rooms. There is insufficient combustible material within the treatment rooms to promote a serious fire.

(ii) *Radium containers.* The radium is hermetically sealed within metallic tubes or 'needles' which may be from 12·5 mm to 75 mm long and may have individual strengths of up to fifty millicuries. When not in use they are stored in a specially adapted safe having massive lead walls. There is therefore a negligible risk of disruption or exposure of the containers as a result of a fire. Their use in treatment is subject to rigid control and patients having radium containers within or on their bodies are treated under special day and night supervision by the nursing staff. Occasionally radium containers may be kept for a few hours in a ward safe, which is usually a small box with thick lead sides kept in the vicinity of the radium ward.

(iii) *Liquid radioisotopes.* Radioactive liquids may be administered to patients under treatment, the principal agents being radioactive isotopes of iodine, gold and phosphorus. Iodine probably accounts for the highest strengths likely to be stored in hospitals in this form and it is only at the storage sites of these liquids that any risk additional to those of normal fire-fighting is likely to be encountered. Isotope storage safes are fully protected by concrete or other enclosures designed to reduce the external radiation hazard to a negligible level. The liquids are stored in glass vessels which may have rubber or cork stoppers. The contents could therefore be discharged from their containers if subjected to only moderately high temperatures.

(b) Fire dangers

Fire risks may be considered low in the case of teletherapy installations and in the vicinity of main radium storage and liquid isotope storage safes. A higher fire risk may be encountered in the vicinity of radium wards, although it is usual to place a ward radium safe remote from readily combustible material.

(c) Fire-fighting procedure—before the call

Hospitals equipped for teletherapy treatment maintain a twenty-four hour service of trained personnel who are qualified to advise and guide fire brigade personnel in fire operations.

Liaison should exist with such hospitals with a view to (i) setting up personal contact between officers of the fire brigade and teletherapy staff, (ii) making predetermined plans for fire-fighting in the light of the position and type of treatment rooms and storage safes and (iii) the use and placing of radiation warning notices.

(d) Fire-fighting procedure—at the time of the call

Rooms marked with the B.S. trefoil radiation warning notices (*see* Plate 31) should only be entered with the authority of, or by prior agreement with, teletherapy officials, whose guidance should be followed in all respects. Since the treatment heads of teletherapy units are composed entirely of non-combustible materials with the exception of a small quantity of electrical insulation, there is no question of the treatment machinery burning, and fire-fighting should be confined to other materials. Where a caesium unit is involved, there is some advantage in the use of CO_2 since there is a risk of the caesium source disintegrating if the case splits due to water accidentally reaching it in quantity while it is at a high temperature. (With cobalt units this consideration does not apply.) Spray branches rather than jets should be used, and, unless the hazard is known to be slight, breathing apparatus should be worn. Gloves and other protective apparel should be worn if necessary, and the teletherapy staff will be in a position to monitor personnel for radiation dangers if this is necessary. Should a fire occur in or near to a radium ward, the ward radium safe, if known to contain radium, should if possible be carried to a place of safety. If, however, the fire precludes access to a ward safe known to contain radium, its immediate vicinity is liable to become contaminated, and this area should be treated as if it were a contamination control zone. Should a serious fire occur in close proximity to a liquid isotope storage site, breathing apparatus should be worn unless advice to the contrary has been given by a radiation safety officer.

5. TRANSPORT

(a) Radiation hazards

Transport of radioactive materials may be by road, rail, sea, or air, and very weak sources may be sent by post: most shipments involve relatively small quantities of radioactivity, these quantities being controlled by the transport authority concerned. In future, however, moderately strong sources may be conveyed by public transport, particularly by rail, but shielding containers for these sources have been designed to prevent damage in the event of collision, fire or other accident.

In addition to the regular shipment of isotopes, and of radioactive materials of the same order of activity, the United Kingdom Atomic Energy Authority occasionally transports large radioactive sources by road. These journeys are planned as special operations: the driver and his mate are trained in the necessary emergency procedure and special shipping containers of very massive construction, designed to withstand impact and fire, are employed. The vehicle is escorted by another which carries a qualified person and radio inter-communication is usually maintained. Fire-fighting apparatus is included in the emergency equipment carried in both vehicles.

(b) Fire hazards

The presence of radioactive sources does not increase the inherent fire risk associated with the particular vehicle involved, unless the source material itself is flammable or pyrophoric. Materials in these categories are transported in special sealed containers designed to minimise the risk. A vehicle fire which involves a small radioactive source would not present a radiation hazard to firemen during fire-fighting, even if the container was damaged or destroyed. Larger sources will be carried only in containers which are designed to withstand accident and fire conditions and should therefore present no hazard to those engaged in fire-fighting. The very strong sources transported by the United Kingdom Atomic Energy Authority are provided with massive shielding (250 mm of steel, and lead, where they emit *gamma* radiation), and a qualified person is always present to give guidance on safety procedure.

The spread of radioactive contamination is unlikely under any circumstances to give rise to hazard during fire-fighting, but salvage should not be undertaken unless a qualified person is present. To assist identification of radioactive material, the containers of isotopes dispatched from the Atomic Energy Research Establishment, Harwell, and the Radio-Chemical Centre, Amersham, carry a written indication of radioactivity, and bear orange markings. All Atomic Energy Authority vehicles transporting radioactive substances carry the B.S. trefoil warning notice in a prominent position in the driver's compartment (or in the interior, in the case of cars).

(c) Fire-fighting procedure

When any information coming to the notice of the officer in charge of any incident leads him to suspect that radioactive sources are present as part of the load of a vehicle, other than one operated by the U.K.A.E.A., he should carry out fire-fighting operations as required. He should adopt cautionary measures as outlined for radioactive sources in industry, except that, if a fire involving a small source is in the open air, breathing apparatus will normally be necessary only if approach has to be made through the down-wind area.

Similar precautions should be taken in the event of a vehicle of the U.K.A.E.A. being involved. In the event of a fire involving a large source, either the crew of the vehicle or that of the escort vehicle will give guidance on safety procedure. The presence and position of radioactive materials may be determined from the warning plate in the driving compartment.

In any case of fire in a vehicle in which the presence of radioactive material is suspected, an informative message should be sent back including the words 'Radiation suspected.' On receipt of this message at Brigade Control the Medical Officer of Health should be immediately informed. In the event of the crew of the U.K. A.E.A. vehicle being incapacitated and unable to send a message to the Authority, this should be done at the first opportunity and in accordance with instructions displayed in the vehicle.

Part 6c, Chapter 45, Section 12

Refrigeration plant risks

REFRIGERATION plants present the fireman with two distinct types of risk. The first of these is the personal danger from poisoning or asphyxiation in the irrespirable atmosphere likely to be produced when a fault or accident causes a leakage of refrigerant. This emergency is likely to occur with all types of refrigeration plant from the fairly small installation of a retail shop (quite commonly found in the basement) to the very extensive and possibly rambling cold storage warehouse found in dock or market areas. Such incidents are not commonly associated with fires, but the fire brigade is frequently called to them because of their ability to work in dangerous atmospheres through the possession of breathing apparatus. (It should be noted, however, that fires in the motors of refrigerators, even in dwellings, are not uncommon, though in these cases leakage of refrigerant is not usually a factor).

The second type of risk is the difficult and dangerous fire which may occur in an extensive cold storage warehouse of old construction, in which the congested and inaccessible interior may produce a smoky fire which is hard to locate and extinguish and in which fire-fighting conditions may well be at their worst (*see* Plate 39).

In both these types of incident it is essential for the fireman to know something of what to expect and of the peculiarities of the work and the situation with which he may be concerned. A similar problem, with added difficulties of its own, exists in insulated ships, and these are described in detail in Part 7 of the *Manual.*

Refrigeration plants

Refrigeration plants vary greatly in design and construction. It is therefore impossible to describe a typical layout. The principles and the physical laws which underlie refrigeration can be found in any textbook on physics. This section is, therefore, confined to a description of the principal refrigerating systems and some of the steps which the fireman may take at a fire or accident involving refrigerating machinery.

1. PRINCIPLES OF REFRIGERATION

Refrigeration plants may be of the non-continuous or continuous types.

(a) Non-continuous

Two examples of this type were the domestic ice-box and the ice-cream tricycle.

(i) *The domestic ice box*. The materials to be cooled were placed in a suitable container surrounded by ice, and the whole was insulated from the atmosphere. The ice absorbed heat from the material in the container; in so doing, the ice melted and had to be renewed from time to time.

(ii) *The ice-cream tricycle*. In this case carbon dioxide snow was used instead of ice. The snow, which evaporated as it absorbed the heat from the material to be cooled, also had to be renewed.

(b) Continuous

In mechanical refrigeration the refrigerant is made to continue its service as a cooling medium by causing the refrigerant to absorb heat from the material to be cooled and then extracting the heat from the refrigerant, which can thus be made to serve continuously. The refrigerants which enable this to be done are certain gases which have a very low boiling point at normal pressure, the most usual of which are ammonia, sulphur dioxide (now obsolescent), methyl chloride, carbon dioxide and dichlorodifluoromethane ('Freon'). In all cases the general principles of the operating plant are the same.

The characteristics of the first four of these refrigerants are discussed in Section 16 on chemicals. Freon is asphyxiating, but non-toxic and non-flammable.

2. PARTS OF A REFRIGERATION PLANT

The essential parts of a refrigeration plant, excluding the piping system, are the regulator, the evaporator, the compressor, and the condenser.

Commencing at the regulator with the refrigerant in liquid state, at high pressure and normal temperature, the liquid flows through the regulator (at a rate determined by the degree and quantity of cooling required), whence it reaches the evaporator, which consists of a series of coils placed in the space or medium to be cooled. Here a change of state occurs, the liquid boiling under the influence of reduced pressure into a vapour and absorbing heat in the process. The refrigerant continues its circuit in vapour form, having gained a certain amount of heat, and reaches the compressor. Here the vapour is compressed to a higher pressure and as a result its temperature is also increased; the gas therefore leaves the compressor at high temperature and high pressure. It then reaches the condenser in this state, where the heat in the vapour is absorbed by the cooling medium, which circulates over the coils of the condenser. This medium may be either air or water. On being cooled the refrigerant which is at high pressure liquefies, and returns to the regulator to begin the circuit again. In some installations a liquid receiver is placed between condenser and regulator to act as a reservoir.

3. METHODS OF REFRIGERATION

There are four main methods of applying these principles to actual refrigeration systems:

(a) Direct expansion

By this method the refrigerant is evaporated in coils of piping placed in the space to be cooled, and the evaporator is, therefore, actually in the cooled space. In this case the refrigerant is encountered throughout the whole of the cooling system either as a liquid or a vapour. Plate 33 shows the principles, in diagrammatic form, of the direct expansion method.

(b) Brine circulation

In this case the cooling is effected by circulating brine through the spaces and then carrying the brine over the coils of the evaporator where heat is extracted from it. The brine therefore conveys heat from the spaces to be cooled to the refrigerant circulating through the evaporator. Brine is used as the conveyor because of its low freezing point, the density being varied according to the degree of cold required. A separate pumping system provides the brine circulation.

(c) Air-cooled

Air-cooled refrigeration is used in the case of commodities which do not require a high degree of cold, such as fruit, vegetables and butter.

In general, air is circulated through the space to be cooled by means of a power-driven fan, this air being at some point forced over a series of coils in which it becomes chilled. If the coils are direct expansion in type, they will themselves contain the refrigerant. If, on the other hand, brine circulation is employed, the coils will carry brine which has been chilled by the evaporator in the refrigerating machinery room.

The air used in an air-cooled system may be either dry or moist as follows:

(i) *Dry*. If the cooler is of the dry type, the air is passed over the coils. Since the moisture-carrying capacity of the air decreases with its temperature, moisture condenses on the coils. If the temperature is low enough, snow or ice is formed. This necessitates periodical de-frosting.

(ii) *Wet*. If the cooler is of the wet type the coils are mounted in a bath filled with brine. A pump picks up brine from the bath and lifts it to a suitable distributor above the coils, whence it trickles down over the coils as the air is forced or drawn over them. The moisture deposited by the air as it cools is absorbed by the brine, which for this reason gradually decreases in density.

(d) Absorption

This is the method used in the domestic gas refrigerator, and in some electric types, and will seldom be met with on a large scale.

Liquid ammonia is evaporated in the evaporator, thus providing the refrigeration. The ammonia then passes to the absorber, which is a receptacle containing water. Sufficient quantities of hydrogen gas are present in the system between the evaporator and the water to maintain a slight pressure. It has no effect on the ammonia, which is absorbed by the water, but the pressure of the hydrogen causes the water to flow to the generator. Heat is applied by means of the gas flame, which drives the ammonia off as a gas from the water. Some of the water in the generator is returned by thermosyphonic action to the absorber to continue the action. The ammonia gas, at high temperature, goes from the generator to the condenser, where the heat gained in evaporation and from the gas flame or electric heating element is dissipated and the ammonia is re-liquefied for return to the evaporator. In this type of refrigerator there are no moving parts.

The principles of the absorption method of refrigeration are shown diagrammatically in Plate 34.

4. ACCIDENTS INVOLVING REFRIGERATION PLANT

An accident involving a refrigeration plant may result in the refrigerant escaping, either in liquid form or directly as gas. If it escapes in liquid form, it will immediately begin to vaporise. This gas or vapour, which is sometimes toxic, makes the use of breathing apparatus essential: heavy concentrations of ammonia can attack the skin: contact with a liquid refrigerant may also cause severe burns.

(a) Small commercial plants

This type of plant is found in butcher's shops, dairies, hotel kitchens, etc. If refrigerant vapour is leaking into the atmosphere surrounding the plant the first action is to break the main electrical circuit in order to put the compressor out of action. The fact that the compressor is not running does not necessarily mean that it is out of action, as the control is thermostatic and may operate at any time. All the valves on the system should be closed to prevent any further circulation, and immediate steps should be taken to ventilate the premises. In general, the use of water from jets or sprays should be avoided, owing to the further damage that may result.

If liquid refrigerant has escaped and it is necessary to use water to swill it away, care should be taken to see that it is well diluted before it is put into the sewers: certain refrigerants, such as sulphur dioxide, can produce a most unpleasant and corrosive liquid.

Fatalities are unlikely to occur, but the fire brigade officer should inquire whether people have been injured or affected by the escape of the refrigerant, so that ambulances can be sent for, if necessary, and first-aid treatment can be given before their arrival.

(b) Large commercial installations

This type of installation is found in large cold stores, meat safes, etc.; on docksides; in large markets; and in such processes as brewing. Fatalities are more likely to occur. A member of the engineering staff is usually available when an installation of this kind is operating, and the fire brigade officer should consult the engineer before attempting to close the valves, giving orders for the laying out of a charged line of hose with diffuser nozzle at the same time. In the case of a heavy escape of gas, the point of escape should be sought. Ammonia is generally used in all large compression-type refrigeration plants: due the the lightness of the gas and its irritating effects upon the eyes and skin, a small leakage will usually soon make itself apparent by convection through the building. People are therefore forewarned, and are able to make a quick escape if necessary. In a major breakdown, however, involving a severe leakage of ammonia gas, they may be overcome before they can escape and may be asphyxiated if they are not immediately removed. The fire brigade officer should therefore always enquire *immediately* on arrival (at a plant using ammonia or sulphur dioxide as the refrigerating medium) whether all persons are out of the building and accounted for, and if not, where they are most likely to be found and what is the quickest route to them.

If a joint is merely leaking, it should be possible to tighten the flanges, fit a new bolt, or take any other necessary steps with the assistance of one of the engineers. Where the leak is due to a fracture or burst, the damaged part of the circuit should be isolated by closing the appropriate valves under the direction of an engineer. The valves should not be operated in the absence of an engineer. Much damage may result from indiscriminately closing the valves; the plant need not necessarily be closed down as a whole merely because damage has occurred at one point. In some cases it would be advantageous to keep the compressors working.

The premises should be ventilated as soon as possible. The staff will advise upon the use of the permanent ventilating system. Fire brigade personnel should, wherever possible, wear protective clothing and full facemasks with breathing apparatus when dealing with severe leakages of ammonia, to prevent burning to the face and body by the fumes. Water should not normally be used if gas has escaped, unless the necessity for using it outweighs the damage it may do. Should liquid refrigerant escape, however it can be diluted with water spray. Before the solution flows into any drain however, it should be thoroughly diluted, and if there is a large quantity the Local Authority should be notified.

5. FIRE INVOLVING REFRIGERATION PLANTS

The following points should be noted:

(**a**) The heat generated by a fire reaching the installation will cause a rapid change in the state of the refrigerant, and an increase of pressure both in machinery and in gas lines. This increasing pressure may cause joints to blow out, or pipes and vessels to burst, thereby releasing the refrigerant. Some refrigerants are flammable and/or explosive, and present a risk to personnel working on the fire. Danger arises when certain parts of the plant, notably the condensers, liquid receivers and separators, are located on the roof. There is also a danger in the machinery space itself, where electrical apparatus or live steam pipes may be found.

(b) Fire is unlikely to break out in the cold spaces if the plant is in operation, though every effort should be made to prevent fire which starts elsewhere spreading to them. The fire brigade officer should find out whether the system depends on brine circulation or direct expansion, as refrigerant is more likely to be released into the fire than brine; this consideration may effect the placing of branches. With the brine system, the main hazards will probably be confined to a fairly compact machinery space unless condensers, etc. are situated on the roof.

(c) Any part of the plant which is affected by the heat should be cooled with water spray even though it is not directly involved in the fire.

(d) Breathing apparatus should be used as the atmosphere may possibly become unbreathable. Danger still exists even if the gases are carried away by convection currents created by the fire itself (see 4(b) above).

Fires in cold stores

1. CHARACTERISTICS OF BUILDINGS

The inherent characteristics of the older type of cold storage buildings can produce very difficult problems in fire-fighting, though in stores of more recent construction the situation is not likely to be so bad.

As far as the fire problem is concerned, the characteristics of these older buildings may be summarised as follows:

(a) Difficulty of access;

(b) Facilities for fire spread;

(c) Inadequate ventilation.

(a) Difficulty of access

In many refrigerated stores of older construction the accommodation consists of a number of relatively small compartments in order to assist in the maintenance of low temperatures. These compartments may be difficult of access through the absence of windows and through very high and over-full stacking, sometimes of combustible containers, and sometimes also severely restricting movement in corridors and doorways. In smoke-logged conditions the task of making one's way through corridors and doorways, of identifying one room from another and of discovering the seat of a fire may be difficult and dangerous in the extreme.

(b) Facilities for fire spread

The compartmentation mentioned above, moreover, does not always involve adequate fire separation between one compartment and another, and the stacking of flammable materials in corridors, etc., provides further opportunity for fire spread throughout the whole area. Furthermore, there may in older buildings be much timber in the construction, and both the insulating linings (*see* Fig. 43) and the wooden or hardboard sheets which cover them may tend to a rapid spread of fire. This will also be aided by ducting and

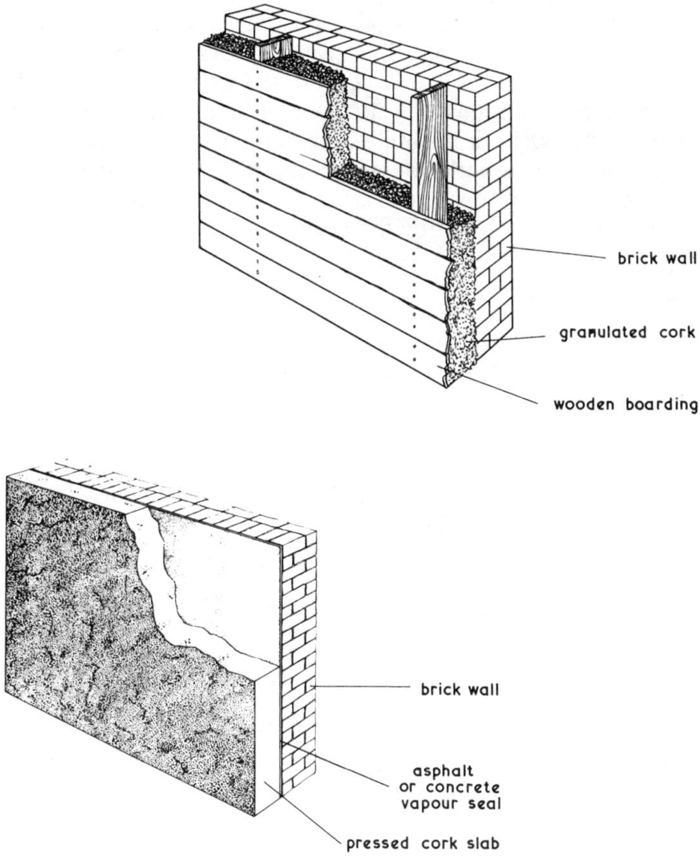

Fig. 43. (ABOVE) *In many converted and old refrigerated stores the wall insulated lining is granulated cork enclosed in wooden boarding.* (BELOW) *In a building of modern construction pressed cork slabs may be used with a vapour seal of asphalt or cement.*

trunking, much of which may itself be constructed of combustible materials in this type of building (*see* Plate 40).

(c) Lack of ventilation

It is essential to the function of a compartmented cold storage building that there should not be unlimited free transfer of fresh air from one part of a building to another or through the building as a whole. Access of fresh air, must, indeed, be strictly limited, and this is often so much a feature of the building's construction that it is virtually impossible to achieve ventilation at the time of a fire. Trunking and ducting provide for such a transfer of air as is required for the normal working of the building, but this will nearly always be on a very limited scale, and the fireman or fire officer will usually have to resign himself to working in breathing apparatus in conditions of very poor visibility and often of high temperature. When fire-fighting operations have reached the point when ventilation is both safe and necessary, it will frequently only be achieved by improvisation, i.e., by cutting or breaking openings in walls and ceilings and by the use of pavement lights, etc.

2. RISKS

It is frequently impossible to ascribe a fire in a cold store to any particular cause: surreptitious smoking and electrical faults are, more often perhaps than any other, thought to be the cause of such fires, but circumstances usually prevent investigations from being conclusive. The fire risk in a well-maintained store, especially in those of newer construction, cannot be said to be high; these fires are therefore comparatively rare—they are more common on board ship than on land—but once a fire has attained a hold it is often exceedingly difficult to effect a stop. In meat and poultry stores the fat content of the goods produces a fierce, tenacious and smoky fire accompanied by considerable heat; in types of storage such as vegetables or fruit, moreover, in which case the goods themselves are not readily combustible, the containers may be highly flammable and may constitute a considerable fire load.

The risks to firemen arising during fire-fighting are (a) that much work will have to be done in an irrespirable atmosphere in conditions of bad visibility and great heat, with the attendant dangers of becoming lost and of losing touch with other personnel; (b) that, in the older buildings, concealed spread of fire may lead to individuals or parties of those engaged in fire-fighting to be cut off, and (c) that, even in parts of the building not seriously affected by smoke or heat, the unexpected release of refrigerant vapour due to damage to the plant or piping may produce casualties. In the larger commercial plants, the commonest used refrigerants are ammonia, carbon dioxide or sulphur dioxide: ammonia is considered by some authorities to be flammable in relatively heavy concentrations but neither of the other two are flammable at all. The toxic and/or asphyxiating qualities of all three are referred to in the list of chemicals in Section 16.

3. FIXED FIRE PROTECTION

Neither sprinklers nor any other form of fixed fire protection are frequently found in cold storage buildings. This is no doubt partly due to the practical difficulties of installation and operation in sub-normal temperatures, but no doubt partly also to the general impression that the risk from this type of occupancy does not justify such installations.

4. FIRE-FIGHTING

Cold stores are pre-eminently a type of risk in which fire-fighting cannot be safely undertaken on a basis of an immediate attack with anything available, such as might be possible and right in a dwelling-house fire or in many of the less complicated industrial fires. A cold store fire must, more often than not, be conducted as a planned campaign, based on the fullest possible knowledge and reconnaissance, and not developed as a result of minute-by-minute assessments of progress and results, since both progress and results may be extremely difficult to judge in practice. The following points are among those to which particular attention should be paid at a fire in a large cold storage depôt, especially in one of older construction:

(a) Liaison should exist between the management and the fire brigade before the fire, and this should embrace the fullest possible knowledge of the plant by the fire brigade and the laying down of agreed procedures as to calling the brigade, access points, control of the plant machinery, ventilation and so on. One of the first actions on the outbreak of fire should be the cutting off, in consultation with the engineer, of the refrigeration machinery and any mechanical ventilation in the section of the building affected.

(b) Stage I or stage II of the Breathing Apparatus Control Procedure should be instituted immediately on arrival, and the highest possible standard of discipline and adherence to procedure should be observed by the breathing apparatus men themselves. In assessing the need for assistance at the fire, officers in charge should bear in mind the punishing nature of this type of fire and take steps to have relief crews available in reserve.

(**c**) Whether or not prior liaison has been established with management so that fire-fighting can follow predetermined lines, the officer in charge is likely to find a confused position in the early stages of the fire, in which it is very unlikely that the whereabouts and direction of spread of the fire can be easily and quickly ascertained. In this situation it is essential that reconnaissance should precede the attack on the fire, and it may often be necessary to await sufficient reinforcements of breathing apparatus men for organised reconnaissance purposes before the attack on the fire can be started. Every effort should be made from the outset to obtain plans, etc. from the management, or at least full information as to exits and entrances, communicating and fire stop doors and the layout of the building generally. The picture so obtained should be supplemented by information brought out by breathing apparatus men on reconnaissance, and the whole should be recorded on sketch plans if circumstances permit.

(**d**) Although the laying out of equipment in readiness, and the covering of surrounding risks, will obviously take place from the first moment of the arrival of the fire brigade, it is not until the officer in charge has, by means outlined above, obtained a reliable survey of the layout of the plant and the situation prevailing that he should attempt an all-out attack upon the fire, and it may be necessary to delay ventilation until the attack can take place. In the course of this, it should be possible to ensure that stops are made in the most suitable places, that crews working in separate places know where and with what object other crews are working and that no unexpected development can take any personnel by surprise. A particular watch should be kept for the obvious dangers from self-closing insulated doors between compartments.

(**e**) The nature of the fire itself will, of course, govern the method of extinction to be employed: where a large body of fire is discovered, hard-hitting jets may well be the best means though sprays have advantages for the absorption of ammonia vapour and the reduction of heat and smoke to improve fire-fighting conditions. If the fire is in a confined compartment giving off dense volumes of smoke, the use of high expansion foam should be considered. Where the fire is unable to develop fiercely due to lack of oxygen jets or sprays will be necessary according to circumstances, and it is certain that much cutting away will have to accompany the work of extinguishment; progress in these circumstances can only be made by ensuring that slab insulation etc. is extinguished in hidden internal pockets as well as on the surface. At the same time, care must be taken to ensure when cutting away that no damage is done to refrigerant lines. Mobile smoke extraction plant may have considerable value in a fire of this nature, though of course, in common with any other form of ventilation, it must not be used without a sound knowledge of the likely effect on the fire of the air currents which it produces.

Part 6c, Chapter 45, Section 13

Fires in resins and gums

VARIETIES OF RESINS AND GUMS

1. **Natural and synthetic resins**
2. **Hard and soft resins and gums**
3. **Origins of resins and gums**

NATURE, PROPERTIES AND RISKS

1. **Nature and properties**
2. **Risks**
3. **Fire-fighting**

LINOLEUM, OILCLOTH AND TARPAULIN MANUFACTURE

1. **Linoleum manufacture**
2. **Oilcloth manufacture and waterproofing processes.**
3. **Tarpaulin manufacture**

THE use of natural resins, which include gums, has been for a very long time widespread in the manufacture of paint and varnish, linoleum, oilcloth, furniture making and other products. They have contributed considerably to the degree of risk which these industries offer; nearly all of them can be called highly flammable, and, since they have the property of melting and flowing at elevated temperatures, they present special problems to the fireman which are similar to those of fats and oils. Fortunately, the use of resins is more or less confined to the industries mentioned above in the sense that they are only likely to be found in bulk in these industries. They are used in relatively small quantities in a much wider variety of industrial processes, but, since the amounts involved are relatively small, these latter industries, which include block-making and process engraving, cannot be said to offer serious risks from resin fires.

With the introduction shortly before the Second World War of synthetic resins on a large scale, the use of resins as a whole has enormously increased. In many cases, notably in the paint and varnish industry, synthetic resins have tended to replace natural resins; on the other hand synthetic resins have tended to create their own new uses (among them are resin-bonded plywoods and surfaced boards of many kinds) and, especially in the more old-fashioned industries, the use of natural resins and gums is still common. Since the properties and manufacturing processes of synthetic resins are largely different from those of natural resins, the former are not dealt with in this Section and have been described fully in Section 10 'Fires involving Plastics'.

Varieties of resins and gums

1. NATURAL AND SYNTHETIC RESINS

Natural resins are products derived from the secretion of the sap of certain plants and trees. Most of the natural juices obtained from the tree contain a number of elements in addition to the pure resin itself, and in nearly all cases they require purification, distillation and other processes before the resin itself is available.

Synthetic resins, on the other hand, are artificial resinous compounds made from synthetic materials by chemical process.

2. HARD AND SOFT RESINS AND GUMS

Some resins are naturally hard at normal temperatures; they issue from the tree in their natural form in a glutinous state, and some of their volatile constituents are evaporated on contact with the air, whereupon they become relatively stiff and, in some cases, brittle. Others are soft; because of the constitution of the natural product, and the smaller proportion of volatile elements in this type, these remain more soft and waxy than hard resins.

Gums are mixtures of resins and essential oils with a minimum of volatile constituents.

3. ORIGINS OF RESINS AND GUMS

The following are typical examples of the main groups of natural resins:

(a) Copals

Congo copal, a yellow or brown natural gum, soluble in benzene, turpentine, chloroform and ether, used for varnishes and linoleum; *amber*, a fossil resin found in the Baltic Sea area, containing resin acid and volatile oils, used as a base for photographic varnishes; *dammar*, white or black, and including *kauri gum*, used in making special varnishes for photographic work.

(b) Lac

These are excretion products of coccid insects in jungle trees, used in the manufacture of *shellac*, a purified product of lac, and a thermo-plastic moulding material which is used in varnish making, photographic work and gramophone record manufacture.

(c) **Balsams**

Canada balsam (*fir balsam*, *Canada turpentine*) is a yellowish liquid, smelling of pine, soluble in ether, chloroform and benzene, used for lacquers, varnishes and adhesives; *Peruvian balsam*, an oleoresin containing esters of benzoic and cinnamic acid; *Tolu balsam*, a similar product to the above.

(d) **Gums**

Gum arabic (*acacia gum*), a fine yellow or white powder, soluble in, and heavier than, water, obtained from some types of acacia tree, used in pharmacy and for adhesives; *gutta-percha* is the coagulated latex of certain trees found in Malaya and the East Indies; it is usually used after removal of the resinous elements, and has particular applications for waterproof electrical insulation; *ester gums* comprise rosin or gums which have been esterified with glycerine. (Esterification is the direct action of an acid on an alcohol resulting in the formation of esters).

(e) **Miscellaneous**

Benzoin is a condensation product of benzaldehyde, which is the oil of bitter almonds, and is used only as a chemical solvent or re-agent. *Dragon's blood* is a resinous exudation from the fruit of palm trees; it is red and normally solid, being soluble in organic solvents; it is used as a pigment in colouring and in process engraving.

Nature, properties and risks

1. NATURE AND PROPERTIES

As has been shown above, resins are in general secretion products from trees, and are either hard or relatively soft; they are in all cases fusible when heated. They are mostly either just heavier than water or just lighter and almost never soluble in water, gum arabic being an exception to this; they are generally soluble in organic solvents. They consist of resinous matter, i.e., highly polymerised acids and neutral substances mixed with terpene derivatives. From their character they are one of the few natural thermo-plastic materials, and this provides the key to their uses in industry as well as to their behaviour in fire.

Natural resins are all flammable, and all are roughly similar in fire risk. When they are heated, essential oils of low flash point (about 220°F. (104°C.) and over) are distilled, and the vapours produced are highly flammable and explosive.

Synthetic resins (*see* Section 10, 'Plastics') are similar in character but are made by condensation or polymerisation of phenol and formaldehyde, formaldehyde and urea, glycerol and phthalic anhydride, etc.

2. RISKS

(a) Burning characteristics

Natural resins are strongly carbonaceous and weak in oxygens they therefore burn very smokily but, if provided with a generous air supply, very fiercely. They will melt and run at temperatures between 110°F. (43°C.) and approximately 500°F. (260°C.), according to the resin concerned.

(b) Oxidisation

While there is no serious risk of spontaneous heating of resins in air, they tend to absorb oxygen and can be dangerous in contact with oxidising agents, such as hydrogen peroxide. Spontaneous ignition is possible in a mixture of resins and oxidising agents.

(c) Dust explosions

Resins are sometimes used for manufacturing processes in finely divided form; a dust cloud could in these circumstances be highly dangerous in air and dust could be a severe hazard in the presence of an oxidising agent.

(d) Industrial processes

Two processes in purification and other treatment of resins have particular hazards: gum running, in which resins and gums are melted prior to distillation, and solvent treatment, in which the resins are dissolved for purposes of purification and the separation of individual constituents. Both these processes will produce flammable or explosive vapours, or both.

(e) Risks of synthetic resins

See Section 10 'Plastics'.

3. FIRE-FIGHTING

In general, fires involving resins and gums, especially in bulk, require the application of foam. This is particularly the case when the fire is well developed when it is first tackled. Since in this case it will be glutinous or flowing, the application of water, while likely in any case to produce violent scattering, may tend to cause the molten material to float and assist the spread of fire.

In some circumstances, however, water sprays may be effective; if by spraying a cooling effect can be produced without the application of excessive water, the fire should be easily extinguishable before it has fully developed and high temperatures are reached. The cooling effect will tend to solidify the surface of the molten mass and prevent spread of fire through flowing. Small fires in resins and gums, if caught in their early stages, can be successfully dealt with by foam, carbon dioxide, or dry powder extinguishers.

Due to the amount of smoke likely to be produced by this type of fire in its early stages at least, breathing apparatus is very likely to be required. There is, however, no likelihood in the normal course of events that toxic vapours will be given off by burning natural resins and gums though this danger is much greater in the case of synthetic resins. (Most of the aminoplastics, within which many synthetic resins are grouped, have either toxic or narcotic fumes associated with their burning. See Section 10).

Industrial processes in linoleum, oil cloth and tarpaulin manufacture

1. LINOLEUM MANUFACTURE

Linoleum is very extensively used as a floor covering. It consists of a layer of ground cork, sawdust, etc., held together by cementing materials, with a backing of coarse canvas (often made of jute).

(a) Raw materials

The raw materials include linseed oil, cork, wood flour, resins and gums, whiting and colouring pigments. Cellulose lacquers and paints, used for printing a pattern on the linoleum, may also be found on the premises.

(b) Process

Large quantities of oxidised linseed oil are obtained by allowing boiled oil to run down sheets of fabric scrim (*see* Plate 49) until a solid layer is formed. These skins are taken down and stored for use. Alternatively, the oil may be 'smacked' by a paddle (*see* Fig. 44), which aerates and oxidises it. In this case it emerges as a liquid. The various fillers such as wood and cork, bonding agents such as gums and resins, and other materials are ground to powder, mixed with the liquid, and heated to form a dough or 'cement' (*see* Plate 50), which is rolled out and spread on the canvas. The rough material produced in this way is heated and pressed in various machines. It is then hung or placed on flat racks in special drying chambers.

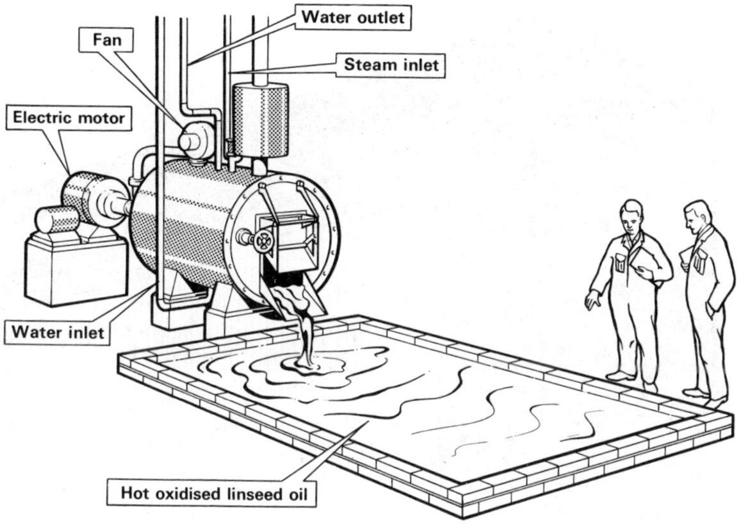

Fig. 44. Oxidising of linseed oil by the 'smacker' process in linoleum manufacture. The cylindrical chamber contains rotating paddles. It is water cooled and presents a fire hazard if the water supply should fail.

(c) Special risks

(i) There is a risk of a dust explosion in both cork and gum grinding (*see* Plate 48). Wood flour is not usually made on the premises, but can also give rise to dust explosions in suitable circumstances.

(ii) Large linseed oil storage tanks or wells will be found on the premises, and large quantities of oil are generally being processed in the oxidising houses.

(iii) The drying rooms (*see* Fig. 45) present a serious fire risk. The canvas edges of the linoleum are flammable and convey fire more readily than linoleum itself.

(iv) Spontaneous heating may occur in connection with the linseed oil or in the drying rooms, especially when jute is used as the backing material.

(d) Fixed fire-fighting installations

Linoleum factories are usually sprinklered. The linseed oil boiling pots are often protected by a CO_2 installation.

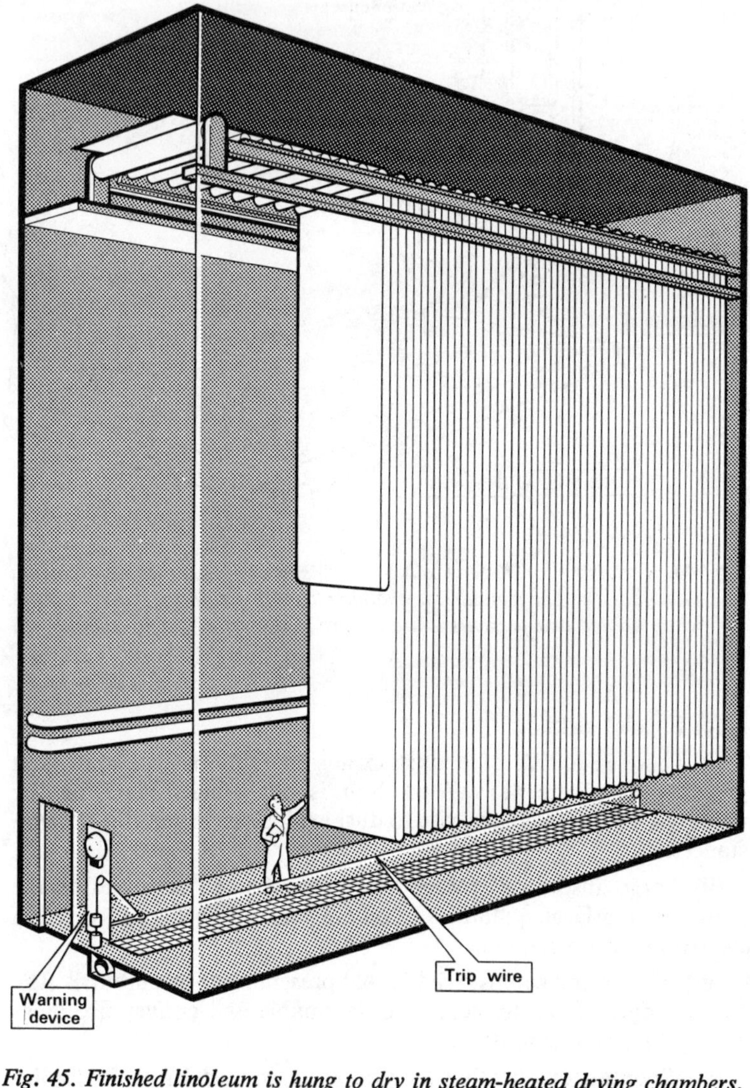

Fig. 45. Finished linoleum is hung to dry in steam-heated drying chambers. An alarm trip wire is provided to give warning in case of linoleum falling to the ground, which could create a serious fire hazard.

(e) Fire-fighting

Jets should be used with care in departments where dust explosion, have taken place or are likely to occur. The use of a hose reel, spray or diffuser branch is effective in most cases. Breathing apparatus will probably be needed. Water can cause considerable damage to finished linoleum by streaking and buckling it, especially when it runs under the end of standing rolls. Salvage operations should aim at protecting the end of linoleum rolls and guarding against an accumulation of water on the floor. If necessary, the rolls of linoleum should be removed to prevent them being damaged in this way.

Should fire occur in a drying room which has a concrete roof, it will often be wise to close the room and let the sprinklers (with which most drying rooms are fitted) deal with the fire. A careful examination should afterwards be made. If the roof is made of wood the fire should immediately be attacked from above.

2. OIL CLOTH MANUFACTURE AND WATER-PROOFING PROCESSES

There are many varieties of waterproofed cloth. Sometimes a material is waterproofed on one side only; in other cases the entire material is treated to make it waterproof.

(a) Raw materials

These may include cellulose nitrate and collodion cotton, celluloid scrap, solvents such as methylated spirit, acetone, naphtha and benzole, castor and linseed oils, soap, paraffin wax and bitumen and cotton or jute fabric.

(b) Process

This may consist of obtaining a plastic material by dissolving celluloid scrap, cellulose nitrate (itself used in celluloid manufacture), and collodion cotton as solvents, mixing the result with oxidised castor and linseed oils and spreading it on cotton cloth. The cloth is then hung up to dry. In the case of proofing processes the cloth is passed through a proofing bath and into drying chambers. These are attended by a high fire risk, as the atmosphere contains much solvent vapour.

(c) Fire-fighting

The dangers and difficulties lie chiefly in the raw materials. Breathing apparatus should be used. A fierce fire of the type found in the case of burning celluloid should be expected.

Part 6c, Chapter 45, Section 14

Rubber fires

NATURE AND PROPERTIES OF RUBBER
INDUSTRIAL PROCESSES AND RISKS

1. **Rubber manufacture**
2. **Manufacture of rubber articles.**
3. **Materials used**
4. **Risks**
5. **Fixed fire protection**
6. **Fire-fighting**

RUBBER as a commodity is typical of the way in which industrial fire hazards have increased during the twentieth century. As a finished product it burns fairly readily and very tenaciously, offering difficulties and dangers to firemen; during manufacture it presents grave fire hazards in many respects and has given rise to many serious outbreaks. At the same time, the use and manufacture of rubber have increased—and are increasing—continually since before the beginning of the century, and now raw rubber, finished products or scrap rubber are liable to be found in almost every branch of industry in ever-increasing quantities.

An additional complication since 1945 has been the manufacture of synthetic rubber on a large scale. It is estimated that upwards of 370,000 tonne of synthetic rubber were produced in Western Europe in 1960 and the estimated world consumption of synthetic rubber in 1956 was 1,500,000 tonne. In Britain the present trend in the tyre manufacturing industry is towards a complete change to synthetic rubber. While the scope of this Section of the *Manual* is sufficient to enable the main risks connected with rubber manufacture to be dealt with, it is not practicable to deal in detail with all the substances and processes involved in the production of synthetic rubber. This is a new industry, expanding on a large scale and still engaged in the discovery of new processes and the search for still newer ones; this brings an ever wider and wider field of synthetic chemical substances into use. It is inevitable that the fireman or fire officer must regard this branch of the industry as too highly specialised to be studied from books, and must leave the finer points of fire-fighting techniques to be settled by liaison with works' chemists and engineers.

Nature and properties of rubber

Natural rubber is a polymerisation product of isoprene, and is obtained from rubber latex, which is the juice of the rubber tree. It is imported into this country in the form of ribbed smoked sheets, crêpe, or plantation shoe sole crêpe. In the normal commercial product these natural rubbers are compounded with other substances to make commercial rubber for the manufacture of a wide variety of articles; the principal materials used for compounding are mentioned in paragraph 3 and include carbon blacks, oils and sulphur. In compounding, which involves temperatures of the order of 300°F. (149°C.) and high pressures, the process of vulcanisation occurs. This causes the rubber mixture to lose its thermoplasticity. Some rubber products are based directly on the use of rubber latex in their manufacture and do not depend on the vulcanisation process.

Synthetic rubbers are artificial materials similiar or superior to natural rubber in certain of its characteristics. G.R.S. is the most widely used of these synthetic polymers, and is made by the co-polymerisation of butadiene with styrene. Also in use throughout the rubber industry are the following synthetic polymers: neoprene (polychlorophene), nitrile rubbers (which are co-polymers of butadiene and acrylonitrile,) and butyl rubbers. These are co-polymers of isobutylene with a small amount of isoprene.

Reclaimed rubber is produced from waste manufactured products such as old tyres and inner tubes. The product is used as a constituent of new rubber in the mixing process.

The calorific value of rubber, 39,540 kilojoules/kilogram is roughly twice that of other common combustible materials. All types of rubber soften and flow when burning (though vulcanised rubber softens but does not flow) and may thus contribute to rapid fire spread; it decomposes slowly when heated, until at about 450°F. (232°C.) or upwards it begins to decompose rapidly, releasing dense volumes of smoke and decomposition products which are noxious and flammable and may give rise to explosion. The ignition temperature of these vapours is approximately 500°F. (260°C.).

Synthetic rubbers behave similarly, though temperatures at which decomposition becomes rapid may be somewhat higher, 660°F. (349°C.) for G.R.S., 750°F. (399°C.) for butyl and 725°F. (385°C.) for nitrile and in individual cases the behaviour may be slightly different. For example, neoprene will not sustain combustion when the source of ignition is removed, and decomposes rapidly at 470°F. (243°C.).

Unless tackled effectively in its early stages a rubber fire is extremely difficult to extinguish, water being the most suitable extinguishing agent. There is no evidence that natural, synthetic or reclaimed rubber is liable to spontaneous heating in the ordinary way, though it is believed to occur in some circumstances in rubber dusts and granules. This may be due to the residual heat from the buffing process.

Latex, being a water-based emulsion, does not present any fire hazards.

Industrial processes and risks

1. RUBBER MANUFACTURE

In most rubber factories, the processes include the production of the rubber compounds themselves from the raw materials in addition to the moulding and fabricating processes which produce the finished rubber articles. The production of rubber compounds involves the preparation of raw rubber (which may be natural or synthetic) by cutting up in machines, by 'masticating' or plasticising in suitable equipment and subsequent mixing.

These processes are carried out in buildings of very different kinds, the industry having been established for approaching 100 years. Many modern buildings are now used, however, and in these instances it is customary for manufacture to be carried out in single-storey factory spaces of wide extent and reasonably incombustible construction. The grinding and mixing processes, however, are often housed in two and three storey buildings of modern construction. It is common also to find a multi-storey construction for the warehousing of finished stocks and for other activities not directly connected with processing rubber. However, many older buildings are still in use and these may be of widely varying construction, not always satisfactory for housing the high fire hazard of rubber manufacture.

Where reclaimed rubber is the basic raw material to be used, this is reclaimed or 'regenerated' by cutting and grinding of the scrap tyres and tubes, etc., in special grinders or 'crackers'. Separation of the rubber from the extraneous materials incorporated in it may be done by mechanical or flotation means, and this may be followed by treatment with oils, water or chemicals under high steam pressures. After regeneration, the rubber is subjected to admixture with solvents, oils or other substances to make the reclaim conform to a particular specification.

Synthetic rubber is not normally produced in the same factories as manufacture the rubber articles, but in chemical works. The majority of this plant is extremely new, having been established since the Second World War, and much of the machinery and vessels used are of modern design and placed out of doors in the same way as in oil refineries and other chemical works. Materials employed are for the most part liquids, and the majority are flammable, some with low flash points and dangerous properties. Butadiene, styrene and acrylonitrile (*see* Section 16) are prominent among them. Processes include purification, mixing, reaction with a catalyst, blending, drying, etc., and are shown diagrammatically in Plate 35. Research laboratories, storage tank farms and other ancillary buildings and plant are to be found associated with synthetic rubber manufacture.

2. MANUFACTURE OF RUBBER ARTICLES

After the initial stages of production, compounded rubber is treated in various ways, depending to some extent on the function of the finished article. The semi-processed material is 'calendered', or squeezed between rigidly-held rotating rolls to produce sheeting; alternatively, it may be 'extruded', i.e. forced through an accurately made orifice to produce tubing, etc. Rubber may be applied to fabric either by calendering or by spreading a 'dough' (rubber/solvent mixture) on to the material. The solvent in the applied dough is then removed by passing it over steam chests. All these materials are finally 'cured' or vulcanised, or are moulded by the application of both heat and pressure.

Curing is a term denoting the reaction between rubber and sulphur or other chemicals, and this can be carried out by several methods: in a heated press, for example, in an autoclave or steam pan, in hot water, or in cold sulphur chloride solution or vapours.

3. MATERIALS USED

(a) Raw rubber of different kinds is the material of which largest stocks are kept, many thousands of tons being held by a large factory. (For storage, *see* below under par. 4 'Risks').

(b) Cotton was originally used in large quantities as a constituent of tyres, waterproof sheeting and other goods. However, it has been replaced by rayon and nylon in tyres, and synthetic fibres are becoming increasingly used in other applications. This tends to reduce fire hazards from these stocks in some respects, though the static electricity dangers may be increased.

(c) Sulphur is added in powder form at the mixing stage. The quantities involved will be very considerable where large-scale rubber production is done.

(d) Carbon and other blacks, which are combustible though slow-burning, are used in large quantities as 'fillers' to give reinforcement and to produce additional toughness and resistance to wear (*See* Section 9 'Paint and Varnish') under 'Carbon Black').

(e) Pine tar (flash point 130°F. (54·4°C.)) or coal tar (flash point 175°F. (79·4°C.)), and paraffin wax (flash point 395°F. (202°C.)) are all added to certain types of rubber to achieve particular properties.

(f) Rapeseed or colza oil is used as an 'extender' or diluent (flash point 325°F. (163°C.)). (*See* Section 8).

(g) Solvents are used in very large quantities. The commonest is petroleum naphtha, which has a flash point (about 40°F. (4·4°C.)) well below ordinary room temperatures and is thus very dangerous. Coal tar naphtha (flash point 75°F. (23·9°C.)) and benzene (flash point 12°F. (—11°C.)) are also used, as are small quantities of many solvents including carbon tetrachloride, ethylene dichloride and tetrachlorethane.

Note: Much research work is in progress in the rubber manufacturing industries, and many more materials than those shown above are liable to be found, normally in smaller quantities, as new processes are experimented with and developed.

4. RISKS

(a) Overheating of machinery

Many of the mechanical processes, especially cutting and grinding, develop a large quantity of heat, and this has to be dissipated to avoid the risk of overheating and consequent ignition of the product. In many instances the risk is added to (such as in the mechanical destruction of scrap rubber) by the foreign metallic bodies being included in the product to be ground; the smaller tyres are, in fact, ground whole, without the removal of the metal bead wire, in the early stages.

(b) Rubber dust

In some processes the grinding and milling operation produces a dust which is liable to ignite or explode if favourable circumstances are accidentally produced. The machinery itself may produce static charges (*see* (d) below) or sparks may be struck from metallic objects. Overheated machine bearings are a similar possible cause. Dust may be ignited in the machine itself, or in trunking; accumulations of the dust in the vicinity of the process area may be ignited by electrical or other faults. Additionally, if the dust or granules are not effectively cooled after a grinding process, spontaneous heating and ignition are possible with the presence of moisture.

(c) Solvents

The greatest risks are associated with the use of solvents. These are normally handled in the works in wheeled 'bowsers' (*see* Plate 51) containing about 136 litres which are manoeuvred into fixed positions and sometimes electrically earthed by flexible leads. The liquid is then normally dispensed into containers of about 570 millilitres capacity and applied in the fabrication processes. This is usually done by means of flannel swabs, and these may experience excessive accumulation of static electricity. To avoid spark discharges occurring, employees are asked not to wear rubber footwear which would prevent them being properly earthed. Precautions also have to be taken against sparks from footwear and from ferrous tools. In some processes flameproof electrical gear is desirable. Spreading machines are subject to frequent outbreaks of fire, caused usually by static charges igniting the flammable vapour produced from dough. Fortunately, the spreading process has been superseded in most cases by the faster calendering process, which does not produce the same degree of risk.

Where solvent recovery is used it also is productive of hazard. After coating, spreading or other processes, the article is passed over steam chests which vaporise the solvent this vapour being removed by fans via a canopy and thence in ducts to the solvent recovery plant. Fires can start in the ducting, and water traps and flame traps are usually provided against the risk of a flashback to the solvent condensers and storage tanks.

(d) Static electricity

Due to the non-conductive nature of most commercial rubbers, the majority of processes in the industry are liable to produce static accumulations in the product and on machine rollers, and also in vessels and piping containing liquids and vapours. Voltages of the order of 5,000 are commonplace. In association with solvent vapours, this produces obvious hazards, but the static risk permeates nearly all branches of the works and demands stringent precautions.

Earth connections are made where applicable, and rubber surfaces are discharged by means of wire brushes or trailing wires.

(e) Sulphur

Sulphur ignites at about 450°F. (232°C.) and is thus a relatively high fire hazard. When burning, it gives off sulphur dioxide, which is highly poisonous. In addition, the vapour of sulphur chloride may be found in rubber works, and this is also highly toxic and injurious to the respiratory organs. (Sulphur chloride is readily soluble in water and the risk can be reduced by means of a water spray).

The risks of a dust explosion in powdered sulphur must also be not forgotten.

(f) Rubber storage

Either natural or synthetic rubber may be stored indoors or outdoors on the premises at a rubber works. In these storages, imported raw rubber may be packed in bales covered with hessian, straw matting or rubber sheet, or in wooden cases. Bales and cases weigh about 100 kilograms. Reclaimed rubber may be in sheets 12·5 millimetres thick and stored without packing. Alternatively, many large 'dumps' of scrap rubber articles (mainly tyres) exist, generally out of doors, which may cover several acres (hectares) and contain thousands of tons (tonnes).

Although the risks involved in satisfactory storage are not high, this is not the case when inadequate precautions are taken, and many extensive fires in rubber storages have occurred which have proved dangerous, costly and difficult to extinguish. Recommendations for rubber storage, both in the open and in buildings, were made in 1953 by an interdepartmental working party set up by the Board of Trade in conjunction with a Committee of the Fire Offices' Committee Fire Protection Association, and were published by that Association and reproduced by the Home Office as Fire Prevention Note No. 5/1953. They dealt, among other points, with exposure hazards, heating, lighting and power, cleanliness and tidiness, naked lights, open fires and smoking, vehicles, the use of petroleum, etc., first-aid fire-fighting and accessibility for fire appliances.

5. FIXED FIRE PROTECTION

Fixed fire extinguishing installations of various types are commonly found. In some works there is a full provision of sprinkler systems, while in others the portion of the buildings used for warehousing may be sprinklered. Special risks connected with the use of

Plate 46. One of the pots used for making varnish by the synthetic resin process. (Section 9).

Plate 47. Fixed varnish pots used in 'oil boiling'. Ducting is provided for the removal of fumes. (Section 9).

Plate 48. Cork-grinding machinery once commonly used in the manufacture of linoleum. (Section 13).

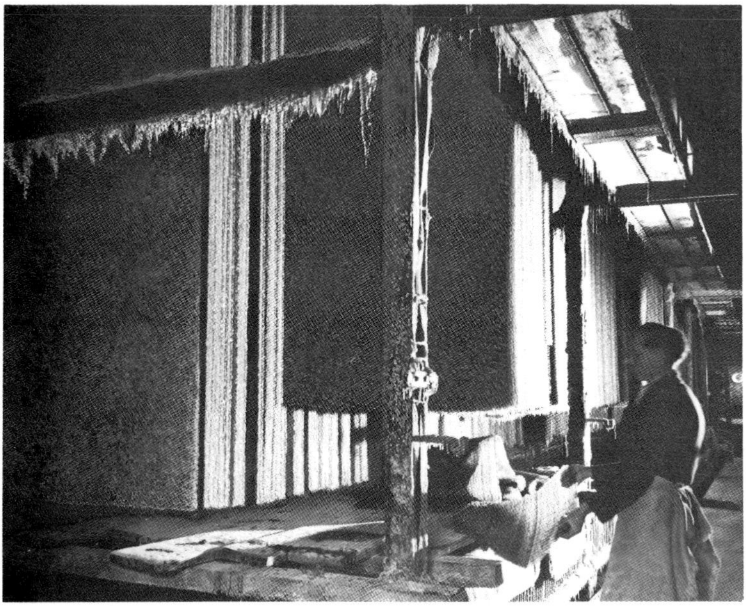

Plate 49. The time-honoured method of oxidising boiled linseed oil by allowing it to run down sheets of scrim until a solid layer is built up. Note the oil-impregnated wooden beams and platform. (Section 13).

Plate 50. Another process in the manufacture of linoleum. The oxidised linseed oil is mixed with resins, etc., to form a 'cement'. This is run into large trays to cool. (Section 13).

Plate 51. A solvent dispenser used in a rubber factory, showing, on the left, the earthing connections. Note also the flame-proof light fitting at the rear. (Section 14).

PHOTO: *Dunlop Rubber Co. Ltd.*

solvents may be covered by carbon dioxide, vaporising liquid or foam.

6. FIRE-FIGHTING

(**a**) At a temperature above about 400°F. (204°C.) rubber will begin to flow in a sticky, slippery, hot mass and in these conditions it is dangerous to walk on it. Rubber liquefied in this way does not afterwards re-solidify. Street drains, etc., must be protected against becoming blocked.

(**b**) When active decomposition sets in (at 450°F. (232°C.) or upwards) heavy white flammable vapours are given off and these are liable to flash at 500°F. (260°C.) and upwards. Pockets of these flammable vapours may be trapped in the molten rubber, and generate great internal pressure while they remain trapped; they will ultimately escape and catch fire with almost explosive force.

(**c**) When burning, rubber produces dense black oily smoke which has some toxic qualities. Breathing apparatus will normally always be necessary on a rubber fire inside a building.

(**d**) When fighting a rubber fire it is necessary to use, from the closest possible range, the hardest-hitting and largest-diameter water jets available; no good will be done by standing back and attempting to flood a rubber fire. The object should be to introduce as much water as possible into the pockets formed in the bulk rubber. In the open, and when direct access to the face of the fire is possible, fixed or portable monitors may be of value. The mass must be pulled apart to expose burning surfaces. Diffuser or spray branches will be found invaluable for clearing away smoke.

(**e**) Foam can seldom be used upon bulk stocks of burning rubber, owing to the lack of a smooth exterior surface; a pile of rubber bales or tyres is usually uneven and filled with air vents. In view of these practical difficulties, however, the use of high expansion foam on rubber fires may prove efficacous as it can be applied in bulk.

(**f**) Water spray is most generally useful at fires involving rubber dust or machines. Water should be used with great care on rubber-and-solvent mixtures ('dough') owing to the danger of scattering them: foam is preferable in such a case. Where the fire is in the mixing or spreading department, the supply of solvent should be cut off and any drums or pockets of solvent should be removed.

(**g**) In some rubber factories a special moulding lubricant called 'facemol' is used. This preparation is a sulphonated fatty alcohol and it minimises the water-repellent characteristic of rubber. It is considered by some to have good qualities as a fire extinguishing medium, and can be used in soda-acid or gas-expelled extinguishers or in a foam-making branch system for producing foam. Steel drums of the liquid are kept at strategic points about some premises.

(**h**) Great difficulties may be caused by fire-fighting equipment, especially hose, becoming contaminated with molten rubber, and every effort should be made to avoid this. The methods of cleaning hose contaminated by rubber and other substances is dealt with in Book 2 of the *Manual*, Chapter 2.

Part 6c, Chapter 45, Section 15

Sugar fires

THE fire hazards associated with sugar are not severe, but they have their own individuality and they have contributed to many serious warehouse and factory fires in the past. The industrial processing of sugar itself, and much of confectionery manufacture, is nowadays carried on by large firms in extensive plants, in which many of the buildings are modern and working conditions good; there is still however, a legacy of poor buildings and storage warehouses from past decades of the confectionery industry which are a high hazard from the point of view both of the building construction and industrial process. The possibility of serious fire and explosion in the industry generally are sufficient to justify a study of sugar fires by the fireman.

Nature and properties of sugar

1. PHYSICAL AND CHEMICAL PROPERTIES

Sugar is produced naturally in vegetable matter (and in some animal products) to a greater or lesser degree; in the case of plants, the process is one of synthesis in which many factors such as light, moisture, carbon-dioxide and chlorophyll play a part. Sugars vary slightly in composition according to the plant in which they are found, and there are thus many varieties. The two forms of vegetable matter from which it is commercially practicable to extract sugar in quantity are sugar cane and sugar beet, though other varieties are extracted for specialised purposes. Chemically, sugar is complex. It is a carbohydrate (composed, therefore, of carbon, hydrogen and oxygen), and is soluble in water. It has the alternative names of 'saccharose' and 'sucrose' and sugars as a whole may be divided into three groups, mono-saccharides (*e.g.* $C_6H_{12}O_6$), di-saccharides (*e.g.* $C_{12}H_{22}O_{11}$) and poly-saccharides (aggregations of molecules similar to those mentioned but aggregated into a larger molecule.) Cane and beet sugar belong in the second group (di-saccharides) and are both sucrose: sugars made from cane or beet are indistinguishable after manufacture.

As a manufactured product sugar varies in two ways. Its constituents may vary by the inclusion of substances such as molasses or colouring matter to produce brown and other sugars, and this has no appreciable effect on its properties from a fire-fighting point of view. Its form is also varied—it takes the form of lump, granulated, icing sugar, etc.—and, this by affecting the specific surface area of the particle, can have a marked effect on its behaviour in fire.

Syrup, treacle and other viscous forms of sugar are solutions in water, and this also affects the burning characteristics of the product.

2. BURNING CHARACTERISTICS

Sugar melts at about 320°F. (160°C.) and then decomposes readily on the application of further heat. It is not easily ignited, but once alight it burns willingly and fiercely if oxygen is allowed free access to it; it bubbles and chars in the process of burning, and the charred coating tends to exclude oxygen and reduces the fire. It will emit quantities of thick smoke, being rich in carbon. It also reacts violently with concentrated sulphuric acid.

In circumstances where the vapours resulting from decomposition can become trapped and fail to ignite, subsequent explosion of these vapours is possible, though the results are unlikely to be severe. Burning sugar can flow when alight, though it does so only sluggishly and solidifies again immediately on cooling or on encountering a cool surface or object.

Syrup and treacle are still less readily ignited, and require the application of heat until the water in which the sugar is dissolved is driven off. After this, the sugar will burn normally, depending on the supply of oxygen.

3. RISKS

In normal circumstances there is no special hazard attached to burning sugar, though it should be remembered that, as with all carbonaceous materials, there is a probability of carbon monoxide being liberated and a certainty of smoke being produced. Finely ground sugar and sugar dust, can, however, very easily give rise to dust explosions of great severity when dispersed in air, and they behave similarly to starch in this respect.

Spontaneous combustion in sugar is not considered a probability in normal circumstances, though in packed sugar storages other factors such as packing materials may affect the hazard.

4. FIRE-FIGHTING

A fire involving lump, granulated or a crystallised mass of sugar requires the application of water. Whether jets or sprays are required will depend upon the form taken by the burning sugar but where finely ground sugar or sugar dust is present precautions should be taken to avoid the risk of dust explosion. This will involve using sprays to avoid stirring up the mass as far as possible: for other precautions against this risk reference should be made to Section 1—'Fires in Dusts'.

Breathing apparatus is likely to be required if space is confined and ventilation inadequate.

Industrial processes and hazards

1. CANE SUGAR REFINING

About three-quarters of the sugar used in the United Kingdom is imported in the form of raw cane sugar and refined in this country.

Large stocks of sugar will always be found on refinery premises, as well as at chocolate and confectionery works, bakeries and similar establishments.

(a) Process

(i) The raw sugar is emptied from bags, and washed to remove impurities.

(ii) Carbon or charcoal in then added to remove impurities, the carbon or charcoal being subsequently re-activated by treatment with chemicals and heating in kilns.

(iii) The mixture is put through filter presses to remove the carbon or charcoal.

(iv) The sugar is then boiled in a vacuum and dried in rotary driers.

(v) It may then be bleached (this is not universal), after which it is sifted and stored in silos until it is bagged and placed in a warehouse.

(vi) It may be ground to make icing or a similar sugar.

(vii) 'Invert sugar' may be obtained by treatment of cane sugar with acids.

(b) Fire risks and special hazards

(i) When sugar is bagged the sugar dust and lint on the bags, and the crevises between the bags, assists the rapid spread of fire. The height of sugar stacks (anything up to 9 metres) makes fire-fighting difficult.

(ii) It is doubtful whether sugar will normally ignite as a result of spontaneous combustion, but fires, which may possibly have originated in this way, have been known to occur where sugar has been stocked in jute bags. Similarly, combustion in empty sugar bags has been known to occur.

(iii) Owing to the risk of a dust explosion, one or more dust extraction systems are fitted, and explosion windows are installed.

(iv) Sugar dust and jute fluff will be found in the receiving, sifting, grinding and bagging departments as well as in the warehouse. The quantities present will depend on standard of housekeeping in the works.

(v) The silos are similar to those used for grain (*see* Section 6) but are somewhat smaller. They are generally made of wood.

(c) Fixed fire protection

Refinery warehouses are not normally sprinklered owing to the great damage which can be done by water. Other buildings may however, be protected in this way if the buildings themselves present a high fire risk.

(d) Fire-fighting

(i) Sugar works are usually kept fairly free from dust, but there is a risk of a dust explosion in the dry processes, especially when sugar is being sifted or ground. In a fire, the partial collapse of the interior of the building may cause a dust explosion. Jets should be used with care.

Dust is not likely to be met with in the processes involving the use of carbon, as these processes are mostly wet ones.

(ii) Sugar is difficult to ignite, and a small fire in the store, for example, is often put out by the collapse of sugar on to the fire when the jute bag has been burnt through. Once alight as the result of the application of external heat, however, it burns with a black smoke and great heat. Stacks are best fought with spray, as burning sugar forms a caramel-like crust when cooled; this crust blankets the fire by keeping out the oxygen. At the same time, as much sugar as possible should be salvaged from the back of the stack. A jet is liable to dissolve and waste sugar and should therefore be used only if it is absolutely necessary to do so.

(iii) Sugar and syrup will flow while hot and solidify again when cooled; they will float upon water. Drums of heated syrup may burst and spread the fire.

(iv) Heated sugar is liable to froth and boil over. It may also give rise to gas explosions.

(v) The chemicals usually found are sulphuric and hydrochloric acids (*see* Section 16.)

2. BEET SUGAR MANUFACTURE

(a) Processes

About a quarter of the sugar used in the United Kingdom is manufactured from sugar beet, which is principally grown on the eastern side of the country, though there are important areas in the West Midlands. The process also produces sugar beet pulp (used as a cattle food) and molasses. Most factories produce granulated sugar, though some produce only raw sugar for refining elsewhere; lump and icing sugar are not normally produced in British factories.

The sugar beet is first washed, weighed and sliced. The juice is then extracted by dissolving the sliced beet in revolving drums of hot water. From this point onwards until the final process of centrifuging the sugar product is a water solution. The beet pulp from which the juice has been extracted is pressed, dried, treated and bagged for distribution to farms. The complete process is shown in diagrammatic form in Plate 11.

Buildings are comparatively modern and usually purpose-built.

(b) Risks

No hazardous processes or materials are involved in most stages of sugar beet preparation, though quicklime (calcium oxide) is used in quantity and is irritating to the eyes, lungs and skin. However, the dust from dried sugar beet pulp has an explosion hazard and may cause troublesome fires in ducts and cyclones. Sugar dust may also be found associated with the final granulating and bagging operations. The majority of fires in the industry are warehouse and storage fires.

3. COCOA, CHOCOLATE AND CONFECTIONERY MANUFACTURE

(a) Construction

Most cocoa and chocolate factories are of modern construction, though this cannot always be said of the smaller confectionery firms.

Confectionery works often contain many wooden partitions dividing the compartments used for different processes, while stairs, wood-enclosed hoists and, especially in chocolate factories, numerous spouts and air-ventilating trunks may accelerate the spread of fire.

(b) Process

(i) *Cocoa.* Cocoa beans arrive at the factory in sacks. They are cleaned, sorted, roasted, and then cooled suddenly in air. The next process is winnowing; the shells are cracked, and the light shell is separated from the kernels or cocoa nibs by means of fans.

The cocoa nibs are then ground (which presents one of the main fire hazards of the plant) and converted into a liquid from which a certain amount of the rich cacao butter is removed. This is done by pouring the cocoa mass into circular steel pots, the tops and bottoms of which are loose perforated plates lined with felt pads. The pots are arranged one above the other, and by means of hydraulic pressure (about 415 bar) the required amount of cacao butter is pressed out. On releasing the pressure each pot is found to contain a hard, dry cake of compressed cocoa. This cake is again ground, sieved, and the powder is packed in tins.

The shell has its fat content extracted by means of a solvent, which is usually trichlorethylene, though exceptionally a flammable solvent may be used. (Trichlorethylene may affect breathing apparatus—*see* Section 16).

(ii) *Chocolate.* The cocoa beans are cleaned, roasted and winnowed as above. The cocoa nibs are then ground to a paste with sugar in a large revolving pan fitted with a granite bed on which run granite rollers. The paste is fed to a refiner, consisting of a series of water-cooled rollers between which the chocolate runs as a thin film. After these processes, which may be repeated more than once, the chocolate goes as a powder to the couching machine, an arrangement of rectangular troughs with granite beds in each of which a heavy granite roller is pushed backwards and forwards on the end of an arm. Finally, the chocolate is moulded, cooled, and wrapped.

(iii) *Confectionery.* Some of the processes described in (ii) above occur in the manufacture of many kinds of confectionery. Other processes differ too widely for any general description to be of value.

(c) Fire-fighting

(i) The dust of sugar, cocoa, ground cocoa bean shell, and starch (used for moulds for gums, etc.) is liable to dust explosion.

(ii) Considerable quantities of sugar and glucose are stored on the premises. Sometimes a large storage tank holding four or five tons of glucose, will be found on the top floor of a confectionery works. This tank supplies the various boiling pans. In a serious fire, the feed pipes may be fractured, in which case the glucose will feed the fire. A stop valve is often fitted immediately beneath the tank, and should be turned off in the case of fire. The co-operation of the management should normally be sought in this matter.

(iii) Cocoa smoulders, and a spark will travel along it in the same way as along sawdust. A spark may, therefore, be conveyed some distance until it comes into contact with, and ignites, woodwork or other combustible material. Burning chocolate and cocoa give off dense volumes of smoke. This both hampers the firemen in locating the seat of the fire, and damages other stocks not yet involved in the fire. Smoke arising from burning cocoa is not in itself dangerous, and the use of breathing apparatus is not necessarily essential. A spray should be used, and the powder should be washed down to locate the buried 'bulls-eyes'. The technique is one of a little water in the right place. Cocoa which has been affected by fire or smoke contaminates good powder, and it is necessary, therefore, to clean out the whole plant after a fire.

(iv) Chocolate, cocoa and some confectionery contain cacao butter (*see* Section 3) in varying proportions. It is sometimes stored in block form in large quantities, and is usually kept at a factory in a cooling room or cool basement. Cacao butter is not easy to ignite, but once alight, burns readily. It melts into an oil at blood-heat and will continue to burn while floating on water. It picks up the taste of smoke very easily. Fabrics and wooden floors easily become saturated with it.

(v) Drying rooms in both chocolate and confectionery factories operate at temperatures of 100°—200°F. (38°—93°C.) for long periods. The woodwork, therefore, becomes very dry.

(vi) Essences used for flavouring are usually very flammable and the vapours of some of them are explosive. They are, however, used in small quantities and do not generally present an exceptional fire risk, although they may add to the fierceness of a fire or complicate the attack upon it. Alcohol is used to dissolve these essences.

(vii) The shells of cocoa beans burn heavily and smokily if they catch fire before their fat content has been extracted. Once alight, they leave many bulls-eyes which have to be dealt with separately.

(viii) The fire hazards connected with the use of many of the principal solvents are dealt with in Section 16 of this Chapter.

(ix) Boiled sweets and toffees are often prepared on coke stoves: jets used in the neighbourhood of these might cause the same troubles as upon burning coke generally (*see* Section 5 'Coal and Coke').

(x) There are considerable quantities of paper, wood fibre, straw, boxes and crates in the packing department. It is important for salvage squads to protect the printed and often highly decorated boxes and wrappings used for confectionery and chocolate, as the destruction of these stocks may dislocate the manufacturer's business even though the confectionery or its production is not itself seriously affected by the fire. The wrapping and bag-making machines should, if possible be protected from damage by water.

(xi) Chocolates normally maintain their shape only at temperatures below 90°F. (32°C.), the melting point of cacao butter. They are greatly affected by heat, smoke, damp or steam, even when, for instance, they are stored in a warehouse adjacent to the building on fire. This applies also to stocks of bottled sweets which may stick together. Salvage work should, therefore, as far as possible aim at removal rather than protection on the spot. As much ventilation should be carried out as is consistent with safety in fire-fighting.

Part 6c, Chapter 45, Section 16

Alphabetical list of dangerous chemicals

Abbreviations used:

F.P.	Flash point
S.G.	Specific gravity
V.D.	Vapour density (Air = 1)
S.I.T.	Spontaneous ignition temperature
M. Pt.	Melting point
B. Pt.	Boiling point
E.L. (L)	Explosive limits (lower)
E.L. (U)	Explosive limits (upper)
C.T.	Critical temperature
C.P.	Critical pressure

Acetaldehyde

Chemical formula CH_3CHO

SUMMARY

Colourless fuming liquid, very strong fruity smell.

Flammable, explosive, toxic.

Also known by the name of acetic aldehyde or ethyl aldehyde.

PHYSICAL DATA

F.P.	– 36·0°F (–37·8°C)
S.G.	0·78
V.D.	1·52
S.I.T.	365°F (185°C)
M. Pt.	– 189·4°F (– 123°C)
B. Pt.	69·4°F (20·8°C)
E.L. (L)	4·0%
E.L. (U)	57·0%

Industrial application

Used in the manufacture of plastics.

Storage and transport

Store away from all possible fire hazards and oxidising agents, and treat as any other flammable liquid.

Dangerous properties

Very dangerous when exposed to any form of heat or flame, and when in such a condition it is very liable to explode. Reacts with oxidising agents.

Fire-fighting

Breathing apparatus must be worn. Water in the form of fog or spray, carbon dioxide, dry powder, vaporising liquid can be used.

Medical effects and treatment

Inhalation of the fumes will cause a feeling of intoxication with violent headaches, the fumes can also seriously affect the eyes.

Acetic acid

Chemical formula CH_3COOH

SUMMARY

A liquid.
Slightly heavier than water.
Vapour heavier than air.
Miscible with water.
Flammable.
Clear, colourless.
Smells strongly of vinegar.

PHYSICAL DATA

F.P.	113°F (45°C)
S.G.	1·05
V.D.	2·07
S.I.T.	1050°F (566°C)
M. Pt.	61·8°F (16·6°C)
B. Pt.	244°F (118°C)
E.L. (L)	4·0%

Industrial application

Used in the manufacture of metal acetates, dyes, textiles, pharmaceuticals, rubber, foodstuffs and cellulose acetate plastics and rayon.

Storage and transport

In bottles (all sizes), demijohns (22·5 litres), carboys (54·5 litres), barrels (180 litres). Also in aluminium or steel drums and road and rail tankers of 11,360 to 20,450 litres capacity.

Dangerous properties

The strong acid burns readily. The fumes are dangerous in enclosed spaces. The acid will blister, and, if in strong solution, destroy the skin.

Fire-fighting

Breathing apparatus may be necessary. Dilute with large quantities of water by means of sprays or open-ended hose lines.

Medical effects and treatments

The fumes have an irritant effect. Remove a sufferer from concentration. Magnesia or chalk in water may be given as an antidote to the effects of fumes. Any affected parts of the skin should be thoroughly washed off with water.

Acetone

Chemical formula CH_3COCH_3

SUMMARY

A colourless liquid with a pleasant 'minty' smell.
Highly volatile and flammable.
Miscible with water.
Vapours heavier than air and poisonous.
Also known as:
dimethyl ketone, ketone propane, propanone.

PHYSICAL DATA

F.P.	0°F (– 17·8°C)
S.G.	0·792
V.D.	2·00
S.I.T.	995°F (535°C)
M. Pt.	– 137·5°F (– 94°C)
B.Pt.	133°F (56°C)
E.L. (L)	2·5%
E.L. (U)	13%

Industrial application

Used extensively as a solvent for nitrocellulose, celluloid, waxes and fats. Also in the manufacture of leather cloth, lacquers, cordite, smokeless powders, paint removers, artificial silk, safety glass, chloroform and acetylene gas.

Storage and transport

In road and rail tankers	2270 – 22700 litres
,, steel drums	200 – 400 litres
,, ,, cans	22·75 – 70 litres

and also in carboys.

Dangerous properties

Very low flash point and explosive limit. Poisonous vapours.

Fire-fighting

Breathing apparatus may be necessary. Water in the form of sprays is the best extinguishing method; CO_2 may also be used.

Medical effects and treatment

Prolonged inhalation of the fumes should be avoided as they cause irritation to the eyes and mucous membrane. They also act as a narcotic, and induce symptoms similar to drunkenness. Remove to fresh air and keep warm, seek medical aid.

Acetylene

Chemical formula HC:CH

SUMMARY

A colourless gas with an ether-garlic smell.

Flammable.

Explosive.

Soluble in its own volume of water and very soluble in acetone.

For cylinder marking colours, see plate 43.

PHYSICAL DATA

F.P.	0°F (– 17·7°C)
V.D.	0·91
S.I.T.	635°F (335°C)
M. Pt.	– 113°F (– 81°C)
B. Pt.	– 119°F (– 84°C) (Sublimes)
E.L. (L)	2·5%
E.L. (U)	80%

Industrial application

Widely used in oxy-acetylene cutting and welding. Also used in plastics manufacture. It forms the basis of the vinyl group.

Storage and transport

In metal cylinders in which it is dissolved in acetone absorbed on porous material. Cylinders must be stored in an upright position.

Dangerous properties

Compressed acetylene is liable to explode (unless in a solution). Acetylene has a very low explosive limit and a very wide explosive range. It forms explosive mixtures with air and oxygen. It explodes violently if mixed with chlorine. Dissolved acetylene cylinders are extremely dangerous in a fire since once they have been heated decomposition is liable to start within the cylinder and to continue even after the cylinder has cooled resulting in explosion (See Section 7 'Fires involving Metals'). When pure, acetylene is only moderately toxic, but where it contains arsine, phosphine or carbon monoxide it is very toxic.

Fire-fighting

For treatment of dissolved acetylene cylinders in fire see Section 7 'Fires involving metals'.

Medical effects and treatment

Can cause shortness of breath and headache, then vertigo, mild gastric symptoms and semi-asphyxia. At a concentration of 40 per cent or more the patient may collapse. Remove to fresh air. Apply artificial respiration. Give oxygen if available. Call a doctor.

Acrylonitrile

Chemical formula CH_2CHCN

SUMMARY

Colourless, mobile liquid.

Mild odour.

Highly toxic when heated, emitting fumes of cyanide.

Highly flammable.

PHYSICAL DATA

F.P.	32°F (0°C)
S.G.	1·39
V.D.	1·83
S.I.T.	898°F (481°C)
M. Pt.	– 115°F (– 82°C)
B. Pt.	174°F (78·9°C)
E.L.(L)	3·05%
E.L. (U)	17·0%

Industrial application

Used in the manufacture of synthetic rubber and plastics.

Storage and transport

Store in a cool, well-ventilated place, away from all means of ignition or persistent heat.

Dangerous properties

When heated or in contact with acids and/or vapours, gives off highly poisonous fumes of cyanides. Reacts violently with oxidising agents. Vapours can cause explosions when mixed with acids at temperatures down to 32°F (0°C).

Fire-fighting

Breathing apparatus should be worn. For extinguishing fire use water spray, carbon dioxide, vaporising liquid, dry powder or alcohol-resistant foam.

Medical effects and treatment

Dangerous to the eyes and can be absorbed through the skin. Extremely toxic preventing the tissue cells from absorbing oxygen. If overcome, carry to fresh air, keep warm and obtain medical attention. Administer oxygen.

Ammonia

Chemical formula NH_3

SUMMARY

A colourless, pungent smelling gas.

Liquefies at 15°C at pressure of 7·4 atmospheres.

Slightly lighter than air.

Very soluble in water.

Not readily flammable but can form explosive mixture in air if heating and/or compression takes place.

For cylinder marking colours, see Plate 43.

PHYSICAL DATA

S.G.	0·817
V.D.	0·771
S.I.T.	1204°F (651°C)
M. Pt.	– 108·5°F (– 77·7°C)
B. Pt.	– 28°F (– 33°C)
E.L. (L)	16%
E.L. (U)	25%

Industrial application

Used extensively as a refrigerant, also in the chemical, explosives, petroleum refining, fertiliser, metallurgical and many other industries. Produced synthetically from atmospheric nitrogen and also as a by-product of coal gas manufacture.

Storage and transport

In glass Winchesters and steel drums as a 33 per cent ammonia/water solution. In steel cylinders as a pressurised liquid. Store away from heat and sunlight. B.S. specification for ammonia cylinders is yellow and red bands on a black background.

Dangerous properties

Dangerously high internal pressures in storage cylinders can be developed as a result of a comparatively slight increase in temperature. Although not readily flammable the gas will burn vigorously in oxygen. Under certain conditions ammonia/air mixtures can be exploded by an electric spark.

Fire-fighting

Breathing apparatus with protection should be worn. High concentrations may require fully-protective clothing. If unable to remove cylinders from source of heat keep them cool by the use of spray jets played along the full length of the cylinders. High level ventilation as soon as fire situation tactics permit. As the gas is very soluble in water spray jets should be used to clear the atmosphere.

Medical effects and treatment

An asphyxiating gas causing pain to eyes, nose and throat. Destroys the mucus membrane and can prove fatal. May cause coughing, sweating and urination. If affected return to fresh air and keep warm whilst recovering. Wash affected parts of eyes and skin with acidulated water. The antidote is lemon juice or vinegar. Clothing upon which liquid ammonia has been spilled should be removed immediately and the body thoroughly drenched with water.

Ammonium hydroxide

Chemical formula NH_4OH

SUMMARY

A water-ammonia solution (usually approximately 25% ammonia).

Dangerous fumes can be fatal.

Has usual characteristics of ammonia vapour.

Vapours slightly lighter than air.

PHYSICAL DATA

(See ammonia.)

Industrial application

Found in works producing ammonia; also in trades producing insecticides, cleansing agents and the textile industry.

Storage and transport

Usually kept in drums or glass carboys.

Dangerous properties

High expansion rate when heated: risk of bursting drums and carboys. Reacts violently with acids.

Fire-fighting

Breathing apparatus should be worn. If unable to remove containers from heat, keep cool with sprays. Dilute liquids and vapours with copious supplies of water. High level ventilation.

Medical effects and treatment

As for ammonia (see page 289).

Ammonium nitrate

Chemical formula NH_4NO_3

SUMMARY

A white crystalline solid.

Soluble in water.

Can cause explosions.

Decomposes when heated to produce poisonous, heavier than air vapours.

PHYSICAL DATA

M. Pt.	337·3°F (169·6°C)
B. Pt.	Decomposes at 410°F (210°C)

Industrial application

Used in explosives (to supply oxygen), in dyes, fertilisers etc. Produced in synthetic ammonia works.

Storage and transport

In iron drums or canisters.

Dangerous properties

May cause explosions if in contact with red-hot substances. Drying out of solution on combustible substances may be followed by spontaneous ignition.

Fire-fighting

Breathing apparatus should be worn as protection against nitrous fumes. Flood with water but avoid scattering crystals or solutions. No high pressure jets. Remove unaffected containers to cool, dry position. Small quantities can be extinguished with water sprays.

Medical effects and treatment

Fumes corrosive on exposed skin. Administer oxygen if breathing is difficult. Keep warm. Remove to hospital.

Special notes

It is considered in some circles that ammonium nitrate is explosive in itself, but the point has not been definitely established. If after-fire operations include removal or breaking up of large masses of ammonium nitrate, tools must be used with great care. Thoroughly wash any items of clothing contaminated by the solution—preferably before they dry out. Do not dry at high temperatures.

Amyl acetate

Chemical formula $CH_3COO(CH_2)_4CH_3$

SUMMARY

A colourless liquid, smelling strongly of peardrops.
Lighter than water, vapours heavier than air.
Slightly soluble in water.
Flammable.
Explosive.

PHYSICAL DATA

F.P.	77°F (25°C)
S.G.	0·88
V.D.	4·5
S.I.T.	750°F (399°C)
M. Pt.	– 109°F (– 78·5°C)
B. Pt.	298·4°F (148°C)
E.L. (L)	1·1%

Industrial application

Used widely as a solvent in the lacquer and paint industry, in the purification of penicillin and in the perfumery and confectionery trades.

Storage and transport

In steel drums of 205 – 410 litres capacity also in steel cans of 22·75 – 70 litres capacity. Rail and shipping regulations state that amyl acetate must be consigned as 'flammable'.

Dangerous properties

Very low explosive limit coupled with wide explosive range.

Fire-fighting

Breathing apparatus may be necessary. Water in the form of sprays to effect rapid cooling. CO_2 or vaporising liquid to extinguish fire.

Medical effects and treatment

A high concentration has a narcotic effect and may produce headaches and sickness. Remove to fresh air and keep warm.

Amyl alcohol

Chemical formula $CH_3(CH_2)_3CH_2OH$

SUMMARY

A colourless liquid.

Flammable liquid with flammable and explosive vapour.

Liquid and vapour dangerous to the eyes and toxic.

PHYSICAL DATA

F.P.	100°F (37·8°C)
S.G.	·817
V.D.	3·04
S.I.T.	700°F (371·1°C)
M. Pt.	– 110·2°F (– 79°C)
B. Pt.	280·4°F (138°C)
E.L. (L)	1·2%

Industrial application

Used in distilleries.

Dangerous properties

Flammable, low flash point and very low explosive limit. Liquid and vapours are dangerous to the eyes and are very toxic.

Fire-fighting

Breathing apparatus and protection for the eyes should be worn. Use water spray, alcohol-resistant foam, carbon dioxide, vaporising liquid, dry powder.

Medical effects and treatment

Vapours irritate the upper respiratory tract and in small amounts cause vertigo and headache. Remove patient to fresh air. Apply artificial respiration if breathing has stopped. Remove any contaminated clothing and wash patient with soap and water. If the chemical has reached the eyes wash them immediately with copious water for at least 15 minutes. Obtain medical attention.

Aniline

Chemical formula $C_6H_5NH_2$

SUMMARY

A colourless oily liquid.
Flammable.
Toxic.

PHYSICAL DATA

F.P.	168°F (75·6°C)
S.G.	1·02
V.D.	3·22
S.I.T.	1000°F (538°C)
M. Pt.	21·4°F (– 6·2°C)
B. Pt.	364°F (184·4°C)

Industrial application

Used in the preparation of plastics, in aniline dyes and in some types of soldering flux.

Dangerous properties

Moderately flammable. Subjected to great heat it will give off flammable and toxic vapours. Poisonous by absorption through the skin or by inhalation of the dust or vapours. Dangerous to the eyes.

Fire-fighting

Wear breathing apparatus, eye protection and protective clothing. Do not use water. Use carbon dioxide, vaporising liquid, foam or special dry chemicals.

Medical effects and treatment

Affects nervous system and blood. Symptoms are a feeling of well-being, then headache, fatigue and vertigo. The speech is affected. *Carry* patient to fresh air. Apply artificial respiration. Take off contaminated clothing *quickly* and wash patient with plenty of warm water and soap. If eyes affected wash them with copious warm water for at least 15 minutes. Obtain medical attention.

Arsine

Chemical formula AsH_3

SUMMARY

A colourless gas with a smell of garlic.

Extremely poisonous.

Not soluble in water.

A moderate fire and explosive hazard.

PHYSICAL DATA

V.D.	2·66
M. Pt.	– 177°F (– 116°C)
B. Pt.	– 67°F (– 55°C)

Industrial application

Liable to be liberated in certain industrial processes involving metals, e.g. in cleaning out steel tanks which have contained sulphuric acid, due to the very dilute acid causing a liberation of hydrogen which reacts with any arsenic present.

Dangerous properties

Extremely poisonous. A moderate fire and explosion hazard.

Fire-fighting

Breathing apparatus should be worn. Use carbon dioxide, dry powder.

Medical effects and treatment

Causes arsenic poisoning. Carry patient shoulder high to fresh air. Keep warm—obtain medical assistance.

Barium

Chemical formula Ba

SUMMARY

A silvery-white malleable metal.

Many compounds soluble in water and alcohol.

Spontaneously flammable in air.

Poisonous if inhaled as a dust or ingested.

Barium salts are found in many forms.

All except barium sulphate should be treated as poisonous.

Non-combustible.

PHYSICAL DATA

S.G.	3·5 at 68°F (20°C)
M. Pt.	1562°F (850°C)
B. Pt.	2084°F (1140°C)

Industrial application

Widely used in one form or other in the explosives, insecticides, chemicals, dyes, paints and linoleum industries.

Storage and transport

In metallic form the substance is stored under paraffin or some similar oxygen-free petroleum derivative. As a salt it may be found in bottles, kegs or boxes.

Dangerous properties

Large quantities of hydrogen are given off when barium is in contact with water or alcohol.

Fire-fighting

Breathing apparatus and gloves should be worn. Remove from contact with heat, water or alcohol.

Medical effects and treatment

Barium can cause injury to the heart as well as being a poison. Avoid contact in any form by wearing protective clothing, especially gloves as dermatitis is a possible result of skin contact. Obtain medical attention. Remove to hospital.

Benzene

Chemical formula C_6H_6

SUMMARY

A clear colourless liquid.

Flammable.

Poisonous vapours which are heavier than air.

Not soluble in water.

Also known as:

- Benzole (Commercial blend)
- Phenyl hydride
- Coal naphtha

PHYSICAL DATA

F.P.	12°F (– 11°C)
S.G.	0·88
V.D.	2·77
S.I.T.	995°F (535°C)
M. Pt.	41·9°F (5·5°C)
B. Pt.	176°F (80°C)
E.L. (L)	1·5%
E.L. (U)	8%

Industrial application

As a solvent for oil extraction from seeds etc., in the dye industry, gasworks, tar distilleries, explosives manufacture, dry cleaning and in motor spirit blending processes.

Storage and transport

In steel tanks on site. Transport in steel drums.

Dangerous properties

Very low flash point. Narcotic and poisonous fumes. Emits heavy black smoke when burning.

Fire-fighting

Breathing apparatus may be necessary. Avoid the use of water. Foam, CO_2, vaporising liquid or similar extinguishing agents should be used. Use sprays to cool containers.

Medical effects and treatment

Prolonged inhalation should be avoided. Effects of inhalation are tightening of the leg muscles, dizziness, breathlessness, perhaps hysteria. Obtain medical attention. Remove to fresh air at once. Keep warm, lie down flat, restrain any energetic movements. Remove to hospital.

Benzine

Chemical formula C_nH_{2n+2}

SUMMARY

A clear colourless liquid (mixture of several of the lighter constituents of petroleum).

Highly flammable liquid with flammable and explosive fumes.

Also known as petroleum ether.

PHYSICAL DATA

F.P.	0°F (– 17·8°C) or lower
S.G.	0·64
V.D.	4·5
B. Pt.	100 – 160°F (38 – 71°C)
E.L. (L)	1·1%
E.L. (U)	4·8%

Industrial application

Used in the manufacture of insecticides and solvents.

Storage and transport

Store away from heat, flame and oxidising agents.

Dangerous properties

Highly flammable: low flash point and explosive limits.

Fire-fighting

Use foam, dry powder, etc.

Medical effects and treatment

Fumes cause headache, nausea and vomiting, irregular respiration and drowsiness. Vapour will also affect the mucous membranes. Remove patient to fresh air.

Bromine

Chemical formula Br_2

SUMMARY

A dark red liquid easily recognised by the characteristic fuming and choking, irritant smell.

Very heavy in liquid and vapour form.

Non-flammable.

Very poisonous and corrosive.

PHYSICAL DATA

S.G.	3·12
V.D.	5·5 at 100°C
M. Pt.	19°F (– 7·2°C)
B. Pt.	138°F (58·9°C)

Industrial application

Used in the preparation of photographic materials, fire chemicals, also in the production of drugs and dyes.

Storage and transport

In blue-glass bottles (approx. 1·15 l) with ground glass stoppers sealed with clay. Bottles packed in Kieselguhr or similar non-combustible substance, in small wooden cases.

Dangerous properties

A very poisonous substance more powerful than chlorine. May re-act with organic materials to produce heating and possibly fire.

Fire-fighting

Breathing apparatus with eye protection and gloves should be worn if bromine is involved in a fire situation. Ventilate thoroughly as soon as possible. (Note vapours are extremely heavy.) Immerse leaking (liquid or vapour) containers in water, or cover with wet sand. Do not attempt to open cases. Liquids can be neutralised with sodium bicarbonate or lime.

Medical effects and treatment

Eyes, breathing organs and skin affected by vapours. Liquid will cause severe painful burns. Remove from concentration and obtain medical attention. Keep warm. Wash burns with a bicarbonate of soda solution or copious quantities of water. Remove to hospital.

Butadiene

Chemical formula CH_2CCHCH_3

SUMMARY

Colourless gas.

PHYSICAL DATA

F.P.	Below 20°F (– 6·7°C)
V.D.	1·87
S.I.T.	842°F (450°C)
B. Pt.	23·9°F (– 4·5°C)
E.L. (L)	2·0%
E.L. (U)	11·5%

Industrial application

Used extensively in synthetic rubber industry.

Storage and handling

When exposed to air may form peroxides that are very explosive. Store containers, which should be tightly closed, in a cool place, preferably out in open, away from fires, flame and any other fire hazards.

Dangerous properties

Narcotic effect in high concentrations. Liquid can produce severe burns on skin and damage to eyes.

Fire-fighting

Breathing apparatus should be worn in heavy concentrations. Use water spray, CO_2, dry powders.

Medical effects and treatment

Splashes on the skin or eyes should be washed with water at once.

Butane

(See Section 5: 'Fuels')

Chemical formula $CH_3(CH_2)_2CH_3$

SUMMARY

A colourless gas; liquid under pressure.
Flammable and explosive.

PHYSICAL DATA

F.P.	− 76°F (− 60°C)
S.G.	0·6
V.D.	2·046
S.I.T.	806°F (430°C)
B. Pt.	31·1°F (− 0·6°C)
E.L. (L)	1·5%
E.L. (U)	9·0%

Industrial application

For heating purposes. Also used domestically for heating and cooking where main gas is not available.

Storage and transport

In cylinders, cylindrical or spherical tanks as a liquid under pressure. Store in a cool place (preferably in the open) away from all forms of heat.

Dangerous properties

Highly flammable with a very low flash point and a low explosive limit. In sufficient concentration causes unconsciousness.

Fire-fighting

Breathing apparatus may be necessary. Also see Section 5: 'Fuels'.

Medical effects and treatment

Causes drowsiness followed by unconsciousness. Remove patient to fresh air. Apply artificial respiration and if necessary give oxygen.

Butyl acetate

Chemical formula $CH_3COOC_4H_9$

SUMMARY

A water-white liquid of characteristic fruity smell.

Lighter than water.

Slightly soluble in water.

Flammable

Explosive vapours.

PHYSICAL DATA

F.P.	73·4°F (23°C)
S.G.	0·88
V.D.	4·00
S.I.T.	790°F (421°C)
B. Pt.	258·8°F (126°C)
E.L. (L)	1·7%
E.L. (U)	15%

Industrial application

Widely used as a solvent of nitro-cellulose lacquer in both cold and hot spray systems and the lacquer industry.

Storage and transport

Stored in cans of 22·75–70 litres capacity, in steel drums of 205–410 litres capacity and in steel road and rail tank cars of 2275–17,275 litres capacity. Must be consigned as 'flammable'. The *Cellulose Solutions Regulations* apply to solutions in which butyl acetate is a constituent.

Dangerous properties

Low flash point of vapours.

Fire-fighting

Breathing apparatus may be necessary. Water in the form of heavy sprays for cooling, CO_2 or vaporising liquid to extinguish fire.

Medical effects and treatment

The vapour has a narcotic effect. Long exposure to fumes may cause giddiness and sickness. Remove to fresh air and keep warm until recovered.

Butyl alcohol

Chemical formula C_4H_9OH

SUMMARY

A water-white liquid with a characteristic smell.
Flammable.
Lighter than water.
Explosive vapours.
Slightly soluble in water.
Also known as butanol.

PHYSICAL DATA

F.P.	84°F (28°C)
S.G.	0·81
V.D.	2·5
S.I.T.	653°F (345°C)
B. Pt.	244°F (118°C)
E.L. (L)	1·7%
E.L. (U)	18%

Industrial application

Used extensively as a solvent or added to other solvents to make them more effective. Widely used in polishes, oil and spirit varnishes, paint removers and similar products.

For other headings see BUTYL ACETATE.

Cadmium

Chemical formula Cd.

SUMMARY

A silvery-white malleable metal.

Extremely poisonous.

PHYSICAL DATA

S.G.	8·64
M. Pt.	610°F (321°C)
B. Pt.	1415°F (768°C)

Industrial application

Used in metallurgy.

Dangerous properties

The metal and its fumes and dust are very poisonous, being analogous to arsenic and mercury. An added danger is that the effects may not be felt for some time.

Fire-fighting

See Section 7: 'Metal Fires'. Breathing apparatus is essential. Use dry sand—Do not use water or any form of liquid or gas extinguisher.

Medical effects and treatment

Symptoms are nausea, diarrhoea and abdominal pain. This poison seriously affects the lungs. It can cause death. Medical attention must be obtained at once.

Calcium carbide

Chemical formula CaC_2

SUMMARY

Dark grey lumps or crystals or colourless transparent crystals.

Dangerous when it contacts moisture.

PHYSICAL DATA

S.G.	2·22
M. Pt.	4172°F (2300°C)

Industrial application

Used in the manufacture of acetylene.

Storage and transport

In drums and tins. Keep away from moisture. Covered by statutory regulations.

Dangerous properties

It reacts with water or moisture to produce acetylene and calcium hydroxide. Contact with the skin will enable calcium carbide to absorb enough moisture to react in this way and skin burns are caused by the resulting calcium hydroxide. Calcium carbide also reacts with metals such as copper or silver to form acetylides which produce small explosions. These are likely to ignite the large amount of acetylene which has been liberated.

Fire-fighting

Breathing apparatus may be necessary. Use absolutely dry sand, powdered talc or asbestos.

Medical effects and treatment

Burns from calcium hydroxide occur where the calcium carbide has contacted any part of the body. If the dust has been inhaled the mucous membranes and upper respiratory tract will be irritated. Wash the affected parts with plenty of water. If the chemical has reached the eyes flush them freely with water for at least 15 minutes and see that the patient sees a doctor.

Calcium cyanamide

Chemical formula $CaCN_2$

SUMMARY

Colourless crystals.

Non-flammable.

Toxic.

PHYSICAL DATA

M. Pt.	2372°F (1300°C)
B. Pt.	2174°F (1190°C)

Industrial application

Used in fertilisers.

Dangerous properties

Poisonous and caustic. Vapours can cause severe irritation of skin and mucous membranes.

Fire-fighting

Use breathing apparatus. Use CO_2 or dry powder.

Medical effects and treatment

Damages the skin if touched. Toxic vapours cause transitory redness in the face, irritation of the nose, throat and skin and headache. Remove patient to fresh air and wash affected parts.

Calcium oxide

Chemical formula CaO

SUMMARY

Colourless crystals or a white powder.

Very caustic.

Also known as quicklime.

PHYSICAL DATA

S.G.	3·4
M. Pt.	4676°F (2580°C)
B. Pt.	5162°F (2850°C)

Industrial application

Used in laboratories as a drying agent. It may be used as a furnace lining and in making crucibles.

Storage and transport

Store away from moisture.

Dangerous properties

Very caustic. As dust it irritates the skin, mucous membranes and upper respiratory tract and is most dangerous to the eyes. It reacts with water to yield calcium hydroxide and in the process evolves a great deal of heat.

Fire-fighting

Breathing apparatus and protective clothing and eye coverings are needed. Use dry lime, absolutely dry sand.

Medical effects and treatment

Affected areas, especially the eyes, should be washed immediately with large quantities of water. Call a doctor.

Camphor

Chemical formula $C_{10}H_{16}O$

SUMMARY

A white translucent crystalline mass with a penetrating smell.

Toxic.

Flammable.

Vapours flammable and explosive.

PHYSICAL DATA

F.P.	150°F (65·6°C)
S.G.	0·99
V.D.	5·24
S.I.T.	871°F (466°C)
M. Pt.	345°F (174°C)
B. Pt.	399°F (204°C)

Industrial application

Used as a plasticiser in making celluloid and other plastics.

Storage and transport

Store in a cool, well-ventilated area away from fire hazards and oxidising agents.

Dangerous properties

A flammable material. Vapours form flammable and explosive mixtures with air. Toxic.

Fire-fighting

Breathing apparatus and eye protection should be worn. Use water spray: foam, carbon dioxide, vaporising liquid or dry powder.

Medical effects and treatment

Causes blurred vision and dizziness. Remove patient to fresh air.

Carbon black

(See Section 9, 'Paint and varnish fires')
Chemical formula C

SUMMARY

A fluffy, velvety black substance consisting of finely divided carbon.

Burns very slowly without smoke or flame.

Toxic.

PHYSICAL DATA

M. Pt.	Above 6332°F (3500°C)
B. Pt.	7592°F (4200°C)

Industrial application

Manufactured separately or, more often, as a by-product. Used in printing inks, polishes and blacking, black paint, carbon paper, gramophone records, brushes for dynamos, carbons for arc lamps, electrodes, insulating materials, imitation leather and other articles.

Dangerous properties

When black is burning the fire is difficult to detect. Dust explosion risk. If black has absorbed water, the water will evaporate and be replaced by air or gas with evolution of heat. Carbon black can irritate the eyes.

Fire-fighting

Breathing apparatus and eye protection needed. Fires in carbon black should be attacked with a fine spray or be flooded through an open-ended hose, care being taken to turn over the heap thoroughly.

Medical effects and treatment

Burning carbon black can cause irritation and conjunctivitis. Remove patient from exposure.

Carbon dioxide

Chemical formula CO_2

SUMMARY

A colourless gas with no smell.

Non-flammable.

Heavier than air.

Soluble in water.

For cylinder marking colours, see plates 42 *and* 43.

PHYSICAL DATA

V.D.	1·53
M. Pt.	– 72°F (– 57·7°C)
B. Pt.	– 109°F (– 78·2°C)

Industrial application

Used in sugar refining, aerated water manufacture, manufacturing sodium carbonate and as a fire extinguisher.

Storage and transport

In steel cylinders. As a liquid in refrigerated tanks and tankers at high pressure.

Dangerous properties

High concentrations cause asphyxiation and death. As liquid carbon dioxide exists only at very low temperature or at high pressure, in normal temperatures containers of it are liable to disintegrate violently if dropped or heated. Solid carbon dioxide causes burns if it touches the body.

Fire-fighting

Breathing apparatus should be worn in the presence of free carbon dioxide. Containers of liquid or gas under pressure should be kept cool, the danger of increase in pressure and violent disintegration being the only risk. Carbon dioxide is non-flammable.

Medical effects and treatment

Exposure to the gas results in laboured breathing followed by collapse. Remove patient to fresh air and if necessary apply artificial respiration and oxygen.

Carbon disulphide

Chemical formula CS_2

SUMMARY

A yellow or colourless liquid with an obnoxious 'coal-gas' smell.

Highly flammable.

Toxic vapours given off.

Liquid heavier than water.

Vapours much heavier than air (note vapour density).

PHYSICAL DATA

F.P.	– 22°F (– 30°C)
S.G.	1·29
V.D.	2·5
S.I.T.	257°F (125°C)
M. Pt.	– 162°F (– 108°C)
B. Pt.	114·8°F (46°C)
E.L. (L)	1%
E.L. (U)	50%

Industrial application

Found in a wide variety of industries, e.g. artificial silk, rubber, leather, oil seed extractions, insecticides, fumigants, etc.

Storage and transport

Small glass or earthenware bottles in laboratories etc. Larger quantities in steel drums or tanks on working site. Transport in steel drums and road and rail tankers. Movement governed by Regulations.

Dangerous properties

Extremely low flash point. Vapours can be ignited by hot steam pipe or heat from electric light bulb. Closed containers subjected to even moderate heat (sunlight) may burst due to internal pressures. Liable to shock ignition. Burns to produce large volumes of SO_2.

Fire-fighting

Breathing apparatus should be worn. CO_2, or similar smothering agent can be used. As CS_2 is heavier than water a gentle application of water to form a seal should be attempted. Use earth and sand on spillage.

Medical effects and treatment

Remove affected persons at once. Even short exposure to the fumes can lead to serious illness such as nervous collapse and paralysis. Direct skin contact should be avoided. Keep patient warm and apply oxygen. Use artificial respiration should breathing cease. Obtain medical attention. Remove to hospital.

Carbon monoxide

Chemical formula CO

SUMMARY

A colourless gas, without odour.

Very poisonous.

Flammable. Burns with a blue flame.

Slightly soluble in water.

May form an explosive mixture with oxygen.

For cylinder marking colours, see plate 43.

PHYSICAL DATA

V.D.	0·96
S.I.T.	1204°F (650·5°C)
E.L. (L)	12·5%
E.L. (U)	74·2%

Industrial application

It is liable to be found in blast furnaces, electric furnaces, kilns, charcoal ovens, coal mines, oil distilleries and garages.

Dangerous properties

Extremely poisonous. Flammable when mixed with air. Severe explosion hazard when exposed to flame. Combines with chlorine in sunlight or in presence of charcoal to yield phosgene (*q.v.*).

Fire-fighting

Breathing apparatus should be worn. Use water spray, carbon dioxide or dry powder.

Medical effects and treatment

Symptoms of carbon monoxide poisoning are headache, dizziness and nausea (followed by collapse and death in severe cases). Remove patient immediately to fresh air and make him lie down. Early treatment is most important. Apply artificial respiration. Give oxygen. Obtain medical attention.

Carbon tetrachloride

Chemical formula CCl_4

SUMMARY

A colourless heavy liquid with a sweet smell.

Non-flammable.

Vapours are toxic and heavier than air.

PHYSICAL DATA

S.G.	1·6
V.D.	5·3
M. Pt.	– 8°F (– 22·6°C)
B. Pt.	170·6°F (77°C)

Industrial application

Carbon tetrachloride was used as a fire extinguishing agent but is no longer recommended on account of its toxicity; used in dye works, dry cleaners etc., as a degreasing agent or solvent.

Storage and transport

In drums and small canisters.

Dangerous properties

Reaction of the vapours to heat may produce phosgene gas which can prove rapidly fatal. Combines with water to form hydrochloric acid.

Fire-fighting

Breathing apparatus should be worn in even a weak concentration of vapours in case phosgene gas forms a part of the atmosphere. Ventilate thoroughly as early as possible.

Medical effects and treatment

Remove exposed persons to fresh air. Giddiness and staggering are typical symptoms. Administer oxygen to relieve breathing, and keep patient warm. Serious illness can result from exposure to the fumes and all patients should be removed to hospital.

Cellulose nitrate

Chemical formula $C_{12}H_{14}(ONO_2)_6O_4$
or various according to composition.

SUMMARY

Synonyms: Nitrated cellulose
Nitrocellulose
Pyrocotton
Guncotton
Pyroxylin
Collodion wool

Cellulose, the tough woody part of vegetation is dissolved in nitric acid to form the basis of the compound. Camphor is added to the solution to form celluloid. Very unstable, decomposes with age evolving oxides of nitrogen. Is the basis of various explosives, ammunition, smokeless (propellent) powders. Highly flammable.

PHYSICAL DATA

F.P.	40°F (4·4°C)
S.I.T.	250°F (121°C) (approx.) (celluloid)
M. Pt.	160°F (71·1°C) (celluloid)

Industrial application
The preparation of explosives of many types. Lacquers. Celluloid.

Storage and transport
Strictly controlled by legislation.

Dangerous properties
Possibility of ignition with explosive violence. Toxic vapours—heavier than air.

Fire-fighting
Breathing apparatus should be worn especially in confined spaces or concentration of the fumes. Once ignited most nitrated celluloses burn as an explosion or, at least, with great rapidity. Flooding with water may be a means of extinguishing the slower burning groups but it should be noted that most compounds contain enough oxygen to enable them to support combustion *under water*. Thorough ventilation of fumes is most important.

Medical effects and treatment
Fumes are highly toxic. Administer oxygen if breathing is affected. Keep warm.

Chlorates

Chemical formula

Potassium chlorate $KClO_3$, Sodium chlorate $NaClO_3$, etc.

SUMMARY

Very strong oxidising agents.

Can cause ignition by friction.

Explosive.

Soluble in water.

Toxic.

PHYSICAL DATA

Decompose violently at 752°F (400°C)

Dangerous properties

Chlorates are explosive and can be caused to ignite violently by shock or friction, and by heat at 752°F (400°C). Any organic material mixed with chlorates may ignite through friction or heating, and materials drying out after soaking in a chlorate solution are highly flammable. Chlorates also react violently with strong acids and other chemicals.

Fire-fighting

Breathing apparatus should be worn for protection against inhaling chlorates dust, which is toxic. Water sprays or open-ended hose lines may be used to flush chlorates away to a point where subsequent reignition will do no harm. Containers must be kept cool. Clothing etc., which has been wetted with a chlorate solution should be kept away from heat and flame and thoroughly flushed to remove all traces.

Chlorinated lime

Chemical formula $Ca(ClO)_2 \cdot 2H_2O$

SUMMARY

Commonly known as chloride of lime or bleaching powder.

A whitish powder which when fresh has a distinct smell of chlorine.

NON-flammable.

Liberates poisonous fumes (chlorine) when heated or in contact with acids.

PHYSICAL DATA

Industrial application

Chlorinated lime is used in the paper and textile industries as a bleaching agent. Also found in the disinfectant trade. Used in the preparation of chloroform, and the purification of water (swimming baths).

Storage and transport

Stored in packets, tins and in 200–300 kilogram casks and steel drums. May be marketed under a trade name.

Dangerous properties

Evolves chlorine gas when heated or when in contact with acids. Explosive by pressure and, in certain instances, by contamination.

Fire-fighting

Breathing apparatus should be worn. Use water preferably in the form of sprays to help lay any attendant dust. Avoid wetting undamaged stocks. Ventilate as soon as possible.

Medical effects and treatment

Is injurious to the eyes in powder form. Fumes cause swelling of air passages resulting in difficult and painful breathing. Oxygen may be required. Remove from fumes. Rest and keep warm. Apply artificial respiration if breathing stops.

Chlorine

Chemical formula Cl_2

SUMMARY

A poisonous gas.
Greenish yellow in colour.
Heavier than air.
Moderately soluble in water.
Explosive (see 'Dangerous properties')
An element.

For cylinder marking colours, see plate 43.

PHYSICAL DATA

V.D.	2·49
M. Pt.	– 150°F (– 101°C)
B. Pt.	– 30°F (– 34·6°C)

Industrial application

Chlorine has a very wide application in industry being found in various quantities in the following trades:

Textiles, water purification, metallurgical processes, dyes and chemicals, paper works, petroleum refining, alkali works, tanneries.

Storage and transport

In steel cylinders in the form of a pressurised liquid.
Also in steel drums, coloured yellow.

Dangerous properties

Apart from the extremely poisonous nature of this gas it is liable to explode with hydrogen under the action of light (sunlight) or heat. Chlorine decomposes ammonia gas to form ammonium chloride which is used in the production of nitrogen chloride, a violent explosive. Hazardous in contact with finely divided metals, coal-gas and similar hydrocarbons.

Fire-fighting

Breathing apparatus should be worn. Effect thorough ventilation as early as possible. Water sprays to dissolve gas. Completely submerge leaking containers in water.

Medical effects and treatment

One part chlorine in 1000 parts of air can prove fatal. Administer oxygen, and give artificial respiration if breathing ceases. Obtain medical attention. Keep warm. Remove to hospital.

Chlorobenzene

Chemical formula C_6H_5Cl

SUMMARY

A colourless liquid.

Highly flammable with flammable and explosive vapours.

Narcotic fumes.

PHYSICAL DATA

F.P.	85°F (29·4°C)
S.G.	1·11
V.D.	3·88
S.I.T.	1245°F (673·8°C) and higher
B. Pt.	269°F (132°C)
E.L. (L)	1·8% (at 212°F) (100°C)
E.L. (U)	9·6% (at 302°F) (150°C)

Industrial application

Used as an intermediary in the production of industrial chemicals.

Dangerous properties

Very flammable. Vapours form explosive mixtures with air. Fumes have narcotic effects.

Fire-fighting

Breathing apparatus should be worn. Use foam, carbon dioxide, vaporising liquid, dry powder.

Medical effects and treatment

Administer oxygen and obtain medical attention.

Chloroform

Chemical formula $CHCl_3$

SUMMARY

A heavy mobile colourless liquid.

Heavier than water.

Pleasant sweet smell.

Vapours heavier than air and strongly narcotic.

Non-flammable (see 'Dangerous properties').

Does not readily mix with water.

PHYSICAL DATA

S.G.	1·5
V.D.	4·12
M. Pt.	– 81°F (– 63·5°C)
B. Pt.	142°F (61·2°C)

Industrial application

Chiefly known as an anaesthetic and found in operating theatres, surgeries, laboratories, chemical works, etc. Also used commercially as a solvent.

Storage and transport

In glass, earthenware or steel containers usually well protected to avoid breakage.

Dangerous properties

Although chloroform is generally accepted as non-flammable it does in fact burn at high temperatures (above 1832°F (1000°C)). Care should be taken to avoid fumes which will rapidly induce unconsciousness.

Fire-fighting

Breathing apparatus should be worn. Ventilate thoroughly as early as circumstances permit.

Medical effects and treatment

Prolonged exposure to the fumes can prove fatal. Remove exposed persons and administer oxygen with—if necessary—artificial respiration. Keep warm.

Chlorosulphonic acid

Chemical formula $ClSO_3H$

SUMMARY

This substance is a clear to cloudy, colourless to pale yellow liquid.

It is non-flammable but can cause ignition when in contact with combustible materials.

Toxic in either liquid or vapour state.

Immiscible with water.

PHYSICAL DATA

S.G.	1·766 at 18°C
V.D.	4·02
M. Pt.	- 112°F (- 80°C)
B. Pt.	303°F (151°C)

Industrial application

Used in soap manufacture.

Storage and transport

Store in tightly closed containers, in a well ventilated and cool storage place, away from any water or excessive moisture, and away from all carbonaceous materials and fire hazards.

In transport all containers should carry a warning notice as a corrosive substance.

Dangerous properties

Corrosive to most metals and in the process of corrosion liberates hydrogen. Very reactive in the presence of water and moisture, causing decomposition evolving a great deal of heat and dense white fumes which contain very large quantities of highly toxic hydrogen chloride. The other component is sulphuric acid, also corrosive and toxic. Both these gases can cause severe burns, irritation to eyes, lungs and mucous membranes. Vapour inhaled can cause conjunctivitis.

Fire-fighting

Do not use water, or water-type extinguishers to control fires involving this substance. Breathing apparatus must always be worn, and, if drums or containers have to be moved, rubber gloves should be worn as protection against burns.

Medical effects and treatment

Personnel overcome by the vapour during fire-fighting operations should be removed to open at once and kept quiet and warm but not hot. If coughing develops oxygen should be administered until the coughing is reduced. Do not encourage to vomit. Call a doctor at once.

Chromates and dichromates

Chemical formula

Varies according to compound, most common are:

Potassium chromate: K_2CrO_4, Potassium dichromate: $K_2Cr_2O_7$

Ammonium dichromate: $(NH_4)_2Cr_2O_7$

SUMMARY

Solids, usually crystalline in character.

Dichromates usually dark orange or red.

Chromates usually bright yellow.

Soluble in water.

Powerful oxidising agents.

Toxic vapours.

PHYSICAL DATA

According to compound.

Industrial application

Both are used in the tanning industry also in the manufacture of photographic materials, paints, dyestuffs. In addition chromates are also used in the match industry, for bleaching and in the printing ink industry.

Storage and transport

In casks or in cases; no special transport regulations.

Dangerous properties

Extremely toxic dusts and vapours, supporters of combustion. Toxic fumes are evolved when in contact with acids, with great violence in the case of sulphuric acid.

Fire-fighting

Breathing apparatus should be worn. Flooding with open ends or sprays offers the best method of attack. Avoid the use of sprays which will scatter the substance. Care should be taken to prevent combustibles from becoming impregnated with the solution as on drying out spontaneous ignition may occur.

Medical effects and treatment

Wash any contaminated parts of the body with copious supplies of warm water and soap. Treat clothing in a similar way until all traces are removed. If personnel show signs of skin irritation or other symptoms obtain medical attention.

Cyclohexanol

Chemical formula $C_6H_{11}OH$

SUMMARY

Colourless needles, or viscous liquid, with a smell of camphor.

Flammable.

Toxic.

Vapour harmful, but the liquid is not very volatile.

Also known as 'Hexoline'.

PHYSICAL DATA

F.P.	154°F (67·8°C)
S.G.	0·94
V.D.	3·45
M. Pt.	59°F (15°C)
B. Pt.	322°F (161·5°C)

Industrial application

Used as an intermediary in chemical production.

Dangerous properties

Flammable. Dangerous to the eyes. Skin contact should be avoided. It can be absorbed through the skin.

Fire-fighting

Use breathing apparatus and protective clothing. Use water (fog or spray), foam, carbon dioxide, vaporising liquid and dry powder.

Medical effects and treatment

Irritates the eyes, nose and throat. Severe exposure can paralyse the central nervous system. Remove patient to fresh air. In serious cases obtain medical advice and apply artificial respiration.

Cymag

SUMMARY

An off-white insecticide powder.

Gives off hydrogen cyanide on contact with water.

Smells of bitter almonds.

Industrial application

An insecticide. May be found in chemists and agricultural stores.

Storage and transport

Supplied in clearly marked hermetically sealed tins.

Dangerous properties

Possesses no special combustible properties but may generate a dangerous and highly toxic concentration of hydrogen cyanide if water reaches it. Evolves hydrogen cyanide also, but more slowly on contact with atmospheric moisture.

Fire-fighting

Breathing apparatus should be worn and civilian personnel should be evacuated. Efforts should be made to remove stocks to the open and the building should be opened for ventilation.

Medical effects and treatment

Anyone overcome should be removed to the open air, artificial respiration applied and a doctor sent for immediately.

Cymene, Para-

Chemical formula $CH_3C_6H_4CH(CH_3)_2$

SUMMARY

Also known as isopropyltoluene.

Colourless liquid.

Smells of parsley.

Flammable, explosive.

PHYSICAL DATA

F.P.	117°F (47·2°C)
S.G.	0·86
V.D.	4·62
S.I.T.	921°F (494°C)
M. Pt.	– 90°F (– 68°C)
B. Pt.	349°F (176°C)
E.L. (L)	0·7% @ 212°F (100°C)

Industrial application

A solvent for the manufacture of resins.

Dangerous properties

Flammable. Explosive.

Fire-fighting

Breathing apparatus may be necessary. Use foam, carbon dioxide, vaporising liquid or special dry chemical.

Dichloroethylene

Chemical formula $C_2H_2Cl_2$

SUMMARY

A colourless liquid with a pleasant smell.

Flammable and explosive when subjected to heat or flame.

Toxic.

PHYSICAL DATA

F.P.	43°F (6·1°C)
S.G.	1·28
V.D.	3·35
M. Pt.	– 112°F (– 80°C)
B. Pt.	138·2°F (59°C)
E.L. (L)	9·7%
E.L. (U)	12·8%

Industrial application

A solvent used widely in industry.

Storage and transport

Store in a cool area away from heat or flame.

Dangerous properties

In presence of air or oxygen it forms violently explosive peroxides. Flammable. Slightly toxic.

Fire-fighting

Breathing apparatus may be necessary. Use water (fog or spray), foam, carbon dioxide, vaporising liquid, or dry powder.

Medical effects and treatment

Irritates mucous surface. Has a narcotic effect similar to chloroform. May cause eyes to water. Remove patient to fresh air.

Diethyl ether

Chemical formula $C_2H_5OC_2H_5$

SUMMARY

A colourless liquid with the characteristic 'ether' smell.

Volatile and flammable.

Slightly soluble in water.

Sometimes referred to as 'sulphuric ether'.

Lighter than water.

Vapour heavier than air.

PHYSICAL DATA

F.P.	– 45°F (– 42·8°C)
S.G.	0·714
V.D.	2·56
S.I.T.	356°F (180°C)
B. Pt.	94°F (34·5°C)
E.L. (L)	1·85%
E.L. (U)	36·5%

Industrial application

Used as an anaesthetic in hospitals, also used commercially in the preparation of collodion and smokeless gunpowder, etc.

Storage and transport

In glass bottles or tins in protective boxes. Also in steel drums. May be found in carboys.

Dangerous properties

Highly flammable. One of the most hazardous of chemicals. Wide explosion limits.

Fire-fighting

Breathing apparatus may be necessary. Use water for cooling containers. Extinguish fire with foam, CO_2, dry powder or dry sand. Vaporising liquid may be used on small fires.

Medical effects and treatment

Prolonged inhalation will induce unconsciousness. High concentrations can be fatal. Remove to fresh air and keep warm.

Ethyl acetate

Chemical formula $CH_3COOC_2H_5$

SUMMARY

A colourless liquid with a pleasant fruity smell.
Lighter than water.
Flammable.
Explosive, heavier than air, vapours.
Soluble in water.

PHYSICAL DATA

F.P.	24·8°F (– 4°C)
S.G.	0·9
V.D.	3·04
S.I.T.	799°F (427°C) and higher
B. Pt.	170·6°F (77°C)
E.L. (L)	2%
E.L. (U)	11·5%

Industrial application

Used as a solvent of lacquers and other nitrocellulose substances. Also in medicine, perfumery and as a food flavouring essence.

Storage and transport

In cans of 22·75 – 70 litre capacity, in drums (steel) of 205 – 410 litres capacity and in tank cars of 2,275 – 17,275 litres capacity. Movement of ethyl acetate is controlled by *Cellulose Solutions Regulations*.

Dangerous properties

Highly flammable; low explosive limit.

Fire-fighting

Breathing apparatus may be necessary. CO_2, or vaporising liquid extinguishers can be used. Use water sprays to clear atmosphere and to cool containers. Avoid inhalation of concentrated vapours.

Medical effects and treatment

The vapours have a narcotic effect if in sufficient strength. Remove exposed persons and administer oxygen if necessary. Keep warm.

Ethyl alcohol

Chemical formula CH_3CH_2OH

SUMMARY

Colourless liquid.

Highly flammable with flammable and explosive vapours.

Soluble in water.

Slightly toxic.

Also known as 'ethanol'.

PHYSICAL DATA

F.P.	55°F (12·8°C)
S.G.	0·79
V.D.	1·59
S.I.T.	700°–800°F (371°–427°C)
M. Pt.	– 173°F (– 114°C)
B. Pt.	173°F (78·5°C)
E.L. (L)	3·5%
E.L. (U)	19%

Industrial application

Used in the manufacture of foodstuffs and drink, and in drugs.

Storage and transport

Store away from heat and flame.

Dangerous properties

Flammable liquid and vapours. Vapours form explosive mixtures with air. Very high concentrations can be toxic.

Fire-fighting

Breathing apparatus and eye protection required. Use water spray, alcohol-resistant foam, carbon dioxide, vaporising liquid, dry powder.

Medical effects and treatment

Irritates eyes and upper respiratory tract. In high concentration vapour can cause intoxication. Remove patient to fresh air.

Ethyl chloride

Chemical formula C_2H_5Cl

SUMMARY

A colourless heavier-than-air gas.
Explosive and toxic.
Smells slightly of 'ether'.
Slightly soluble (2 per 100) in water.
Sometimes called 'monochloroethane'.

For cylinder marking colours, see plate 43.

PHYSICAL DATA

F.P.	– 58°F (– 50°C)
S.G.	0·92
V.D.	2·22
S.I.T.	966°F (520°C)
B. Pt.	54°F (12°C)
E.L. (L)	3%
E.L. (U)	15%

Industrial application

Used in the fire chemical industry, in hospitals and doctors' surgeries. An anaesthetic. A refrigerant.

Storage and transport

Easily condensed into a liquid (54°F (12°C)) and usually stored in this form in steel cylinders. When used as a refrigerant methyl bromide is added to reduce its flammability. Sometimes found in small sealed glass bottles packed in boxes.

Dangerous properties

Low flash point, toxicity. Decomposition products may include fumes of hydrochloride acid.

Fire-fighting

Use breathing apparatus. Extinguish fires with CO_2 or vaporising liquid. Keep containers cool with sprays. Ventilate thoroughly.

Medical effects and treatment

Causes irritation to the skin and is harmful to the eyes. Remove splashed clothing to reduce skin absorption of liquid. Keep warm. Apply oxygen if necessary.

Ethylene

Chemical formula $H_2C{:}CH_2$

SUMMARY

A colourless gas with a sweet smell.

Flammable and explosive.

Slightly soluble in water.

For cylinder marking colours, see plates 42 *and* 43.

PHYSICAL DATA

V.D.	0·98
S.I.T.	842°F (450°C)
E.L. (L)	2·7%
E.L. (U)	28·6%

Industrial application

In hospitals and the manufacture of drugs and as a fruit ripener.

Storage and transport

In steel cylinders coloured mauve with a red cap.

Dangerous properties

Flammable and explosive. It explodes spontaneously in presence of chlorine in sunlight. May asphyxiate in an enclosed air space by reducing the oxygen.

Fire-fighting

Use breathing apparatus. Use fine water spray, carbon dioxide or dry powder.

Medical effects and treatment

Remove patient to fresh air. Apply artificial respiration. Give oxygen if necessary.

Ethylene dichloride

Chemical formula $CH_2Cl.CH_2Cl$.

SUMMARY

A heavy oily colourless liquid.
Smells like chloroform.
Not soluble in water.
Vapours heavier than air.
Flammable.
Otherwise known as:
Ethylene chloride
Ethylene-glycol-dichloride
Dichloroethane
Dutch liquid.

PHYSICAL DATA

F.P.	70°F (21°C)
S.G.	1·25
V.D.	3·35
S.I.T.	775°F (413°C)
M. Pt.	– 23°F (– 30·6°C)
B. Pt.	182°F (83·3°C)
E.L. (L)	6%
E.L. (U)	16%

Industrial application

Used as a solvent of oils, fats, waxes, gums and certain resins. Also used as an anaesthetic.

Storage and transport

In steel cylinders or cans.

Dangerous properties

Vapours are anaesthetic. Low explosion limit.

Fire-fighting

Breathing apparatus should be worn. Use water spray, foam, CO_2, vaporising liquid or dry powder.

Medical effects and treatment

Prolonged inhalation will bring about unconsciousness. Keep warm. Administer oxygen if necessary.

Ethylene glycol

Chemical formula $CH_2OH.CH_2OH$

SUMMARY

A clear heavy syrupy liquid.
Slightly heavier than water.
Freely soluble in water.
Flammable.
Vapour heavier than air.
Usually referred to as 'glycol'.

PHYSICAL DATA

F.P.	232°F (111°C)
S.G.	1·1
V.D.	2·14
S.I.T.	775°F (413°C)
B. Pt.	387°F (197°C)
E.L. (L)	3·2%

Industrial application

Used extensively as an anti-freeze agent in internal combustion engines. Also used as a de-icing agent on aircraft, as a solvent for dyes and as a lubricant in the wool spinning industry.

Storage and transport

In cans or drums.

Dangerous properties

Low explosive limit.

Fire-fighting

Dilution would be the most effective method if possible. Use alcohol-resistant foam, CO_2 or dry powder.

Medical effects and treatment

No effects on skin etc. Little ill effects unless prolonged exposure to heavy concentration of vapours.

Fulminates

Chemical formula

Fulminate of silver $Ag_2C_2N_2O_2$, Fulminate of mercury $HgC_2N_2O_2$

SUMMARY

A high explosive.

Very hazardous.

Sensitive when dry to shock and friction.

Explodes violently on contact with flame.

Toxic fumes.

PHYSICAL DATA

Variable according to compound.

Industrial application

May be found in varying quantities and forms in arsenals, military establishments, quarries, mines and other industries where high explosives might be used. Another compound, gold fulminate, is used in the gilding trades.

Storage and transport

Strictly controlled by Regulations, showing explosives labels, etc.

Dangerous properties

Extremely sensitive to flame, impact and friction. Explodes violently. Toxic fumes are produced during manufacture. A solution of fulminates and water or wetted dusts may explode on drying out.

Fire-fighting

Breathing apparatus may be necessary. Fire-fighting in the accepted sense is not normally possible because of the sensitivity of the substances. Complete flooding reduces the risk of explosion. Care must be taken to ensure that ALL stocks in the danger area are so treated.

Medical effects and treatment

Toxic fumes and dusts may be encountered. Avoid handling the raw materials. If nausea or other symptoms appear, obtain medical attention immediately.

Glycerine

Chemical formula $CH_2OH \cdot CHOH \cdot CH_2OH$

SUMMARY

Colourless or pale yellow syrupy liquid, with no smell.
Flammable.
Soluble in water.

PHYSICAL DATA

F.P.	349°F (176°C)
S.G.	1·25
V.D.	3·17
S.I.T.	739°F (393°C)
M. Pt.	64°F (17·9°C)
B. Pt.	554°F (290°C)

Industrial application

Used in the manufacturing of fats, soaps, polishes, drugs, explosives.

Dangerous properties

Flammable. Can ignite spontaneously. Reacts with permanganate of potassium.

Fire-fighting

Use water (fog or spray) to dilute, alcohol-resistant foam, carbon dioxide, vaporising liquid or dry powder.

Hydrochloric acid

Chemical formula HCl

SUMMARY

A water-white fuming liquid when pure. Usually found in commercial form (28—32% strength) when it is straw coloured due to traces of iron dissolved in it. Sometimes called 'spirits of salts' or 'muriatic acid'.

Non-combustible in air.

Corrosive and irritating fumes.

Slightly heavier than water.

Highly soluble in water.

Poisonous.

PHYSICAL DATA

S.G.	1·2
M. Pt.	– 173°F (– 114°C)
B. Pt.	– 121°F (– 85°C) (pure form)

Industrial application

Used extensively in the textile, galvanising, chrome-tanning and wire drawing industries, also metal etching, soldering and the dyestuffs industries.

Storage and transport

Usually in glass carboys. Rubber lined steel tankers are used to transport the acid in bulk. Vessels made of 'Staffordshire blue' bricks bonded with a patent cement may be found in use on static storage sites.

Dangerous properties

Non-combustible in air, as stated, but when in contact with most metals hydrogen gas—which may of course form explosive mixtures—is evolved. Solutions are poisonous and vapours may damage mucous membrane.

Fire-fighting

Breathing apparatus should be worn. Every effort must be made to safeguard containers against damage. Keep acid away from other substances. Use soda ash or slaked lime to dam free liquid. Dilute thoroughly and ventilate.

Medical effects and treatment

Even a short exposure to the fumes may cause congestion of the lungs and possibly death. Remove affected persons at once and obtain medical attention. Keep warm and resting and administer oxygen if necessary. Remove to hospital. Skin burns should be treated with a cloth moistened in bicarbonate of soda solution.

Hydrocyanic acid

Chemical formula HCN

SUMMARY

A colourless liquid smelling of bitter almonds.
Soluble in water.
Liquid and vapours are extremely poisonous.
Liquid highly flammable.
Vapours highly flammable and explosive.
Burns with a violet flame.
Also known as 'prussic acid' or as 'hydrogen cyanide'.

For cylinder marking colours, see plate 44.

PHYSICAL DATA

F.P.	0°F (– 17·8°C)
S.G.	0·69
V.D.	0·932
S.I.T.	1000°F (538°C)
B. Pt.	78·8°F (26°C)
E.L. (L)	5·6%
E.L. (U)	40%

Industrial application

Used in chemical manufacture.

Storage and transport

In lined steel drums, and glass carboys.

Dangerous properties

Flammable liquid and vapours. Vapours form explosive mixtures with air at all temperatures. Exceedingly poisonous. Paralyses the respiratory centre of the brain and acts very quickly. Can be absorbed through the skin.

Fire-fighting

Breathing apparatus and protective clothing needed. Use large volume of water. Make sure the acid is thoroughly diluted. Warn the Local Authority of its presence in the drains. Any clothing that has been splashed with the acid must be changed immediately.

Medical effects and treatment

Symptoms are irritation of the throat, watering of the eyes, difficulty in breathing, weakness, giddiness, then collapse, convulsions and death. Call a doctor immediately. This poison acts very rapidly and death is likely to occur before the doctor can arrive. Get instructions by telephone. Recommended treatments are:—Remove patient to fresh air. At once apply artificial respiration. Inhalations of amyl nitrate can be given.

Patient can be given to drink solutions of hydrogen peroxide in water or potassium permanganate in water. Vomiting can be induced with emetic or by irritating the pharynx.

Special notes

Burning celluloid forms hydrocyanic acid. A small quantity of celluloid can produce a fatal amount of the acid.

Hydrofluoric acid

Chemical formula HF

SUMMARY

Aqueous solution of hydrogen fluoride.

Gas evolved when solution is heated.

Acid miscible with water.

Extremely toxic.

Extremely corrosive.

PHYSICAL DATA

S.G.	0·988
M. Pt.	– 134°F (– 92°C)
B. Pt.	67°F (19·4°C)

Industrial application

Distilleries, glass works (etching), laboratories and chemical stores. It may also be present as fumes in fertiliser works, during the purification of graphite from silica, in the cleaning of iron castings and in the removing of vitreous enamel from sheet steel.

Storage and transport

As a solution, usually 30 to 60 per cent., in lead carboys, gutta-percha bottles, paraffin wax containers (small amounts only) or even in platinum containers. Subject to transport regulations.

Dangerous properties

Extremely corrosive, evolves hydrogen gas when in contact with metals. The presence of this gas calls for 100 per cent protection as a precautionary measure. (An exposure time of 30–60 minutes in a concentration of 50 parts per million is considered to be fatal.)

Fire-fighting

Breathing apparatus *must* be worn with protection for eyes and exposed skin. Neutralise spilt acid with plenty of chalk. Avoid contact with acid.

Medical effects and treatment

Liquid in the eyes can cause immediate blindness, and serious and painful burns on the skin. Wash affected parts of the body in running stream of water. Treat burns with compresses soaked in a solution of bicarbonate of soda. Obtain medical attention for affected cases.

Hydrogen

Chemical formula H_2

SUMMARY

A colourless gas with no smell.

Highly flammable and explosive (note wide limits).

Much lighter than air.

Slightly soluble in water.

For cylinder marking colours, see plate 44.

PHYSICAL DATA

V.D.	0·069
S.I.T.	1085°F (585°C)
E.L. (L)	4·1%
E.L. (U)	74%

Industrial application

Used for hydrogenation of coal, for hardening oils, in the manufacture of ammonia and of hydrogen chloride. Can be liberated in a wide number of metal reactions (see Section 7: 'Fires involving Metals'), for example where heated metals or metal dusts such as aluminium or magnesium contact water or even where unheated metals of the sodium type contact water.

Storage and transport

Store in a cool, well-ventilated place away from all sources of heat.

Dangerous properties

Extremely dangerous fire and explosion risk. Forms highly explosive mixtures with air or oxygen. Can, in very high concentration, cause asphyxiation by replacing the necessary oxygen in the air.

Fire-fighting

Breathing apparatus may be necessary. Use carbon dioxide or dry powder.

Medical effects and treatment

Remove patient to fresh air. Apply artificial respiration.

Hydrogen peroxide

Chemical formula H_2O_2

SUMMARY

A water-white pungent liquid.

Decomposes to water and oxygen.

Non-combustible (see 'Dangerous properties').

Soluble in water.

Similar substances:
Hydrogen dioxide.

PHYSICAL DATA

S.G.	1·67
B. Pt.	305°F (152·1°C)

Industrial application

Used in various solutions with water for bleaching straw (hat trade), hair, ivory, oils, etc., also as a food preservative and as an antiseptic, especially in dental surgery, as a mouthwash.

Storage and transport

In carboys, Winchesters and bottles.

Dangerous properties

The dangers are dependent on the strength of the solution. As a mouthwash (3 per cent) and for similar uses it is comparatively stable but strength over 70 per cent may be hazardous. It is a powerful oxidising agent and although non-combustible, can cause violent combustion to take place when in contact with combustibles or their vapours. Strong light such as sunlight, can cause decomposition which may build up dangerous pressures in containers leading to violent explosions.

Fire-fighting

Breathing apparatus should be worn. Flood with water and, if possible, protection afforded to exposed skin surfaces. Beware of exploding bottles or containers.

Medical effects and treatment

Hydrogen peroxide may cause irritation to the skin and should be flushed clear with plenty of water. If the eyes are affected irrigate for at least 15 minutes with warm water followed by a flushing with boric acid solution. Seek medical advice if eyes are affected.

Iron (powder)

Chemical formula Fe

SUMMARY

A solid.
Heavier than water.
Liable to dust explosion.
Liable to spontaneous heating.
A heavy, dark grey powder.

PHYSICAL DATA

S.G.	7·86
M. Pt.	2795°F (1535°C) (approx.)
B. Pt.	5432°F (3000°C)

Industrial application

May be found in a wide variety of places from school laboratories to engineering works, foundries, munitions factories, etc.

Storage and transport

No special regulations.

Dangerous properties

Is liable to spontaneous heating, a condition which is assisted by the presence of oil and soap residues (lathe emulsion). Steam or water applied to red hot iron or iron powder will liberate hydrogen. Hydrogen is also produced when the powder is in contact with hydrochloric or sulphuric acid.

Fire-fighting

If a prolonged exposure is anticipated breathing apparatus should be worn. Dry sand, asbestos graphite or powdered graphite should be used. Sprays may be used as long as the evolution of hydrogen is allowed for.

Medical effects and treatment

Conjunctivitis may result from exposure to the powder. Seek medical attention if symptoms occur.

Mercury

Chemical formula Hg.

SUMMARY

An extremely heavy liquid. Also known as 'quick-silver'.

The only metallic element which is liquid at ordinary temperatures.

Silvery colour, easily recognised by the formation of characteristic globules.

Extremely toxic vapours.

PHYSICAL DATA

S.G.	13·6
M. Pt.	– 37°F (– 38°C)
B. Pt.	675°F (357°C)

Industrial application

Used in the manufacture of thermometers, barometers, mirrors, certain electrical appliances, drugs, chemical fulminates.

Also in gold extraction processes.

Note: 'Amalgams' are alloys or combinations of mercury with other metals.

Storage and transport

Large quantities in thick-walled iron bottles known in the trade as flasks. Small quantities in heavy stone bottles.

Dangerous properties

Extremely toxic vapours, even at low concentration.

Fire-fighting

If mercury is involved breathing apparatus must be worn. Gloves should be worn when handling containers etc. as poisoning is possible through the skin.

Medical effects and treatment

Symptoms of mercury poisoning are tremors of the extremities, mental disturbances, and soreness of the gums. Serious damage to kidneys or nervous system may result. If mercury has been ingested obtain medical attention immediately.

Special notes

Persons employed in the use of mercury receive careful instructions as to the dangers involved. If the skin is contaminated wash frequently and thoroughly.

Note. At ordinary room temperatures the vapour given off by the exposed mercury is 200 times stronger than the calculated safe concentration. Wet mercury does not vaporise.

Methane

Chemical formula CH_4

SUMMARY

A colourless gas with no smell.

Flammable and explosive.

Slightly soluble in water.

Also called 'marsh gas'.

For cylinder marking colours, see plate 44.

PHYSICAL DATA

V.D.	0·554
S.I.T.	1000°F (538°C)
B. Pt.	– 256·8°F (– 161°C)
E.L. (L)	5%
E.L. (U)	15%

Industrial application

A constituent of town gas.

Storage and transport

In steel cylinders and insulated tanks.

Dangerous properties

Flammable. Forms explosive mixtures with air or oxygen. Also with chlorine.

Fire-fighting

Breathing apparatus may be necessary. Use carbon dioxide or dry powder.

Medical effects and treatment

Slightly toxic but can cause asphyxiation if in sufficient concentration to exclude oxygen. Remove patient to fresh air. Apply artificial respiration.

Methyl Acetate

Chemical formula $CH_3COO \cdot CH_3$

SUMMARY

An ester.

A liquid.

Lighter than water.

Soluble in water.

Vapour heavier than air, irritating.

Flammable.

Explosive vapours.

PHYSICAL DATA

F.P.	15°F (– 9·4°C)
S.G.	0·93
V.D.	2·55
S.I.T.	850–935°F (454–501°C)
B. Pt.	140°F (60°C)
E.L. (L)	4%
E.L. (U)	14%

Industrial application

Used as a solvent for cellulose acetate and similar cellulose solutions, as used in cellulose lacquers, aero-dope, photographic film, perfumery etc.

Storage and transport

In steel drums or cans of 22·75 – 13,600 litres capacity. The *Cellulose Solutions Regulations* apply.

Dangerous properties

Low explosive range coupled with low flash point makes this substance a serious fire hazard.

Fire-fighting

In heavy vapour concentrations breathing apparatus should be worn if exposure is prolonged. Water spray to affect dilution is the best form of attack. CO_2, vaporising liquid or dry powders may be used.

Medical effects and treatment

Irritating to the eyes and, if absorbed into the body, may affect the eyesight. Remove patient to open air. If symptoms persist obtain medical attention.

Methyl alcohol (methanol)

Chemical formula CH_3OH

SUMMARY

A colourless liquid with a slight rather pleasant smell.
Flammable—burns vigorously.
Poisonous.
Soluble in water.
Also referred to as:
- 'Methanol'
- 'Carbinol'
- 'Wood naphtha'
- 'Wood spirit'

PHYSICAL DATA

F.P.	65°F (18·3°C)
S.G.	0·8
V.D.	1·11
S.I.T.	878°F (470°C)
B. Pt.	151°F (66°C)
E.L. (L)	6%
E.L. (U)	36%

Industrial application

Used extensively in industry as a solvent of fats, oils, dyes, perfumes, etc. Also used in the manufacture of spirit varnishes and polishes, and of mineralised methylated spirits.

Storage and transport

In cans, steel drums and steel tanks.

Dangerous properties

Low explosive limit. Extremely poisonous and can be absorbed through the skin and lungs.

Fire-fighting

Breathing apparatus should be worn. Dilution with water should be attempted if ullage space permits. Care should be taken to avoid contact with the liquid. Use water spray, CO_2, vaporising liquid, alcohol-resistant foam or dry powder.

Medical effects and treatment

Vapour affects central nervous system and damages internal organs. Exposure to the fumes and liquid should be avoided. Symptoms are headache, nausea, vomiting, abdominal pains and unconsciousness. Keep warm and rested. Administer oxygen if necessary. Obtain medical attention or remove to hospital.

Methyl bromide

Chemical formula CH_3Br

SUMMARY

A gas at ordinary temperatures.
Usually found as a liquid under pressure.
Heavier than air.
Explosive within very narrow limits.
Sickly sweet smell.
Extremely toxic.
Also known as 'bromomethane'.
For cylinder marking colours, see plate 44.

PHYSICAL DATA

S.G.	1·73
V.D.	3·27
S.I.T.	999°F (537·5°C)
B. Pt.	40°F (4·4°C)
E.L. (L)	13·5%
E.L. (U)	14·5%

Industrial application

At one time used as a fire extinguishing medium but no longer recommended because of its toxicity. Used to render ethyl and methyl chlorides non-flammable.

Storage and transport

Under pressure in steel cylinders.

Dangerous properties

Both liquid and vapour are very poisonous and can cause damage to the eyes.

Fire-fighting

Breathing apparatus must be worn and care taken to protect exposed parts of the skin.

Medical effects and treatment

Half to one per cent. concentration in air can be fatal in 30 minutes. Liquid will cause painful and serious burns to the eyes and skin. Remove immediately to fresh air and if necessary apply artificial respiration with oxygen. Obtain medical attention. Remove all contaminated clothing and wash affected skin surfaces freely. If eyes are affected use copious supplies of water and hold lids apart to ensure thorough washing out. Keep warm.

Methyl chloride

Chemical formula CH_3Cl

SUMMARY

A gas of ethylene odour; liquid under pressure.

As liquid lighter than water.

As gas heavier than air.

Soluble.

Flammable.

Explosive.

Toxic.

For cylinder marking colours, see plate 44.

PHYSICAL DATA

S.G.	0·92
V.D.	1·78
S.I.T.	1170°F (632°C)
B.P.	– 11°F (– 24°C)
E.L. (L)	8%
E.L. (U)	20%

Industrial application

Used in medicine, in the extraction of perfumes and for refrigeration.

Storage and transport

Methyl chloride is often rendered non-flammable by the addition of methyl bromide.

Dangerous properties

Two to two and a half per cent. concentration will produce death or serious injury in two hours.

Fire-fighting

Breathing apparatus may be necessary. Thoroughly ventilate at low level. Use carbon dioxide or dry powder.

Medical effects and treatment

Fumes have narcotic properties, and are dangerous to the eyes. Rescue shoulder high. Unconscious cases should be given oxygen. Keep warm. If breathing stops, apply artificial respiration with oxygen.

Methyl ethyl ketone

Chemical formula $CH_3COC_2H_5$

SUMMARY

A colourless liquid.

Lighter than water.

Vapour heavier than air.

Miscible in water.

Flammable.

Explosive in air.

Also known as 'butanone'.

PHYSICAL DATA

F.P.	21°F (– 6°C)
S.G.	0·8
V.D.	2·4
S.I.T.	960°F (516°C)
M. Pt.	– 123°F (– 87°C)
B. Pt.	176°F (80°C)
E.L. (L)	1·8%
E.L. (U)	11·5%

Industrial application

Used as a solvent in many manufacturing processes and in making smokeless gunpowder.

Dangerous properties

Flammable and explosive in air.

Fire-fighting

Use alcohol-resistant foam, carbon dioxide or dry powder.

Methylated spirits

Chemical formula see 'Special notes'

SUMMARY

Industrial methylated spirits is water-white and has a characteristic smell.

Ordinary grade spirits are dyed with methyl violet to render them unpalatable.

Flammable.

Soluble in water.

PHYSICAL DATA

Varies according to grade: for general characteristics see 'ethyl alcohol'.

Industrial application

Used in hospitals, surgeries, laboratories etc., as a solvent or cleansing fluid. Also in dyes, chemicals, explosives, flavouring extracts, perfumes, paint and varnish as well as for domestic use.

Storage and transport

In bottles, Winchesters, stone or glass carboys. In bulk it is stored in sheet iron tanks.

Dangerous properties

Usually has a low flash point.

Fire-fighting

Breathing apparatus should be worn. Flood with water to dilute if ullage space permits. Water spray or alcohol-resistant foam. CO_2 can be used in a small fire.

Medical effects and treatment

The fumes are highly intoxicating and exposure should be avoided. Affected persons should be removed to fresh air and rested. Obtain medical attention. Keep warm, give coffee to drink.

Special notes

Methylated spirits is made up mainly of ethyl alcohol to which certain other spirits (approved by Board of Customs and Excise) are added. There are many grades of the spirits and many uses. The industrial grade is the 'natural' colour, water white and other grades have different colours. The most common form is the 'toilet' grade easily recognised by the purple colour. This is done by the addition of methyl violet.

Naphtha

SUMMARY

Flammable liquids.

Toxic, explosive vapours.

Includes coal tar naphtha and petroleum naphtha.

PHYSICAL DATA

F.P.	0–100°F (– 17·8° to 38°C)	
S.G.	0·6–0·87	
S.I.T.	900°F (482°C) or lower	
B. Pt.	158–352°F (70°–178°C)	
E.L. (L)	1·1%	Petroleum naphtha
E.L. (U)	4·8%	Petroleum naphtha

Industrial application

Used as a fuel for lamps, in varnish making and as a solvent of rubber, etc.

Storage and transport

In tanks and drums.

Dangerous properties

A severe fire hazard.

Fire-fighting

Breathing apparatus should be worn. Avoid the use of water except for tank cooling purposes. Use foam, CO_2, vaporising liquid, dry powder, dry sand, etc.

Medical effects and treatment

Avoid exposure to the fumes and excessive skin contact which may result in dizziness, vomiting and narcosis or even chronic poisoning. Remove affected personnel to fresh air and keep warm. If symptoms persist seek medical aid.

Naphthalene

Chemical formula $C_{10}H_8$

SUMMARY

White crystalline flakes with a strong smell.

Flammable.

Explosive.

Toxic.

Heavier than water, insoluble in water.

PHYSICAL DATA

F.P.	176°F (80°C)
S.G.	1·145
V.D.	4·42
S.I.T.	1038°F (558°C)
M. Pt.	176°F (80°C)
B. Pt.	424°F (218°C)
E.L. (L)	0·9%
E.L. (U)	5·9%

Industrial application

Raw material in chemical and dyestuffs manufacture.

Storage and transport

Store away from sources of heat and flame.

Dangerous properties

Flammable with explosive vapours. Can cause dust explosions. Not liable to spontaneous heating. Fumes are toxic and the liquid is a skin irritant.

Fire-fighting

Breathing apparatus may be necessary. Use water, carbon dioxide, vaporising liquid or dry powder.

Medical effects and treatment

Wash affected parts of the skin with copious water. Remove affected persons to open air, administer oxygen and obtain medical attention if necessary.

Nitric acid

Chemical formula HNO_3

SUMMARY

Transparent colourless or yellowish liquid.

Fuming, suffocating, caustic and highly corrosive.

Explodes in contact with organic materials.

Fumes very toxic.

Not itself flammable.

Also known as 'aqua fortis'.

PHYSICAL DATA

S.G.	1·5
B. Pt.	186°F (86°C)

Industrial application

Used in the manufacturing of explosives and in metallurgy.

Storage and transport

In carboys.

Dangerous properties

Extremely corrosive. Both liquid and vapour are very dangerous to the eyes. Contact with any part of the body causes severe burns. Fumes are extremely toxic. Reacts very violently with organic materials to cause explosions and fire, giving off very toxic fumes. Explodes in contact with many chemicals e.g. hydrogen sulphide.

Fire-fighting

Breathing apparatus and completely protective clothing must be worn to ensure that the acid cannot reach any part of the body, especially the eyes. Use extinguishing material appropriate to the material burning. Dilute the acid with copious water supplies.

Medical effects and treatment

Where the patient has been splashed with the acid:—Remove all contaminated clothing and flush affected parts, particularly the eyes, *at once* with large quantities of warm water. Obtain medical attention immediately and continue bathing affected parts until available. The speedy application of plenty of water can greatly lessen the severity of the burns. Inhalation of the fumes can cause severe pulmonary damage and death. Call a doctor immediately even if no symptoms are apparent as these can be delayed.

Nitro benzene

Chemical formula $C_6H_5NO_2$

SUMMARY

Bright yellow crystals or a yellow oily liquid.

Flammable with explosive vapours.

Highly toxic.

PHYSICAL DATA

F.P.	190°F (87·8°C)
S.G.	1·2
V.D.	4·25
S.I.T.	900°F (482°C)
M. Pt.	42·5°F (5·7°C)
B. Pt.	412°F (210·9°C)
E.L. (L)	1·8% by volume in air at 200°F (93·3°C)

Industrial application

Used in the manufacturing of aniline.

Storage and transport

Store away from combustible matter and all forms of heat and flame.

Dangerous properties

Flammable liquid and vapours. Vapours form explosive mixtures with air. Very toxic whether absorbed through the skin, swallowed or inhaled in vapour form.

Fire-fighting

Breathing apparatus and protective clothing required. Use foam, carbon dioxide or dry powder.

Medical effects and treatment

Symptoms of poisoning are elation, then headache, sleepiness then sometimes nausea. Lips, tongue and lobes of ears may have a bluish tinge. In acute cases the pupils of the eyes may contract and later dilate with no reaction to light. Quick treatment is important. Carry patient to fresh air. Remove contaminated clothing. Wash affected parts with soap and plentiful luke-warm water. If eyes are involved flush for at least 15 minutes with luke-warm water. If necessary give oxygen. Where fumes inhaled give an emetic (mustard water or luke-warm soapy-water). *Never* give alcohol or drugs. Call a doctor.

Nitrous oxide

Chemical formula N_2O

SUMMARY

A gas.
A colourless liquid, under slight pressure, with a faint sweetish smell.
Gas heavier than air.
Liquid heavier than water.
An oxidant.
Soluble.
Toxic.
Explosive with hydrogen.
Also known as 'laughing gas'.
For cylinder marking colours, see plate 42.

PHYSICAL DATA

S.G.	1·23 (liquid)
V.D.	1·53 (gas)
M. Pt.	– 151·6°F (– 102°C)
B. Pt.	– 129°F (– 89·5°C)

Industrial application

As an anaesthetic in hospitals and dental surgeries.

Storage and transport

Stored in cylinders.

Dangerous properties

Nitrous oxide will tend to intensify the fierceness of a fire. It is explosive with hydrogen. Keep away from organic matter, as with oxygen. Flammable in contact with substances such as phosphorus, sulphur or carbon when these are warm.

Fire-fighting

Breathing apparatus may be necessary. Ventilate the premises at floor level and take the usual precautions for compressed gas cylinders.

Medical effects and treatment

When breathed for a short time induces excitement and in many cases laughter. Not likely to be found at fires without poisonous nitrous fumes. If the man is unconscious with the pure fumes give oxygen after loosening clothing, in sitting position. If the breathing fails, use artificial respiration and administer oxygen.

Octane

Chemical formula $CH_3(CH_2)_6CH_3$

SUMMARY

A clear liquid.

Flammable with explosive vapours.

PHYSICAL DATA

F.P.	56°F (13·3°C)
S.G.	0·7
V.D.	3·86
S.I.T.	450°F (232°C)
B. Pt.	258°F (125·8°C)
E.L. (L)	0·84%
E.L. (U)	3·2%

Industrial application

A constituent of petroleum fuels used in industry as a solvent.

Storage and transport

Store away from heat and oxidising agents.

Dangerous properties

Highly flammable. Vapour forms explosive mixtures with air. Liable to spontaneous combustion.

Fire-fighting

Breathing apparatus may be necessary. Do not use water. Use foam, carbon dioxide, vaporising liquid or dry powder.

Medical effects and treatment

In high concentrations it can be narcotic and irritating. Remove patient to fresh air.

Octel anti-knock compound

SUMMARY

Liquids containing tetra-ethyl lead (T.E.L.) or tetra-methyl lead (T.M.L.).
Compounds dyed orange, blue or green.
Insoluble in and heavier than water.
Soluble in petroleum spirit and kerosene.
Highly toxic and can be made to burn.

Industrial application

Used in the blending of petroleum fuels to improve the anti-knock qualities.

Storage and transport

By road in 205–225 litres steel drums or in 5 or 10 tonne steel containers; by rail in tank wagons. Ocean-going tankers berth at the wharf of the Associated Octel Co. Ltd. on the Manchester Ship Canal, Ellesmere Port.

Dangerous properties

Octel anti-knock compounds are toxic and may enter the body through the mouth, by inhalation of the vapour or by absorption through the skin. The compounds can be made to burn, the difficulty of ignition being comparable with kerosene or heavier hydrocarbon fractions. A vapour explosion hazard can arise following an extensive spill of compound in an unventilated enclosure. There is danger of explosion if the liquid is heated in sealed containers.

Fire-fighting

If anti-knock compound is spilled, or where the material is involved in fire, keep to windward and wear breathing apparatus. Cover any free liquid with foam or a blanket of water carefully applied as spray. Never disperse the liquid by using jets or allow the material to enter drains as this will spread the toxic hazard. Drums, tanks or other containers involved in fire must be cooled with water. Failure to cool effectively may result in the violent rupture of sealed containers. Water spray, foam, dry powder and CO_2 are suitable extinguishing agents; with dry powder and CO_2 re-ignition may occur unless sufficient cooling has been achieved. Water spray is generally considered to be the most effective.

Medical effects and treatment

Compounds can produce systemic poisoning of varying degree depending upon the quantity of material absorbed. If the compound gets on the skin wash immediately with kerosene followed by soap and water. Remove any contaminated clothing as quickly as possible and wash the underlying skin.

Oxidising agents

Dangerous properties

Many compounds which contain a high percentage of oxygen part with it violently to oxidisable substances i.e. those which are prone to take up oxygen. The principal groups of oxidising agents are bromates, chlorates, chromates, dichromates, iodates, nitrates, perchlorates, peroxides and persulphates. Oxidising agents are dangerous because:

(a) They react violently with oxidisable substances, giving off heat which may be sufficient to ignite the substance or its surroundings spontaneously.

(b) They break down with the evolution of oxygen, often when only slightly heated or sometimes even under the influence of friction, and thereby increase the fierceness of a fire.

(c) In the case of potassium and sodium peroxide, they evolve oxygen in contact with water.

(d) Most of them are soluble in water, and in case of fire there is a danger that strong solutions may be carried about the building or into contact with other materials when on evaporation of the water a residue of the oxidising agent is left behind. This residue may ignite spontaneously.

(e) When in the presence of dust, oxidising agents increase the dangers of dust explosion by increasing the heat of combustion, and thus the explosive power of the dust.

Oxidising agents are used in industry both as sources of free oxygen and as bleaching agents. The principal ones used are hydrogen peroxide and sodium peroxide.

Oxygen

Chemical formula O_2

SUMMARY

Colourless, odourless gas.

Liquid at very low temperatures (below the critical temperature in a 'Dewar' vessel), or under high pressure at a very low temperature.

Gas and liquid greatly increase severity of a fire.

Liquid explodes in contact with oxidisable materials.

Liquid dangerous to touch.

For cylinder marking colours, see plates 42 *and* 44.

PHYSICAL DATA

V.D.	1·42 @ 32°F (0°C)
C.T.	– 182°F (– 119°C)
C.P.	50 atmospheres

Industrial application

Used in oxy-acetylene and oxy-propane welding and many other industrial processes and particularly in engineering. Used in hospitals.

Storage and transport

In steel cylinders.

Dangerous properties

Adds enormously to the fierceness of any fire: even heavy types of clothing fabric can burn quickly and fiercely in an oxygen-enriched atmosphere if ignited by a spark or cigarette end. Gas forms an explosive mixture in contact with oxidisable materials such as oil. Oxygen is at high pressure in the cylinders and these are liable to explode violently if they are jarred or heated. Liquid oxygen exists only at very high pressure or very low temperature and in either case causes severe damage to the tissues if it contacts the body.

Fire-fighting

Use extinguishing media appropriate to the material burning.

Paraffins

Chemical formula C_nH_{2n+2} (general)

SUMMARY

Paraffins is a term given to the whole series of saturated aliphatic hydrocarbons of the general formula above.

Varies from a solid (wax) to a thin liquid (kerosene).

Prepared from shale oil by distillation.

Of varying degrees of flammability.

PHYSICAL DATA

F.P.	81 – 165°F (27 – 74°C)
S.I.T. M. Pt. B. Pt.	Variable according to composition.

Industrial application

Used extensively in industry in all its forms as a solvent in medicine, cosmetics, candle making, waxed papers, polishes, lubricants and the water-proofing industries.

Storage and transport

In drums or tanks mainly but varies greatly according to state.

Dangerous properties

Mainly a high fire hazard especially as a melted 'running' fire.

Fire-fighting

Avoid the use of water which is likely to cause explosive scattering of the liquid. Use foam or vaporising liquid. Use sand to dam up running liquid. Use water sprays to cool tanks and containers.

Medical effects and treatment

No serious effects are likely to arise from a short exposure to the fumes which may arise from a paraffin fire. Fresh air or the administration of oxygen will usually lead to complete recovery.

Parathion (trade name)

Chemical formula $C_{10}H_{14}NO_5PS$

SUMMARY

A dark-brown liquid resembling oil

Highly toxic.

Faint odour.

PHYSICAL DATA

B. Pt. 707°F (375°C)

Industrial application

An insecticide. Found at agricultural stores and farms. Used absorbed on a wettable powder, or mixed with diacetone alcohol or similar flammable solvents.

Dangerous properties

Highly toxic either by absorption through the skin or by inhalation of dust or fumes carrying the chemical.

Fire-fighting

Breathing apparatus should be worn in fighting a fire indoors in which 'Parathion' is involved. Out of doors no protection is necessary.

Medical effects and treatment

Any personnel feeling unwell during or after fire-fighting operations should be given medical attention. All personnel who have been in fire-fighting operations involving 'Parathion' should, as a precautionary measure, take a bath and change their clothing as soon as possible.

Pentane

Chemical formula $CH_3(CH_2)_3CH_3$

SUMMARY

A colourless liquid.

Highly flammable.

Vapours highly flammable and explosive.

PHYSICAL DATA

F.P.	Below – 40°F (– 40°C)
S.G.	0·63
V.D.	2·48
S.I.T.	588°F (309°C)
B. Pt.	96°F (36°C)
E.L. (L)	1·4%
E.L. (U)	7·8%

Industrial application

A constituent of motor fuels.

Storage and transport

Store away from heat or flame.

Dangerous properties

Flammable liquid and vapours. Vapours form explosive mixtures with air.

Fire-fighting

Breathing apparatus may be necessary. Do not use water. Use foam, carbon dioxide, vaporising liquid or dry powder.

Medical effects and treatment

Can act as an asphyxiant in very high concentration by replacing the necessary oxygen. Remove patient to fresh air. Apply artificial respiration.

Petroleum spirit

SUMMARY

A clear colourless liquid.

Highly flammable.

Vapours highly flammable and explosive.

Toxic.

PHYSICAL DATA

F.P.	– 50°F (– 45·6°C)
S.G.	0·635 to 0·660
V.D.	2·50
S.I.T.	475°F (246°C)
B. Pt.	104° to 158°F (40 to 70°C)
E.L. (L)	1·4%
E.L. (U)	5·9%

Industrial application

Universally as a fuel.

Storage and transport

Controlled by *Petroleum Consolidation Act* (1928) and *Dangerous Substances (Conveyance by Road) Regulations*, 1981.

Dangerous properties

Extremely flammable liquid and vapours. Vapours form explosive mixtures with air. Vapours slightly toxic and can be narcotic.

Fire-fighting

Breathing apparatus may be necessary. Use foam, carbon dioxide, vaporising liquid or dry powder. See Section 5, 'Fires in fuels'.

Medical effects and treatment

Remove patient to fresh air. If necessary apply artificial respiration.

Phenol

Chemical formula C_6H_5OH

SUMMARY

A white crystalline mass, pink or red if impure.

Strong carbolic smell.

Liquid or solutions very corrosive.

Vapours flammable.

Liquid and vapours very toxic.

Soluble in water and in alcohol; also very soluble in ether.

Also known as 'carbolic acid'.

PHYSICAL DATA

F.P.	175°F (79·4°C)
S.G.	1·072
V.D.	3·24
S.I.T.	1319°F (715°C)
M. Pt.	106°F (41°C)
B. Pt.	360°F (182°C)

Industrial application

Used in plastics manufacture, and as a disinfectant.

Storage and transport

In 205 litre drums. Store in a cool, dry atmosphere away from all forms of heat and flame.

Dangerous properties

Flammable vapours. Extremely corrosive if it touches the body, causing severe burns. Liquid, solutions and vapours (which can be evolved from the solid material or from solutions) are all highly toxic and can cause death. Can be absorbed easily through the skin, mucous membranes, gastronomical tract or lungs. Very dangerous to the eyes (whether liquid or solid).

Fire-fighting

Breathing apparatus, eye protection and fully protective clothing are necessary. Use water spray. Carbon dioxide, vaporising liquid or dry powder may also be used. Ensure that any phenol which is liable to run down drains has been thoroughly diluted.

Medical effects and treatment

The toxic effects first attack the central nervous system. Death can take place within a few hours of exposure. Symptoms of absorption are great muscular weakness, headache, dizziness, ringing in the ears, rapid irregular and difficult breathing and a weak pulse. Remove patient to fresh air. Obtain medical attention immediately. At once apply artificial respiration if breathing has stopped. If patient has been splashed remove contaminated clothing and flush affected parts (body or eyes) with quantities of luke warm water for at least 15 minutes. In the case of the body, soapy water or a large amount of either alcohol or glycerine can be used as the washing agent. Do not apply any salves or ointments. Keep patient warm (not hot).

Phosgene

Chemical formula $COCl_2$

SUMMARY

A colourless gas; heavy liquid under slight pressure.
Choking smell of musty hay.
Gas much heavier than air. Liquid heavier than water.
Extremely toxic.
Hydrolysed by water.
Correctly known as 'carbonyl chloride'.

For cylinder marking colours, see plate 44.

PHYSICAL DATA

S.G.	1·39
V.D.	3·5
M. Pt.	– 155°F (– 104°C)
B. Pt.	46°F (8°C)

Industrial application

Used as a chlorinating agent in the fine chemical industry. Solutions of the gas with petrol, benzole and ethyl acetate are used in the coal-tar dye industry, for exterminating vermin, and for making synthetic essence of violets. Used as a poison gas in warfare.

Storage and transport

As a liquid in large tanks, under slight pressure.

Dangerous properties

The gas is highly toxic and containers which may release their contents should be handled only under expert supervision.

Fire-fighting

Breathing apparatus must be worn. Thoroughly ventilate at floor level premises in which gas is suspected. Spray copiously with water, and preferably with caustic potash or caustic soda solution. Water hydrolyses and destroys phosgene, but hydrochloric acid (q.v.) is formed, which corrodes metals and rots clothing.

Medical effects and treatment

In fairly high concentrations, will cause coughing, choking and watering of the eyes. If breathed in any quantity it is highly lethal, causing water-logging of the lungs. It is not unusual, after slight initial symptoms, to have a period of well-being before the onset of more serious symptoms. In a mild case these are flushed face, rapid breathing and cough, drowsiness. In more severe cases, colour becomes bluish. In the latter case, give oxygen at once. Don't take clothes off, but loosen at neck, keep very warm. Hurry to hospital. Do not allow patient to walk or to exert himself. Do not give alcohol or artificial respiration.

Phosphoretted hydrogen (phosphine)

Chemical formula PH_3(gas), P_2H_4(liquid)

SUMMARY

A colourless gas; liquid under pressure.
Offensive garlic-like smell.
Gas heavier than air.
Liquid fractionally heavier than water.
Flammable. Toxic.
Spontaneously flammable at room temperature when impure. Also known as 'hydrogen phosphide' or 'phosphine'.

PHYSICAL DATA

S.G.	1·01 (liquid)
V.D.	1·185 (gas)
S.I.T.	Below atmospheric temperature (when impure)
M. Pt.	– 209°F (– 134°C)
B. Pt.	– 125°F (– 87·9°C)

Formed (a) when red phosphorus is gently heated in a current of hydrogen gas; (b) by the action of a hot solution of sodium hydroxide on white phosphorus; (c) by the action of water on calcium phosphide; (d) encountered as an impurity of acetylene; (e) by the action of moisture on aluminium phosphide.

Industrial application

The gas is not used in commerce but in addition to the above modes of formation, it may be found at fires in marine stores through sea markers becoming involved. (Sea markers usually contain a canister filled with a mixture of calcium phosphide and calcium carbide. Water applied to this mixture produces acetylene and phosphoretted hydrogen and the latter ignites the acetylene.) Sometimes is given off in small quantities by damp sawdust.

Dangerous properties

A highly toxic gas which in the impure state (as generally prepared), ignites spontaneously at room temperature. When it is formed indirectly, as in the case of sea markers, both liquid and gas are given off. The liquid vaporises in air at ordinary temperatures and usually ignites the gas given off with it. Phosphoretted hydrogen is spontaneously flammable with chlorine and explosive with oxygen.

Fire-fighting

Breathing apparatus should be worn. Thoroughly ventilate the premises. Avoid the use of water if phosphine is being evolved by calcium phosphide.

Medical effects and treatment

Symptoms are fainting, slowing of pulse, lowered blood pressure, nausea and vomiting. In serious cases convulsions ensue. Give oxygen and secure urgent medical attention.

Phosphorus

Chemical formula P_4

SUMMARY

White Phosphorus—a colourless to yellow waxlike solid. Ignites spontaneously when exposed to air. Causes severe skin burns. Toxic. Scarcely soluble in water. Shines in the dark.

Red Phosphorus—a reddish brown powder. Flammable. Toxic. Not soluble in water.

PHYSICAL DATA

	White	*Red*
F.P.	Spontaneously in air	572°F (300°C)
S.G.	1·82	2·2
V.D.	4·42	4·77
S.I.T.	86°F (30°C)	500°F (260°C)
M. Pt.	111°F (44°C)	1094°F (590°C)
B. Pt.	536°F (280°C)	536°F (280°C) (ignites)

Industrial application

Used in the manufacture of matches and pyrotechnics.

Storage and transport

Stored under water at all times.

Dangerous properties

White phosphorus is exceptionally dangerous since it ignites and burns vigorously when exposed to air. It readily starts fires and explosions and if in contact with the skin causes severe burns. It is especially dangerous to the eyes. Although red phosphorus has a higher ignition temperature, it is still very dangerous, being highly flammable. Both types give off poisonous vapours.

Fire-fighting

Breathing apparatus and fully protective clothing (including eye protection) must be worn when fire is indoors. Keep the phosphorus wet with water sprays. When it dries it will flame again, and it should therefore be removed to a safe place where it can burn out. Buildings that have been involved should be examined in the dark for any traces of phosphorus (white) which will glow with a greenish gleam. Phosphorus must not be dealt with by anyone who is not fully protected.

Medical effects and treatment

Keep any part of the skin which has contacted phosphorus wet (preferably under water) until a doctor can remove the phosphorus.

Phthalic anhydride

Chemical formula $C_6H_4(CO)_2O$

SUMMARY

White crystalline needles.

Flammable.

PHYSICAL DATA

F.P.	305°F (151°C)
V.D.	5·10
S.I.T.	1083°F (584°C)
M. Pt.	268°F (131°C)
B. Pt.	545°F (284°C)

Industrial application

Used in the manufacture of plastics.

Dangerous properties

Flammable. Fumes cause coughing and choking and are irritating to the mucous membranes. Dangerous to the eyes, causing corneal burns.

Fire-fighting

Breathing apparatus and eye protection needed.
Use water spray, carbon dioxide, vaporising liquid or dry powder.

Medical effects and treatment

Remove patient to fresh air. If eyes are affected obtain medical attention.

Picric acid

Chemical formula $(NO_2)_3C_6H_2OH$

SUMMARY

Yellow crystals.

Can be highly explosive.

Flammable.

Very toxic.

PHYSICAL DATA

F.P.	302°F (150°C)
S.G.	1·76
V.D.	7·9
S.I.T.	572°F (300°C)
M. Pt.	252°F (121·8°C)

Industrial application

An explosive and a dye.

Storage and transport

Frequently kept under water.

Dangerous properties

Can be a powerful explosive if mixed with impurities. Flammable Picric acid will stand warming up to 300°F (149°C) but will detonate if heated rapidly to 572°F (300°C). Very toxic. Liable to cause strong irritation to the skin, eyes, lungs and other organs.

Fire-fighting

Breathing apparatus and eye protection should be worn. Use copious water spray or open hose lines to keep container cool.

Medical effects and treatment

Irritation of the lungs can lead to bronchitis (through inhaling dust or vapours). Contact with the eyes or skin can cause conjunctivitis and dermatitis respectively. Remove contaminated clothing and flush affected parts thoroughly with plenty of water.

Polytetrafluoroethylene

SUMMARY

Known also by the initials 'PTFE' and the trade name 'FLUON'.

A white plastic moulding powder.

Non-combustible.

Gives off toxic vapours when heated.

PHYSICAL DATA

Decomposes at temperatures above 752°F (400°C).

Industrial application

As a fabricated plastic product in the chemical, electrical and engineering industries.

Dangerous properties

PTFE is incombustible, but at temperatures over 752°F (400°C) it decomposes, i.e. it slowly depolymerises into its monomer and other fluorine-containing compounds. Toxic vapours are given off. (See Section 10 'Plastics').

Fire-fighting

Breathing apparatus must be worn. No smoking in the possible presence of the fumes.

Medical effects and treatment

Inhalation of the vapours may cause a severe feverish chill and necessitate a few days in bed. Persons suspected should be medically examined as soon as possible.

Potassium

See Section 7. 'Metals'.

Chemical formula K

SUMMARY

A soft silver-white lustrous metal which burns with a lilac-coloured flame.
Reacts with water.
Lighter than water.
Flammable.
Explosive.
Causes burns.

PHYSICAL DATA

S.G.	0·87
M. Pt.	144°F (62°C)
B. Pt.	1396°F (758°C)

Industrial application

In alkali works, fine chemical works, chemical stores. (Its commercial use is not extensive.)

Storage and transport

In glass bottles, tin cans or iron drums, stored beneath oil or kerosene, or in airtight containers.

Dangerous properties

Decomposes explosively with water due to the condensation of water vapour: hydrogen is released, and sufficient heat is liberated to ignite the hydrogen. Explosive by pressure or friction. Ignites in warm air, oxidises rapidly in moist air. Potassium is explosive with carbon disulphide, hazardous with chlorine, sulphuric and nitric acids and all oxidants. Solutions containing water have the same effect as pure water. Do not touch potassium particles with bare hands, as deep burns will be caused. Flying particles are dangerous to the eyes.

Fire-fighting

Use dry sand, talc powder, asbestos or asbestos-graphite powder to smother the fire, but *not* water, CO_2, soda-acid or foam. Sand, etc., should be applied carefully and from a distance to avoid scattering. Scattered particles may cause fires, and should be collected and disposed of safely. Breathing apparatus, goggles and a dry face cloth should be used. Water should not be used on or near potassium. It is possible in certain circumstances to use heavy oil (*not* kerosene or other spirit) to extinguish potassium and sodium, which do not react in a hazardous way with it. Should the oil catch fire it can be dealt with in the normal way.

Medical effects and treatment

Burns should be dressed with lint soaked in water containing a small amount of acid.

Potassium hydroxide

Chemical formula KOH

SUMMARY

A solid, commonly called 'caustic potash'.

Usually in sticks or flakes or pellets.

Extremely corrosive.

Soluble in water with evolution of heat.

Dangerous fumes.

Strongly alkaline.

PHYSICAL DATA

S.G.	2·04
M. Pt.	680°F (360°C)
B. Pt.	2408°F (1320°C)

Industrial application

Used in alkali works and in the chemicals industry generally.

Storage and transport

Stored in bottles, cans or drums usually in a petroleum derivative.

Dangerous properties

Exposure to moisture—as in air—will cause evolution of heat and possibly fire. Extremely corrosive on skin. Even weak solution in eyes can lead to permanent blindness.

Fire-fighting

Breathing apparatus should be worn. Complete protection of eyes and all exposed parts should be attempted to avoid splashes of solution. Dilute solutions thoroughly and cautiously with water or neutralise with dilute acid such as vinegar.

Medical effects and treatment

Solution attacks all animal and vegetable tissues. Dangerous fumes may be evolved. Remove affected persons to open air and wash burns with a dilute acid such as vinegar. Obtain medical attention.

Potassium nitrate

Chemical formula KNO_3

SUMMARY

Transparent, colourless or white crystalline powder or crystals.

Powerful oxidising agent.

Slightly soluble in water at room temperature and very soluble at 212°F (100°C).

Also known as 'saltpetre', or 'nitre'.

PHYSICAL DATA

S.G.	2·1
M. Pt.	635°F (335°C)
B. Pt.	Decomp. at 752°F (400°C)

Industrial application

Used for pickling meat, as a fertiliser and as a constituent of gunpowder.

Dangerous properties

Very liable to cause fire and explosion. (See 'oxidising agents'.)

Fire-fighting

Breathing apparatus may be necessary as protection against nitrous fumes.

Potassium permanganate

Chemical formula KM_nO_4

SUMMARY

Dark purple crystals.
Powerful oxidising agent.
Soluble in and heavier than water.
Explosive with many substances.
Liable to spontaneous heating with many substances.
Can cause burns.

PHYSICAL DATA

S.G.	2·703
M. Pt.	Decomp. below 464°F (240°C)

Industrial application

Used in nickel refining, as a disinfectant, in oxidising, bleaching and tanning processes, and in explosives manufacture.

Storage and transport

In airtight wooden containers lined with zinc or paper; corrugated iron drums and tins.

Dangerous properties

Liberates oxygen when heated. Very liable to cause fire and explosion. When mixed with sulphuric acid it will ignite all flammable gases and vapours. Ignites spontaneously with glycerine. Dangerous to the eyes. If swallowed it is corrosive and a strong irritant.

Fire-fighting

Breathing apparatus and eye protection essential. Flood with water by means of sprays or open ends; do not use high pressure jets. Care should be taken to prevent flammable materials from becoming soaked in the solution and subsequently drying out; if this occurs in a heated atmosphere fresh fires may be caused. Small fires may be dealt with by carbon dioxide or dry powder.

Potassium sulphides

Chemical formulae K_2S, K_2S_2, K_2S_3, K_2S_4 and K_2S_5

SUMMARY

Reddish brown solids.

Flammable.

Soluble in water.

Toxic.

Liable to spontaneous ignition.

PHYSICAL DATA

Varies according to compound.

Industrial application

Used in tanning.

Storage

In airtight iron drums, cans or glass bottles when in anhydrous form. In wooden casks and cases when fully hydrated.

Dangerous properties

When burning they evolve sulphur dioxide (q.v.). In the freshly prepared condition they absorb oxygen quickly on exposure to air, and especially when in a state of fine division (or when mixed with free carbon, such as lampblack), they may do so sufficiently quickly to ignite spontaneously. In large heaps to which air cannot easily penetrate, the interior may burn away with little exterior indication. They are not dangerous when allowed to absorb moisture and become hydrated.

Fire-fighting

Breathing apparatus must be worn. Water may be used for extinguishment but it should be remembered that hydrogen sulphide (q.v.) is produced as a result of its reaction with the sulphide.

Propane

(See Section 5 'Fuels')
Chemical formula $CH_3CH_2CH_3$

SUMMARY

Gas; liquid under pressure.

Flammable.

Narcotic.

Explosive in suitable concentration.

A petroleum derivative, which can also be produced synthetically.

The pure gas is odourless but a stenching agent is usually added to make detection easier.

PHYSICAL DATA

F.P.	– 156°F (– 104°C)
V.D.	1·52
S.I.T.	870°F (466°C)
B. Pt.	– 43·8°F (– 42°C)
E.L. (L)	2·4%
E.L. (U)	9·5%

Industrial application

Used as a refrigerant; as a substitute for acetylene in flame cutting and welding; as a source of gas in isolated buildings, yachts, caravans, etc. for cooking and sometimes (with incandescent mantles) for lighting.

Storage and transport

In cylinders as a liquid under pressure; cylinders vary in capacity from a few pounds to some hundreds of pounds and may be four or more feet long. Also in spherical tanks under pressure.

Dangerous properties

As for butane (q.v.).

Fire-fighting

See Section 5. 'Fuels'.

Medical effects and treatment

As for butane (q.v.).

Pyridine

Chemical formula C_5H_5N

SUMMARY

A colourless liquid.
Sharp, burning odour.
Highly flammable.
Explosive.
Toxic fumes.

PHYSICAL DATA

F.P.	68°F (20°C)
S.G.	0·982
V.D.	2·73
S.I.T.	900°F (482°C)
B. Pt.	239·5°F (115·3°C)
E.L. (L)	1·8%
E.L. (U)	12·4%

Industrial application

A product of coal-tar distillation; a constituent of methylated spirits.

Storage and transport

Store away from heat and flame.

Dangerous properties

Highly flammable when exposed to heat or flame: vapours explosive. Emits toxic fumes of cyanides. Reacts with oxidising agents.

Fire-fighting

Use breathing apparatus. Use water spray, foam, carbon dioxide, vaporising liquid or dry powder.

Medical effects and treatment

Irritating to the eyes and mucous membranes. Prolonged exposure can cause serious effects. Remove to open air at once and obtained medical attention.

Sodium

Chemical formula Na

SUMMARY

A soft silvery metal.

Floats in water with violent reaction accompanied by flame.

Very irritating fumes.

Oxidises in air.

PHYSICAL DATA

S.G.	0·97
S.I.T.	Above 239°F (115°C)
M. Pt.	204°F (95·5°C)
B. Pt.	1616°F (880°C)

Industrial application

Used in alkali and other chemical works, wholesale chemists; also in the production of cyanides and sodium peroxides.

Storage and transport

In bottles or jars immersed in a petroleum distillate or in blocks covered with naphtha which are then packed in sheet iron containers holding approx. 50·8 kg, in hermetically sealed drums.

Dangerous properties

Although not as hazardous as potassium (which it resembles in many ways) sodium must nevertheless be treated with care. It decomposes in contact with moisture as in air, liberating hydrogen gas. The heat of the decomposition is usually high enough to ignite the gas.

Fire-fighting

Breathing apparatus should be worn and full protection against contact with exposed parts of the skin should be attempted. Water or any agent containing oxygen must *NOT* be used. Use only *DRY* sand or dry soda-ash to smother and do so at a safe distance as scattering of molten metal may result.

Medical effects and treatment

The fumes given off by sodium burning in air, viz.:-sodium oxide are extremely irritating to the skin, nose, and throat. If fragments of the molten metal come in contact with the body they may cause deep burns. Brush off quickly *before* applying water then treat burns with dilute acetic acid (vinegar). Obtain medical attention.

Sodium chlorate

Chemical formula $NaClO_3 \cdot 3H_2O$

SUMMARY

A solid.

Soluble.

Explosive by shock, etc.

Toxic.

Liable to spontaneous heating.

PHYSICAL DATA

S.G.	2·49
M. Pt.	478°F (248°C)

Industrial application

Used for bleaching textiles and paper; also in the manufacture of weed killers.

Storage and transport

In wooden boxes with inside containers of glass, earthenware or metal; if of glass or earthenware the capacity does not exceed 1 kilogram; if of metal, 2·25 kilogram each.

Dangerous properties

It is hazardous with organic material and may explode by shock in such circumstances. Toxic (which is not the case with sodium hypochlorite). In contact with acids, it emits chlorine dioxide, a hazardous and explosive gas. It will ignite if mixed with sulphur and wetted or rubbed.

Fire-fighting

Use breathing apparatus. Flood with water through diffuser branches or open ends. Do not use high-pressure jets. Take care a solution does not wash down on to material which is flammable, or on drying the salts may cause fresh fires in a heated atmosphere.

Clothing

Articles of clothing or other materials soaked in chlorate solution should be thoroughly washed by rinsing and wringing, otherwise they may ignite spontaneously when dry if heated.

Sodium hydroxide

Chemical formula NaOH

SUMMARY

Commonly called caustic soda.

A white crystalline substance.

Soluble in water with the evolution of heat.

Has soapy feel in solution.

Dangerous fumes.

Strong alkali.

Non-flammable.

PHYSICAL DATA

S.G.	2·13
M. Pt.	604°F (318°C)
B. Pt.	2534°F (1390°C)

For other information see potassium hydroxide, to which sodium hydroxide is similar.

The latter is widely used in industry.

Sodium nitrate

(See Section 7, 'Metals, nitrate, baths')

Chemical formula $NaNO_3$

SUMMARY

Colourless transparent crystals with no smell.

Powerful oxidising agent.

Soluble in water.

PHYSICAL DATA

S.G.	2·261
M. Pt.	580°F (304·4°C)
B. Pt.	Decomp. at 716°F (380°C)

Industrial application

For metal treatment baths.

Storage and transport

In sacks.

Dangerous properties

Liable to cause fire and explosion.

Fire-fighting

Breathing apparatus may be necessary as a precaution against nitrous fumes. Water from open ended hoses should be applied copiously but gently. Do not allow water flowing from fire to come into contact with flammable materials. For other information, see Section 7, 'Metals, nitrate baths'.

Sodium peroxide

Chemical formula Na_2O_2

SUMMARY

A yellowish powder.

Reacts violently with water and acids.

Toxic fumes possible.

Similar substances:

'Sodium dioxide'
'Sodium superoxide'
'Sodium binoxide'

PHYSICAL DATA

S.G.	2·805
M. Pt.	Decomposes at 860°F (460°C)

Industrial application

Used in the manufacture of organic chemicals and soaps. Also for bleaching textiles, sponges, bristles etc.

Storage and transport

Packed in tins and metal drums.

Dangerous properties

Reacts violently (sometimes with explosive force) with water or moisture in any form. Very damaging to clothing.

Fire-fighting

Breathing apparatus should be worn. Water must not be used. Use a smothering agent such as dry sand, dry graphite or bicarbonate of soda.

Medical effects and treatment

The dust causes irritation when in contact with the skin, eyes and mucous membrane. Remove all traces by flushing with copious quantities of water. Gargle or spray the nose and throat if affected. If irritation does not disappear rapidly seek medical advice. Contaminated clothing should be thoroughly washed before re-use.

Sodium-Potassium alloy

Chemical formula NaK

SUMMARY

A liquid.

Self igniting in contact with atmosphere (moisture).

Caustic vapours heavier than air.

Very hazardous if inhaled.

Explosive gas (hydrogen) liberated.

PHYSICAL DATA

(mixture containing 44% potassium)

S.G.	0·886 (at 212°F) (100°C)
M. Pt.	66°F (19°C)
B. Pt.	1517°F (825°C)

Industrial application

Used as a heat-transfer medium in processes involving very high temperatures.

Storage and transport

In steel drums (sealed of up to 250 litre capacity. (Weight approx. 160 kilograms)). May be found sealed in a petroleum derivative.

Dangerous properties

(1) Decomposes in contact with water to liberate hydrogen.

(2) The alloy is more re-active than either component.

Evolves large volumes of sodium oxide vapours which are highly caustic and extremely dangerous especially if inhaled.

Fire-fighting

Breathing apparatus and protection should be worn. Use ONLY dry salt or soda ash and beware of scattering of burning metal.

Medical effects and treatment

Vapours cause damage to eyes and mucous membrane. Very irritating. Causes dermatitis if left on skin surfaces. Wash immediately with copious quantities of water, under deluge shower if possible. Irrigate affected eyes for 15 minutes with warm water. Seek medical attention always.

Sodium sulphide

Chemical formula Na_2S

SUMMARY

A reddish-yellow compound.

A solid.

Soluble in water.

Flammable.

Liable to spontaneous ignition.

PHYSICAL DATA

S.G.	1·85 @ 57°F (14°C)
M. Pt.	2156°F (1180°C)

Industrial application

Used in rayon manufacture, in silver and gold treatment, in manufacturing sulphur dyes, in paper making, tanning and the rubber industry. For commercial purposes it is used in crystalline form (30–33 per cent. strength) or fused form (60–65 per cent. strength).

Storage and transport

In iron drums or bottles.

Dangerous properties

When burning, it gives off sulphur dioxide (*q.v.*). In contact with acids hydrogen sulphide (*q.v.*) is evolved.

Fire-fighting

Breathing apparatus must be worn. Water may be used for extinguishment. Hydrogen sulphide (*q.v.*) may be produced as a result of a reaction between the sodium sulphide and the water.

Styrene (monomer)

Chemical formula $C_6H_5CHCH_2$

SUMMARY

A colourless oily liquid, smells strongly.

Flammable.

Toxic vapours.

Also known as 'vinyl benzene', 'phenyl ethylene', 'cinnamene'.

PHYSICAL DATA

F.P.	88°F (31°C)
S.G.	0·9 @ 68°F (20°C)
V.D.	3·6
S.I.T.	914°F (490°C)
B. Pt.	293°F (145°C)
E.L. (L)	1·1%
E.L. (U)	6·1%

Industrial application

A raw material in the manufacturing of plastics.

Storage and transport

Store away from heat and flame.

Dangerous properties

Flammable. Liable to spontaneous ignition on heating. Vapours narcotic. Liquid can cause skin infections.

Fire-fighting

Breathing apparatus must be worn. Keep liquid from the exposed skin. Use foam, carbon dioxide, vaporising liquid or dry powder.

Sulphur

Chemical formula S

SUMMARY

Yellow crystals or powder.

Dust or vapours flammable when mixed with air.

Toxic fumes.

Also known as 'brimstone', 'flowers of sulphur'.

PHYSICAL DATA

F.P.	405°F (207°C)
S.G.	2·07; liquid 1·803
S.I.T.	450°F (232°C)
M. Pt.	235°F (112·8°C)
B. Pt.	831°F (444·6°C)

Industrial application

Used in making sulphuric acid and carbon disulphide; explosives; dyes; rubber.

Storage and transport

In bulk, sacks or drums. As a liquid in heated and insulated road tankers. Store away from heat and flame.

Dangerous properties

Liable to cause fires and dust explosions. Dust or vapour form flammable mixtures with air. Dust can irritate the eyes and skin. When heated emits sulphur dioxide (*q.v.*).

Fire-fighting

Breathing apparatus must be worn. Large fires can be flooded with water sprays or open ends. (Do not disturb the powder or dust explosion dangers arise.) Small fires can be extinguished by smothering with the sulphur.

Sulphur dioxide

Chemical formula SO_2

SUMMARY

A gas, strong smell.

Non-flammable.

Heavier than air.

Toxic and irritating.

For cylinder marking colours, see Plate 44.

PHYSICAL DATA

V.D.	2·264
M. Pt.	– 104°F (– 75°C)
B. Pt.	14°F (– 10°C)

Industrial application

Used as a bleaching agent in the textile industry, also as a food preservative, a disinfectant and in the petroleum refining industry. The liquefied gas is in very common use as a refrigerant in small cold storages and retail shops.

Storage and transport

The gas in solution will be found stored in carboys. More often found as a liquefied gas in steel gas cylinders.

Dangerous properties

Poisonous and corrosive vapours.

Fire-fighting

Breathing apparatus should be worn. Use water sprays to assist in 'laying' the gas. Thorough ventilation as soon as situation permits. Remove unaffected cylinders to safety. Keep heated cylinders cool by sprays. Application along the entire length.

Medical effects and treatment

Dangerous to the eyes and affects the upper respiratory tract and lungs. Remove to fresh air at once. Wash burns with warm water, administer oxygen if breathing becomes painful and difficult. Obtain medical attention.

Sulphuretted hydrogen

Chemical formula H_2S

SUMMARY

A colourless gas—smells of rotten eggs.

Slightly heavier than air.

Very poisonous.

Flammable.

Moderately soluble in water. Also referred to as 'hydrogen sulphide'.

PHYSICAL DATA

V.D.	1·189
S.I.T.	500°F (260°C)
M. Pt.	- 122°F (- 85·5°C)
B. Pt.	- 76°F (- 60°C)
E.L. (L)	4·3%
E.L. (U)	46%

Industrial application

For metallurgical works, gasworks, waste coal bins or stacks, rubber works, etc. More frequently produced accidentally as in sewers and similar locations.

Storage and transport

May be found in cylinders but is usually produced on the spot as, and when, required.

Dangerous properties

An extremely poisonous gas (1 part H_2S in 500 parts air—can prove fatal in a very short time). Forms an explosive mixture with air. When burning in air it forms sulphur dioxide (*q.v.*).

Fire-fighting

Breathing apparatus *MUST* be worn with full eye protection. Use water sprays, carbon dioxide or dry powder. Thoroughly ventilate as soon as possible.

Medical effects and treatment

Remove to fresh air at once. If breathing has ceased, apply artificial respiration and give oxygen until removal to hospital. Secure medical attention immediately.

Sulphuric acid

Chemical formula H_2SO_4

SUMMARY

A heavy oily colourless liquid.

Corrosive.

Non-flammable but can ignite carbonaceous materials.

Toxic vapours.

Miscible with water.

Also known as 'oil of vitriol'.

PHYSICAL DATA

S.G.	1·834
B. Pt.	626°F (330°C)

Industrial application

Used in the manufacture of textiles, explosives, dyes and chemicals, accumulators, fertilisers; also in the metal galvanising, ammonia recovery, petroleum refining and alkali industries.

Storage and transport

Glass bottles, carboys, iron drums and tank cars.

Dangerous properties

Liable to ignite on contact with carbonaceous materials. Dangerously reactive in contact with water and metals, evolving hydrogen gas and heat. May cause extremely deep burns if in contact with the skin.

Fire-fighting

Breathing apparatus may be necessary. Avoid the use of water unless large quantities of water can be applied to dilute the acid. Smother with chalk, soda, or ashes.

Medical effects and treatment

The acid is highly corrosive on the skin, eyes and respiratory tracts. Wash off acid at once with copious quantities of water, remove contaminated clothing. If shock symptoms appear keep warm and as in all cases of contact, seek medical aid. Do not apply oils or ointments unless on doctor's orders.

Titanium tetrachloride

Chemical formula $TiCl_4$

SUMMARY

Colourless or light yellow liquid.

Corrosive.

Very toxic liquid and fumes: gives off thick white smoke.

Reacts with water or moisture to give off hydrogen chloride.

Non-flammable.

PHYSICAL DATA

S.G.	1·726
M. Pt.	– 22°F (– 30°C)
B. Pt.	277°F (136·4°C)

Industrial application

As a smoke producer and as an intermediary in chemical processes.

Dangerous properties

Causes serious burns if it touches the skin. If exposed to moisture of the air it gives off hydrogen chloride fumes. If inhaled it absorbs enough moisture from the tissues to evolve these fumes. (Hydrogen chloride is very toxic and very irritating.) If dissolved in water titanium tetrachloride becomes very strongly acid due to excess hydrogen chloride.

Fire-fighting

Use breathing apparatus, eye protection and protective clothing.

Toluene

Chemical formula $C_6H_5CH_3$

SUMMARY

A volatile, flammable liquid.

Poisonous and irritant.

Explosive vapours.

Not miscible with water.

Also known as:
'Methylbenzene'
'Phenylmethane'
'Toluol'

PHYSICAL DATA

F.P.	40°F (4·5°C)
S.G.	0·86
V.D.	3·14
S.I.T.	1026°F (552°C)
B. Pt.	230°F (110°C)
E.L. (L)	1·3%
E.L. (U)	6·8%

Industrial application

Used in industry as a solvent, in dye-making and the perfumery trade, also as an intermediate, and in the preparation of T.N.T. and other explosives. Obtained as a product of coal-tar distillation.

Storage and transport

In storage tanks on the site and in metal drums.

Dangerous properties

Very low explosive range. Low flash point. Poisonous and irritant fumes.

Fire-fighting

Breathing apparatus should be worn. Avoid the use of water except to cool tanks and containers. Use foam, CO_2, or vaporising liquid.

Medical effects and treatment

Avoid exposure to the fumes which will cause burns to the cornea of the eyes; in addition it has a depressant effect on the central nervous system sometimes affecting the bone marrow. It has the effect of dulling the sense of smell and thus risking dangerous exposures. Remove to fresh air at once. Keep warm; seek medical advice.

Trichlorethylene

Chemical formula $CHCl{:}CCl_2$

SUMMARY

A colourless mobile liquid.

Sweetish smell, like chloroform.

Flammable only at high temperatures.

Heavier than water.

Extremely poisonous.

Also called 'trichloroethylene', and 'ethylene trichloride'.

PHYSICAL DATA

S.G.	1·45
V.D.	4·5
S.I.T.	770°F (410°C)
B. Pt.	188·5°F (87·0°C)

Industrial application

As a general solvent or de-greasing agent in engineering works and the paint and varnish industry, also as the main ingredient in the 'dry' cleaning of fabrics.

Storage and transport

Usually stored in steel tanks, drums or bottles. Transported in drums.

Dangerous properties

An extremely poisonous gas which may, when in contact with red hot metals, decompose to form a gas closely resembling phosgene.

Fire-fighting

Breathing apparatus will give adequate protection against concentration of vapours found in normal circumstances. High concentrations of vapour will however pass through rubber. Care should be taken in the use of breathing apparatus in high concentrations. Use foam, CO_2, dry powder.

Medical effects and treatment

First reaction is irritation of eyes, nose, and throat followed by dizziness, vomiting, etc. Remove to fresh air at once. 3·5 per cent. vapour in air may prove fatal. Keep warm, and secure medical attention. Do not allow smoking near fumes. Administer oxygen if necessary.

Note: Prolonged exposure of the skin will produce ill effects.

Turpentine

Chemical formula $C_{10}H_{16}$

SUMMARY

A colourless flammable liquid.

Explosive vapours.

Irritating fumes.

PHYSICAL DATA

F.P.	95°F (35°C)
S.G.	0·86
V.D.	4·84
S.I.T.	464°F (240°C)
B. Pt.	313–347°F (156–175°C)
E.L. (L)	0·8%

Industrial application

Used as a solvent for oils, fats and resins and as a thinner for oil varnish. Used in the manufacture of synthetic camphor.

Storage and transport

Steel tanks and drums.

Dangerous properties

Highly flammable with dangerous irritating fumes. Susceptible to spontaneous heating.

Fire-fighting

Breathing apparatus should be worn if exposed to strong concentrations of vapour. Avoid the use of water. Use foam, vaporising liquid, CO_2 or dry powder.

Medical effects and treatment

Comparatively low concentrations of the vapour will cause irritation of eyes and nose and higher concentrations can cause headache, vertigo, bronchial inflammation, etc. Continuous contact on the skin may lead to dermatitis. Remove affected persons to open air, where recovery will normally occur quite rapidly.

Turpentine substitute (white spirit)

SUMMARY

A colourless flammable liquid.
Explosive vapours.

PHYSICAL DATA

F.P.	85–95°F (29·4–35°C)
S.G.	<1
V.D.	4·3
S.I.T.	450°F (232°C) approximately
B. Pt.	300–400°F (149–204°C)
E.L. (L)	1·0%
E.L. (U)	6·0%

Industrial application

A blend of hydrocarbon distillates used as a solvent for oils, fats and resins, and as a thinner for paints and varnish.

Storage and transport

Steel tanks and drums.

Dangerous properties

Flammable.

Fire-fighting

Use foam, vaporising liquid, CO_2 or dry powder.

Medical effects and treatment

No serious effects are likely to arise from an exposure to the fumes from a fire involving white spirit.

Xylene

Chemical formula $C_6H_4(CH_3)_2$

SUMMARY

Colourless liquid.
Lighter than water.
Immiscible with water.
Flammable.
Explosive vapour.
Can react with oxidising materials.
Otherwise known as 'Xylol'.

PHYSICAL DATA

F.P.	73°F (22·8°C)
S.G.	0·86
V.D.	3·66
M. Pt.	– 54°F (– 47·9°C)
B. Pt.	282·5°F (139°C)
E.L. (L)	1·1%
E.L. (U)	7·0%

Industrial application

Found in tar distilleries as a derivative of coal tar. Used in the manufacture of dye-stuffs, explosives, synthetic varnishes, etc.

Storage and transport

Stored in drums 225 litres or less.

Dangerous properties

Low flash point. Vapour explosive when exposed to heat or flame. Low explosive limit.

Fire-fighting

Breathing apparatus may be necessary. Use foam, CO_2, dry chemical or vaporising liquid.

Medical effects and treatment

Vapours are toxic. Liquid can cause irritation of the skin.

INDEX

Printed in the United Kingdom for HMSO
Dd300510 4/95 C20 G559 10170